p. 265
selection ≠ hybrids

THE PROCESS OF EVOLUTION

McGRAW-HILL SERIES IN POPULATION BIOLOGY

Consulting Editors
Paul R. Ehrlich, Stanford University
Richard W. Holm, Stanford University

Briggs Marine Zoogeography
Edmunds and Letey Environmental Administration
Ehrlich, Holm, and Parnell The Process of Evolution
George and McKinley Urban Ecology
Hamilton Life's Color Code
Poole An Introduction to Quantitative Ecology
Stahl Vertebrate History: Problems in Evolution
Watt Ecology and Resource Management
Watt Principles of Environmental Science
Weller The Course of Evolution

PAUL R. EHRLICH
RICHARD W. HOLM
Department of Biological Sciences
Stanford University

DENNIS R. PARNELL
Department of Biological Sciences
California State University, Hayward

with illustrations by
Anne H. Ehrlich

THE PROCESS OF EVOLUTION

SECOND EDITION

McGraw-Hill Book Company
New York St. Louis San Francisco Düsseldorf
Johannesburg Kuala Lumpur London Mexico Montreal New Delhi
Panama Paris São Paulo Singapore Sydney Tokyo Toronto

Library of Congress Cataloging in Publication Data

Ehrlich, Paul R
    The process of evolution.

    (McGraw-Hill series in population biology)
    Includes bibliographies.
       1. Evolution. 2. Genetics. I. Holm, Richard W.,
joint author. II. Parnell, Dennis R., joint author.
III. Title. DNLM: 1. Evolution. 2. Genetics.
QH366 E33p 1975
QH336.2.E35 1974   575   74–3030
ISBN 0–07–019133–6

THE
PROCESS OF
EVOLUTION

2 3 4 5 6 7 8 9 K P K P 7 9 8 7 6 5

This book was set in Helvetica Light by Black Dot, Inc.
The editors were William J. Willey and Michael Gardner;
the designer was Anne Canevari Green;
the production supervisor was Sam Ratkewitch.
New drawings were done by B. Handelman Associates, Inc.

To
Edgar Anderson
Joseph H. Camin
Harlan Lewis
Herbert L. Mason
Charles D. Michener
Peter H. Raven
Robert R. Sokal
Robert E. Woodson

CONTENTS

PREFACE

Modern evolutionary theory is the great unifying concept of biology. It represents the major theoretical triumph of the biological sciences—an all-embracing theory which attempts to explain the manifold complexities of biological phenomena. The biochemist attempting to understand the genetic code, the neurophysiologist probing the complex mechanisms of the mind, the embryologist seeking to understand how one tissue affects the development of another—indeed—all biologists are working on problems whose broad *theoretical* significance can be measured primarily by their contribution to our understanding of evolutionary phenomena. The biochemist may be able eventually to cure cancer, the neurophysiologist to understand mental disorders, and the embryologist to discover how the genetic code is translated into an organism. But, without a theory that interrelates all these phenomena, their work would have only applied significance.

The central position of evolution in biology has long been recognized. Nevertheless, most laymen and many biologists are largely ignorant of modern evolutionary theory. This book is an attempt to supply a reasonably concise volume dealing with organic evolution. It has been written for the reader concerned more with the *process* of evolution than with its products per se. There are no pictures of dinosaurs, no taxonomic descriptions of organic diversity, and no discourses on the history of evolutionary thought.

We have assumed that our readers have reached at least that level of biological sophistication attained in university courses in biology and chemistry. We hope that the book will serve as a challenging text for an undergraduate course in evolutionary theory, as a basic text to be supplemented with outside reading for a graduate course, and as general reading for biologists in other fields who may wish a brief review of what is known of the process of evolution. Most of the material presented has been used in either undergraduate courses in population biology or courses in advanced topics in evolution.

An attempt has been made to present evolutionary theory as a unified whole. Because we are assuming some familiarity, at least on a casual level, with phenomena such as selection and mitosis, we have felt free to make passing reference to them before they are treated in detail. Life, meiosis, genetic systems, culture, and the like have not been taken for granted. Rather we have attempted to show how these phenomena are themselves the result of evolutionary processes. Necessarily this involves speculation, which we feel to be rewarding and stimulating. It is, however, important that it be recognized as speculation. In some areas, other evolutionists certainly will find our treatment heterodox. In particular, we have deemphasized taxonomic ideas such as species and subspecies, which we feel have restricted the thinking of biologists about evolutionary problems. The term adaptation has been given the relatively inconspicuous role that we feel it deserves. Our reasons are discussed in the final chapter.

We have tried to make our descriptions and discussions as rigorous as possible, except where it becomes absurdly pedantic to avoid taxonomic concepts or the casual use of words such as selection and adaptation. Lapses into what may be termed teleology we regard as teleonomy (relating structures and activities to the functions served in or for organisms). We hope the reader will agree that a somewhat more unified and logical treatment of evolutionary phenomena is possible if a rigid taxonomic framework is not followed. Scientific names used in this book connote kinds of organisms and carry no implications of genetic attributes or phylogeny.

At the end of each chapter is a list of references chosen in part because of their currency and extensive bibliographies. In the text, reference without a direct citation often is made to scientists closely associated with a particular concept or experiment; direct citations can be found in the bibliographies of general papers listed. An extensive glossary includes common biological terms undefined in the text, as well as those used primarily by evolutionists.

In the 10 years since the publication of the first edition of *The Process of Evolution*, significant advances have been made in the biological sciences. What at that time could be described as important breakthroughs in the fields of molecular genetics, neurophysiology, developmental biology, and population biology have today become the foundations for much of the current and most productive research in these disciplines.

Nevertheless, it is still the unifying concept of organic evolution which makes it possible to interrelate and integrate these developments into a single theoretical framework. It should be made clear, however, that this relationship is reciprocal, since in the last 10 years important insights into the evolutionary process have also been gained as a result of new information from other areas of the biological sciences.

With this in mind, we have attempted in this revision to maintain the basic aims of the first edition. It has been our intention to provide not only an updating and clarification of theoretical considerations but also studies from the recent literature which in our opinion best illustrate the topics under consideration.

Our intellectual indebtedness to a very large number of evolutionists will be obvious. We must specifically acknowledge the writings of Edgar Anderson, C. D. Darlington, Theodosius Dobzhansky, Herbert L. Mason, Ernst Mayr, George Gaylord Simpson, G. Ledyard Stebbins, and Sewall Wright, which have had a profound influence in interesting us in evolutionary problems and shaping our thoughts about them.

We should like to thank the following persons who have helped us in many ways in the task of preparing this book: R. I. Bowman, Joseph H. Camin, Verne Grant, P. H. Greenwood, N. K. Johnson, Alan E. Leviton, Harlan Lewis, L. Mintz, J. A. Moore, George S. Myers, R. Quinta, Peter H. Raven, C. L. Remington, Jonathan Roughgarden, R. G. Schmieder, and Robert C. Stebbins.

One or more chapters of either edition were read by Kenneth B. Armitage, William K. Baker, D. L. Bilderback, Marsden S. Blois, Winslow R. Briggs, Howell V. Daly, Ruth R. Ehrlich, Marcus Feldman, M. M. Green, Robert W. Hull, George B. Johnson, Joan Johnston Kendig, Donald Kennedy, Charles D. Michener, Ashley Montagu, Robert M. Page, John F. Pelton, David D. Perkins, Timothy Prout, Peter H. Raven, David C. Regnery, G. G. Simpson, Robert R. Sokal, Michael E. Soulè, John H. Thomas, Robert P. Wagner, Norman K. Wessells, and Charles Yanofsky. Theodosius Dobzhansky read and criticized the first edition. This generous donation of time and effort on the part of all these individuals is deeply appreciated. Many of the subjects considered here have been discussed in detail with colleagues and students in our Population Biology Seminar. The authors accept full responsibility for all errors of fact and interpretation, as they have not always been able to adopt the suggestions of the reviewers.

Paul R. Ehrlich
Richard W. Holm
Dennis R. Parnell

THE
PROCESS OF
EVOLUTION

PART ONE

ORGANISMS: ORIGIN AND FUNCTION

The Process of Evolution *is divided into four major sections. These deal with (1) the origin and functioning of organisms, (2) the properties of populations of organisms, (3) how differentiation of populations occurs and results in major patterns of variation, and (4) the evolution of man and his culture (the latter of course, includes evolutionary theory).*

This initial section deals to a large extent with subjects that often are taken for granted in discussions of evolution, and the organization may appear unusual. Nevertheless, we believe it to be important to approach the study of evolution in a levels-of-organization framework. The basic properties of life are themselves products of an evolutionary process. In these first four chapters, certain properties of living systems critical to the study of evolution are outlined. Emphasis is given to the ways in which a continuity of information is maintained in the cyclic stream of life and to ways in which this information is elaborated. Where possible, intelligent speculation about ways and means of ancient transformations and origins of ubiquitous mechanisms is included. Such speculation, no matter how inaccurate it may turn out to be, serves to remind us that such things as photosynthesis, DNA, meiosis, dominance, and cellular differentiation did not always exist in their present forms. No attempt has been made to give an encyclopedic account of these major areas of biological thought; rather we have tried to set the stage for the consideration of the process of evolution in organisms as we know them today.

CHAPTER **ONE**

THE ORIGIN OF LIFE

It is axiomatic that all life originates from preexisting life so that the question of how life began in the first place is rarely considered. The ancients solved the problem with the idea of the spontaneous generation of such complex organisms as flies and mice from nonliving matter. But these, as well as more sophisticated ideas, were laid to rest by the experiments of Redi and Pasteur. As a result, however, the basic question—How did life originate?—was brought into focus. Without some type of spontaneous generation, how can the origin of the myriad entities which are called "alive" be explained? Often this problem has been confused by the tendency to equate life with the properties of highly complex organisms. The contrast between a bird and a rock or between a bacterium and an iron filing is self-evident. Indeed, it is so striking that the difference between the living and the nonliving could be misconstrued as one of kind rather than one of degree.

The great majority of biologists believe that there is no significant discontinuity between the living and the nonliving, even though they may not agree on a definition of "life" or even upon the "properties" of life. Many obstacles can be avoided merely by viewing life as a special property of matter at a certain stage of complexity and not attempting a rigorous definition. At least it can be said that living systems handle energy in a regulative manner so as to establish an energy potential between an organism and its environment; certainly one of the most fundamental properties of life is the continuous and directed movement of electrons among the complex molecules of which living organisms are made. It is important to note that these energy transformations are precisely controlled. The regulated release of an amount of energy, which uncontrolled would cause a mild explosion, results in what is thought of as life. In addition, living systems have the property of reproducing themselves. Thus when the problem of the origin of life is considered, answers must be sought to the questions of how the systems that extract and utilize energy from the environment could arise and how they could replicate. There seems to be a sort of twilight zone between the extremes of living and nonliving, an area in which these terms may not be applicable. In this zone of viruses, nucleic acids, and specialized sorts of colloids some of the answers to our questions may be found.

Life as it is known today arose sometime in the $1\frac{1}{2}$-billion-year interval between the formation of the earth (currently estimated to have occurred 4.7 billion years ago) and the formation of the oldest known alga-like fossils dated at 3.2 billion years. It does not seem likely that the spontaneous origination of life could be observed at the present time. If new life did appear spontaneously, it would probably quickly be eliminated by modern heterotrophic organisms even if the present environment were favorable. The situation was quite different under the conditions that probably obtained on the earth billions of years ago.

THE EARLY STAGES

The *sine qua non* of the production of life *as we know it* is the development of certain organic compounds—compounds built around carbon and consisting, in the main, of this element joined in diverse configurations with nitrogen, oxygen, hydrogen, phosphorus, and sulfur. The early stages in the chemical evolution of the earth's surface must have been characterized by the presence of much simpler inorganic molecules. The questions that immediately arise are how these were combined to produce the more complex compounds found in living systems and what the source of energy for such transformations may have been.

There is considerable evidence to support the thesis that the atmosphere of the early earth was chemically reducing, although its exact composition is the subject of much speculation. For example, Oparin has suggested that it was made up principally of methane, water vapor, ammonia, and hydrogen. The behavior of these substances under a variety of conditions has been studied by Miller, who placed mixtures of these gases in an apparatus (Fig. 1-1) in which they could be exposed to electrical discharges. Circulation was produced by boiling water on one side of the apparatus and condensing it on the other. Chromatographic analysis at the end of the experiments revealed the presence of amino acids (the subunits of protein structure), hydroxy acids, and aliphatic acids—three basic

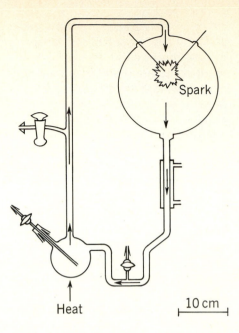

Figure 1-1 Spark-discharge apparatus. *[After S. L. Miller, Ann. N.Y. Acad. Sci., vol. 69 (1957).]*

Heat   10 cm

types of organic molecules. The amino acids included glycine and alanine (the most common amino acids in proteins), aspartic acid, and glutamic acid. It is interesting to note that α-alanine predominated over β-alanine in these experiments; modern proteins contain only α-amino acids (α-amino acids have the NH_2 and COOH groups both attached to the same carbon atom). Similar results have been obtained with reducing atmospheres of differing composition and by employing different discharge techniques.

Miller argues that the same types of compounds would have been produced under the influence of ultraviolet light and electrical discharges if the primitive earth had had a reducing atmosphere. Contending that organic compounds would not be produced if oxidizing conditions were present, he points out that if amino acids (and other organic compounds) are necessary for life, the presence of life on earth is evidence for a primitive reducing atmosphere. The recent discovery in meteorites of organic compounds, including possibly amino acids, suggests that conditions similar to those proposed for the early atmosphere of the earth exist today elsewhere in the universe.

Free oxygen, which first appeared some 800 million to 2 billion or more years ago and which gives the present atmosphere its oxidizing character, came from two sources: part had a photochemical origin (from water undergoing photolysis in the upper atmosphere, with the hydrogen escaping into space); the rest was produced photosynthetically by living organisms (the main source today).

In addition to the formation of organic compounds as outlined above, simple organic compounds are formed by the interaction of water vapor with carbides in magma (molten rock and gases) brought to the surface by volcanic activity ($3Fe_mC_n + 4mH_2O = mFe_3O_4 + C_{3n}H_{8m}$). Calvin and others have shown in experiments with ionizing radiations (of the sort that would be produced by radioactive materials or by cosmic rays) that, in the presence of molecular hydrogen, partial reduction of carbon dioxide can occur. Further irradiation of aqueous solutions of the substances produced (formic acid, formaldehyde) leads to the formation of such compounds as oxalic acid or acetic acid. Eventually molecules of two-carbon compounds (acetic acid) may combine to produce a four-carbon compound (succinic acid). In these experiments amino acids are also produced.

In other experiments, Fox has shown that heating dry amino acid mixtures results in the formation of synthetic polypeptides (proteinoids). Proteinoids in water tend to form spherules of varying size and shape, depending upon their interaction with substances mixed with them. In some respects these spherules resemble coacervates and other cell models. Among other things, the proteinoid spherules retain their integrity for rather long periods and are not destroyed by high-speed centrifugation.

ORIGIN OF SELF-REPLICATING SYSTEMS

Thus it can be seen that there are diverse ways in which organic compounds might have been produced on the primitive earth. It is not unreasonable to assume, therefore, that the primitive ocean was comparable to a thin soup of organic materials. There is little agreement on how the first **self-replicating** systems developed in this soup. Obviously, what was first required was the selective construction of molecules. Calvin has pointed out that the phenomenon of autocatalysis is a selective process. Autocatalysis occurs whenever the *product* of a chemical reaction has the property of influencing catalytically the rate of its *own* formation. There follows a progressive buildup of products in a sequence of increasingly complex compounds formed from simpler ones. An early selection for complexity must have gone on in this way in the organic soup.

Autocatalytic reactions are only partly analogous to a self-replicating living system. No presently known substance, when isolated, will replicate itself. Only *systems* have the ability to replicate. The living systems familiar to us are composed of proteins and nucleic acids, together with some means of energy mobilization. Polypeptide chains, the backbones of protein molecules, are formed by the linkage of amino acids in linear series. The linkage is accompanied by the elimination of water as the amino acid chain lengthens. The bonds between the amino acid units are known as peptide bonds. The spontaneous formation of even a small protein in a solution of amino acids requires outside energy and is a very improbable event. But in the absence of free oxygen and predatory organisms, the life of an amino acid soup could be extremely long, long enough to turn the improbable into the probable. (The chance of being struck by lightning in a 70-year life span is very slight, but if one were to live for 700 million years it would become almost a certainty.)

However, as Wald points out, the spontaneous generation of protein molecules is opposed by their tendency toward spontaneous dissolution. Indeed, the equilibrium point in the reversible, spontaneous protein-generation reaction lies on the side of dissolution rather than synthesis. Wald suggests, nevertheless, that molecules seem to be able to resist dissolution both through large size and through aggregation with other molecules. Proteins may be an unstable midpoint, subject either to dissolution into their component amino acids or to the formation of more stable aggregates. The first "organisms" may well have been the result of the formation of larger and larger aggregates.

ENERGY SOURCES

Ultraviolet light usually is considered the chief source of energy for early synthetic processes. With simple molecules, only very short wavelengths are absorbed, but as more complex molecules appear, absorption of longer ultraviolet wavelengths takes place. As the earth evolved its thick atmospheric layers, including water and ozone, ultraviolet light of short wavelengths could no longer penetrate to the earth's surface to be used as an energy source. The appearance of colored pigments (e.g., porphyrins, mentioned below) made possible the absorption of energy in the visible spectrum. However, whatever the source of energy, there is a considerable gap between the absorption of a quantum of energy and its mobilization for use in biological processes.

The problems of how energy is used for protein synthesis and the conditions under which this may occur are particularly vexing ones. In present-day biological systems the enzymes responsible for the mobilization of energy and for the synthesis itself *are* proteins. Thus to postulate the functioning of such systems in the formation of the first proteins embroils one in a "chicken or egg" dilemma. It has been suggested that, in the absence of proteins, other substances (e.g., clays) may have served as catalysts, since many of the known enzymatic phenomena are fundamentally molecule-surface reactions. This raises the question of how proteins subsequently came to assume this function. At least it can be said that the surface phenomena of clays and surface configurations of proteins have certain aspects in common.

Chemical energy for synthesis in modern biological systems involves **organophosphate bonds.** The energy released upon cleavage or transfer of these bonds is regulated by a complex system of catalysts (enzymes plus their coenzymes); the characteristics of these sets of reactions are unique to living systems. They change rate in response to changes in concentration of product, are dependent upon physical conditions (temperature, pressure) in a fashion distinct from nonliving systems, and demonstrate an efficiency of energy transfer not often equaled in inorganic reactions. Such reactions in living systems are referred to as *biological oxidations*. The transfer of energy needed by a heterotrophic organism is almost universally mediated by a single type of organophosphate bond, which occurs in the "energy-rich" compound called adenosine triphosphate (ATP). ATP mediates energy transfer in a biological reaction by releasing one phosphate group and becoming adenosine diphosphate (ADP). The latter still includes one energy-rich phosphate bond and may be rephosphorylated into ATP. It is this rephosphorylation (called oxidative phosphorylation) which is controlled by the system of enzymes referred to above. Although these enzymes also require the cleavage of ATP bonds to function, more ATP bonds are formed than are used. The extra energy is derived from the energy stored in the glucose (or other) molecule upon which the enzymes are acting directly. By this interlocked series of reactions, chemical energy supplied to the organism as molecules of carbohydrate, lipid, or protein (which the organism cannot use as such) is transformed into ATP-bond energy that the organism can use.

In nonliving systems, energy transfer by molecular degradation yields smaller molecules plus much heat. In living systems, the products are "high-energy" organophosphate bonds, smaller molecules, and surprisingly little heat. In fact, one of the salient features of the living energetic machinery is the efficient coupling between energy-yielding and energy-storing reactions.

Of course the ultimate source of energy in existing organisms (except chemosynthetic bacteria) is that of the sun trapped by photosynthetic organisms and eventually stored in phosphate bonds by a related process involving photophosphorylation in the photosynthetic organism or by synthesis into carbohydrate with phosphorylation. Many important catalysts in both photosynthesis and biological oxidations are colored compounds involving metal ions (Fe, Mg) and the organic substances known as porphyrins. Calvin has diagrammed (Fig. 1-2) how such important biological materials might have arisen in the course of chemical selection involving autocatalysis. In the sequence from simpler to more complex molecules, later stages are catalysts for succeeding stages. Since the use of porphyrins by nonphotosynthetic organisms is widespread, Calvin feels that the presumably random variation involving small changes in the porphyrins led eventually to the construction of chlorophyll and the origin of **photosynthesis.** In addition, Granick believes that all the colored compounds in the sequence that leads to chlorophyll might have had the same function as chlorophyll. In the early stages, metallic ions, present as constituents of minerals, might have served to catalyze the same reactions they now catalyze as constituents of metalloenzymes.

Photosynthesis is the result of a complex series of reactions. Some of these can take place in the dark, whereas others can occur only with illumination. Most of the many reactions usually included under the rubric photosynthesis, in the broad sense, are actually dark

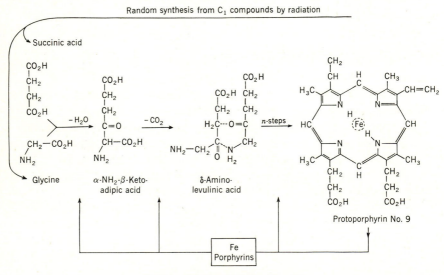

Figure 1-2 Possible steps in the synthesis of porphyrin compounds from molecules produced randomly under the influence of radiation. Iron may act as a catalyst at points marked by arrows but is a much better catalyst when combined with porphyrin. Thus the production of protoporphyrin 9 is facilitated by the presence of protoporphyrin 9. *[From M. Calvin, Evolution, vol. 13 (1959).]*

reactions, involving the addition of carbon from CO_2 to —C—C—C— chains. These dark reactions can be carried out by most cells. It seems likely that many of these reactions evolved independently, perhaps earlier than strictly photosynthetic reactions. The light reaction leads to the production of a reducing agent and high-energy phosphate, adenosine triphosphate (ATP). The reducing agent usually is hydrogen but occasionally is a phosphorus compound. These substances then operate the carbon reduction cycle, and hexose sugar molecules are produced. This energy is mobilized, in ways which are imperfectly understood, so that excited chlorophyll transforms other molecules to produce the reducing agent and the ATP. Thus, typically, photosynthesis involves photophosphorylation, i.e., the transformation of light into the "energy currency" of phosphate bonds.

It probably never will be possible to say with certainty whether the coupling of colored compounds with biosynthetic processes took place before or after the appearance of what today would be called living organisms. Calvin believes that the final step in the development of modern photosynthesis, the production of oxygen, did not take place until relatively late in the sequence of events. Therefore reactions like those of some modern organisms, which are photosynthetic but do not emit oxygen, came first.

ORIGIN OF STRUCTURE

In the light of the above discussion, it is not overwhelmingly difficult to imagine how the substances required for the processes we think of as metabolism could have arisen. However, living systems are not fluid, structureless entities. Generally they have a characteristic and complex organization of the matter constituting this energy-conversion mechanism. Now the factors involved in the evolution of structure as well as of function must be considered. In the sea, the original molecules probably were dispersed as a rather uniform colloidal suspension. However, in colloids of different substances, semiliquid colloidal gels or coacervates are formed, and it might be expected that these arose as the organic soup became increasingly complex. From the work of physical chemists, much is known about the behavior of coacervates. They often do not form as a continuous layer but separate out of the equilibrium liquid (thus left colloid-poor) in the form of discrete droplets.

Recently, by polymerizing the nucleotide adenine, Oparin has made coacervate droplets which selectively absorb and concentrate small molecules from the environment. Once inside, the molecules are changed and incorporated. Such droplets are called **protobionts.** According to Oparin, daughter droplets, each with its parent's attributes, could form by the breaking up of the protobionts, under the influence of wave action. Once this has happened, "prebiological natural selection" could operate and the more successful drops; those with greater catalytic ability to incorporate materials from their surroundings would grow at the expense of the less successful. One can also imagine that accidental fusion of droplets might occur. Should the droplets have different compositions, a sort of protosexual recombination process would occur. Thus one can see in coacervates many properties that would qualify them as links in a chain leading to the structure of life as now known. They are clearly separated from their environment, have internal structuring, absorb matter from their environment, and have sufficient multiplication and "recombination" to permit the operation of natural selection.

It seems clear that, in the vast stretches of geologic time, mechanisms such as those outlined above (and others as yet undiscovered) produced the ancestors of the living systems we know today. Indeed, when one pictures the vast oceans, lakes, and hot springs rich in organic compounds and presenting a wide variety of conditions of temperature, light, salt concentration, and physical substrate (crystals, clays), it is difficult not to believe that living systems developed more than once. It is not unlikely that modern organisms are the descendants of the victor in a fierce energy war among the early "organisms."

ORIGIN OF THE GENETIC CODE

The level of complexity of the hypothetical ancestral organism discussed to this point does not involve a system by means of which the entity could be replicated as a unit. Splitting by fission may or may not result in the formation of equal parts; in fact, it might be imagined that occasionally one of the parts would lack a component essential for the maintenance of life. At this stage there was no system of heredity, no genetics, which would ensure the continued production of functional entities. The first principle of genetics is "like begets like." This is not the result of a great immutable "law of nature" but rather the functioning of a complex system for transmitting genetic "information," the information needed to construct a new organism. Without such a system, it seems certain that life would not have evolved beyond the level of coacervate droplets. The efficiency of the transmission apparatus has been a major factor in determining the limits of intricacy of living entities.

In the $1\frac{1}{2}$ billion years of chemical evolution that preceded the evolution of life, many systems of transmitting information may have arisen and been eliminated in a selectional process. It is clear that the hereditary system must have been coupled to the synthetic and energy-converting systems. Therefore it is no surprise to note that the substances involved in the hereditary system, the nucleic acids, have adenosine phosphate as a building block. The system found in most cellular organisms is based on coding information in two macro-molecular nucleic acids, ribonucleic acid (RNA) and deoxy-ribonucleic acid (DNA).

Oparin has proposed that such a hereditary system might have arisen at the protobiont stage. Even at these early stages, clans of protobionts might have had some ability to organize the polynucleo-tides, later to become DNA and RNA. Such organization might have favored the polymerization of amino acids which could have led to the formation of primitive enzymes. This would represent a far more stable and reproducible condition than random protein synthesis. Such clans of protobionts would have been at a strong selective advantage.

In general outline, the functioning of the genetic mechanisms which eventually evolved utilizing nucleic acids is as follows. The units of information, known as genes, are coded into the structure of giant self-replicating molecules of DNA. These molecules, repro-duced and passed from generation to generation, are the master blueprints from which all living organisms develop. They maintain the continuity of life. Slight changes in these blueprints are also repro-duced and are responsible for the variation that permits evolution. DNA molecules are chains made up of four **nucleotide units:** deoxyguanylic acid, deoxycytidylic acid, deoxyadenylic acid, and deoxythymidylic acid. Similarly, RNA molecules are made up of four nucleotides: guanylic, cytidylic, adenylic, and uridylic acids. Each nucleotide group contains a pentose group. The backbones of the DNA and RNA molecules are made up of the pentose sugars linked by the phosphate groups. Attached to each sugar residue in this regular chain is one of the five bases: either a purine (adenine or guanine) or a pyrimidine (cytosine and thymine in DNA or uracil in RNA). The configuration of the DNA molecule is a double helix of two sugar-phosphate backbones whose bases face each other—they are often spoken of as paired. Because their chemical properties differ, only certain pairs of bases are possible within the space between the two sugar-phosphate backbones: adenine with thymine and cytosine with guanine. This helical model of the structure of DNA has been repeatedly confirmed by the results of x-ray diffraction and other studies of the physical properties of the molecule.

When chromosome duplication takes place in preparation for

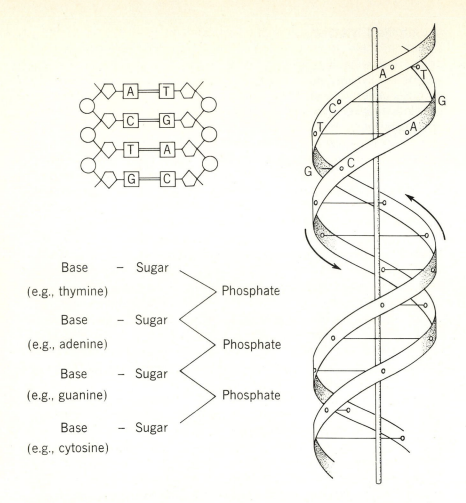

Figure 1-3 The structure of DNA. Upper left, complementary chains of nucleotides. Letters, bases; five-sided figures, pentose sugars; circles, phosphate groups; double and triple lines, hydrogen bonds. Lower left, detail of segment of one strand. Right, model of double helix in which the nucleotide chains are arranged. Ribbons, phosphate-sugar backbone; bars, paired bases joining backbones. This double-helix configuration is known as the Watson-Crick model. *[After Sinnott, Dunn, and Dobzhansky, Principles of Genetics, McGraw-Hill, 1958.]*

cell division, the complementary helices unwind, and each helix serves as a chemical template on which DNA precursors attach to form a new complementary strand and reestablish the double helix. This is the mechanism by which genetic information, coded in terms of the sequence of nucleotides in the DNA strand, is passed from cell to cell and (through the gametes) from organism to organism.

READING THE CODE

DNA serves as a template against which another complementary strand of DNA or one of the three types of RNA can be assembled: ribosomal RNA, messenger RNA, and transfer RNA, which play different roles in protein synthesis. Protein synthesis takes place at the ribosomes of the cell, organelles composed of protein and ribosomal RNA (*r*RNA). The ribosomes, which consist of two unequal, more or less spherical bodies, contain the vast majority of the RNA in the cell. The genetic code is transcribed from the DNA sequence to the RNA sequence of messenger RNA (*m*RNA), which then carries it to the ribosome. Transfer RNA (*t*RNA) functions in the assembly of amino acids into polypeptides.

RNA is synthesized against the DNA template through the action of RNA polymerase, an enzyme with a molecular weight of about $1/2$ million. This enzyme is constructed of five polypeptide chains, four of which form the core enzyme and a fifth σ (sigma) which easily dissociates from the others. The σ chain is responsible for recognizing the initiation sites along the DNA molecule. After the initiation of RNA synthesis occurs, σ dissociates while the core enzyme continues to be operative. Termination of RNA synthesis depends on the presence of another specific protein ρ (rho), whose mode of action remains to be elucidated.

The code, transcribed from the DNA sequence to the RNA sequence, must be able unambiguously to control the sequential positioning of 20 common amino acids into particular linear sequences. It has been demonstrated rigorously that the DNA code consists of nonoverlapping triplets of nucleotide bases called *codons* (Table 1-1). Each codon corresponds to an amino acid. Triplet sequences corresponding to all 20 of the amino acids have been found. In fact, since there exist 64 (4^3) possible triplet sequences, many amino acids are coded by more than one triplet. The code is thus redundant. In all, 61 of the 64 triplets have been shown to code for amino acids. The three remaining triplets are called nonsense codons and are used for punctuation. That is, they designate separation points along the DNA sequence between regions coding for different polypeptide chains. Insertion of these nonsense triplets gives the code its nonoverlapping nature.

How do the amino acids "read" the messenger RNA sequence so that they condense into proteins containing the proper order of amino acid residues (see Fig. 1-3)? Amino acids become bound to the relatively small soluble *t*RNA molecules *before* the acids are linked together into proteins. There are different *t*RNA molecules for each amino acid. In one part of the molecule is a sequence of

TABLE 1-1. THE RNA NUCLEOTIDE TRIPLET CODES SPECIFYING AMINO ACIDS* †

Triplet Amino Acid	Triplet Amino Acid	Triplet Amino Acid
UUU UUC} Phenylalanine	CGU CGC CGA CGG} Arginine	GCU GCC GCA GCG} Alanine
UUA UUG CUA CUU CUG CUC} Leucine	CAU CAC} Histidine	GGU GGC GGA GGG} Glycine
	CAA CAG} Glutamine	
UCU UCC UCA UCG AGU AGC} Serine	AUU AUC AUA} Isoleucine	GAU GAC GAA} Aspartic acid
	AUG} Methionine	GAA GAC} Glutamic acid
UGU UGC} Cysteine	ACU ACC ACA ACG} Threonine	
UGG} Tryptophan		UAA UAG UGA} Termination signal
	AAU AAC} Asparagine	
UAU UAC} Tyrosine	AAA AAG} Lysine	
CCU CCC CCA CCG} Proline	GUU GUC GUA GUG} Valine	

*   A, Adenine; U, Uridine; G, Guanine; C, Cytosine.

†   Adapted from Paul R. Ehrlich, Richard W. Holm, and Michael E. Soulé, *Introductory Biology*, copyright McGraw-Hill Book Company, 1973; used by permission.

nucleotides that determines with which amino acid it can react, and in another part is the sequence (a triplet) that determines the position on the *m*RNA template to be assumed by the *t*RNA unit (Fig. 1-4). The compounds of *t*RNA and amino acids are formed with energy supplied by ATP. The *t*RNA–amino acid units then find their places on the long *m*RNA template. The amino acids are brought into close proximity and condense (with the aid of an enzyme) to form a protein with the proper sequence of amino acid residues.

One can well imagine that this system is a far cry from the first system of transmitting genetic information. For example, the DNA

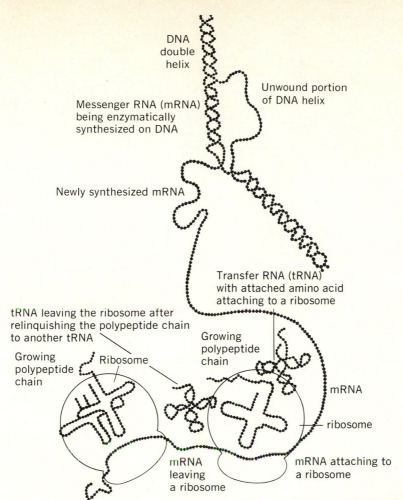

DNA double helix

Unwound portion of DNA helix

Messenger RNA (mRNA) being enzymatically synthesized on DNA

Newly synthesized mRNA

Transfer RNA (tRNA) with attached amino acid attaching to a ribosome

tRNA leaving the ribosome after relinquishing the polypeptide chain to another tRNA

Growing polypeptide chain

Growing polypeptide chain

Ribosome

mRNA

ribosome

mRNA leaving a ribosome

mRNA attaching to a ribosome

Two tRNA molecules attached to adjacent mRNA base triplets (codons) on a ribosome; the growing polypeptide chain is being transferred from left to right

tRNA attached to a triplet (codon) of mRNA bases on the ribosome; at its opposite end, the tRNA has added its amino acid to the polypeptide chain

Figure 1-4 Diagrammatic theoretical representation of the process of protein synthesis. *m*RNA is synthesized on a DNA molecule; then imprinting a portion of its code on a strip of *m*RNA, the molecule attaches to a ribosome, which proceeds to read off the message, codon by codon. At each codon in the *m*RNA strip a *t*RNA molecule plugs in momentarily. The *t*RNA molecule carries a specific amino acid at one end and at the other end an anticodon which fits only the codon for that particular amino acid. Thus, one by one, following the exact order laid down by the DNA code, the proper amino acids are brought into line and formed into a polypeptide chain (protein). As the ribosomes move along the strip of *m*RNA, other ribosomes attach to the disengaged end and start down the strip, making their own strands of protein. The *m*RNA is always read in a specific direction, and proteins are always synthesized starting at their N-terminal ends. *(From P.H. Raven and H. Curtis, Biology of Plants, Worth Publishers, Inc., 1970.)*

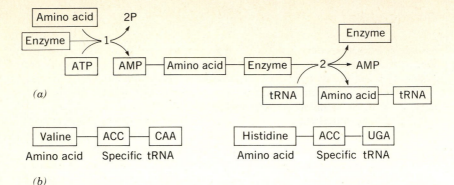

(a)

(b)

Figure 1-5 Amino acid activation. *(a)*. The attachment of an amino acid to a specific *t*RNA molecule begins with the linkage of the amino acid to a specific enzyme. (The energy required for this reaction is obtained from ATP.) Following the activation of the amino acid, *t*RNA specific for both the amino acid and the enzyme attaches to the amino acid, and the enzyme is released. *(b)*. Each amino acid has a separate *t*RNA molecule. One end of the molecule is the attachment site for the amino acid (ACC). The other end of the molecule has a specific triplet sequence (codon) that determines the position on the *m*RNA molecule to be assumed by the *t*RNA unit (CAA for valine and UGA for histidine). *(From P.B. Weisz, The Science of Biology, copyright McGraw-Hill Book Company, 1971; used by permission.)*

remains as a master template, reducing the possibilities for error that would be inherent in a system in which copies are made from copies. Similarly, the short *t*RNA molecules are highly specialized to accomplish the proper positioning of the protein subunits. The manifold interactions of DNA and RNA in the organism are not confined to processes concerning transmission of genetic information and protein synthesis. There is speculation that RNA functions in those higher-organism systems that involve training and memory. The mental properties of primates may eventually be described in terms of fundamental chemical properties recognizable in the simplest cells and organisms.

In viewing the complexities of function found in the cells of present-day organisms—the highly specialized organelles, the very efficient system for utilizing high-energy phosphate bonds, the precise mechanisms for distribution of genetic information and for cell division—one may find it hard to believe that such complexity ever arose from the coacervate stage previously described. It is like looking at a unicellular organism and a man and trying to imagine one as the ancestor of the other without knowledge of any intermediates. It should be remembered that the time available for the evolution of the coacervate into the complex cell was of the same magnitude as that available for the journey from protistan to man.

SUMMARY

Life is a complex energy-matter nexus whose origin can be explained logically in general terms. Important events in the origination of life certainly were the development of organic compounds, their segregation into structural entities, the origin of energy-mobilizing cycles, and the development of systems for self-replication. These events presumably were partially synchronous and controlled by a sort of protoselection. The present system for self-replication utilizes information coded in macromolecules of nucleic acids that control protein synthesis.

Life may be considered to be an aspect of the matter-energy continuum characterized by incessant replication. Perfect replication is impossible, and therefore natural selection is inevitable since some replicates will be more fit than others.

REFERENCES

Barghoorn, E. S.: The Oldest Fossils, *Sci. Am.*, **224**(5):30–42 (1971). A discussion of the nature of the oldest known fossils and the implications these discoveries have for understanding how life originated on earth.

Calvin, M.: Round Trip from Space, *Evolution*, **13:**362–377 (1959). A good brief discussion of the problem of the origin of life.

Miller, S. L.: The Formation of Organic Compounds on the Primitive Earth, *Ann. N.Y. Acad. Sci.*, **69:**260–275 (1957). Gives details on the atmosphere experiment. This article is in a number entitled Modern Ideas on Spontaneous Generation, which contains several interesting papers.

Needham, A. E.: The Origination of Life, *Q. Rev. Biol.* **34:**189–209 (1959). A stimulating discussion in very broad terms.

Oparin, A. I., A. E. Braunshtein, A. G. Pasynskii, and T. E. Pavolvskaya (eds.): *The Origin of Life on the Earth*, Pergamon, New York, 1959. This symposium volume contains a large number of important papers, most of which are highly technical.

Oparin, A. I.: *Genesis and Evolutionary Development of Life*, Academic, New York, 1968. The latest revision of the classic work on the origin of life.

Ponnamperuma, C.: *The Origins of Life*, Dutton, New York, 1972. A well illustrated account by a leading exobiologist.

Sagan, Carl: On the Origin and Planetary Distribution of Life, *Radiat. Res.*, **15:**174–192 (1961). A summary paper with extensive bibliography.

Watson, J. D.: *Molecular Biology of the Gene*, 2d ed., Benjamin, New York, 1970. A comprehensive survey of the state of knowledge in the field of molecular genetics by one of the key workers in the field.

CHAPTER TWO

UNITS OF REPLICATION

One of the most dramatic results of modern scientific technology and the development of the electron microscope has been the revival of interest in cytology. The increased resolution of the electron microscope has revealed structures of amazing complexity where none was known to exist and, indeed, has brought form and function together at the level of macromolecules and their aggregates. From our point of view, these results are particularly interesting; from them we can hypothesize that the membrane systems of which cells are largely composed may have been the inevitable result of the mixture of large and complex molecules, such as lipids and proteins, before the origin of life itself, as suggested in Chap. 1. These macromolecular structures, originating by chance, may be similar to the membranes seen to be combined in cells in a variety of ways. The basic cellular constituents are common to plants and animals, providing a structural ground plan for all life except in the most highly specialized cells or organisms. A brief review of cell structure is given below to provide background for the later discussion of the evolution of genetic mechanisms and systems.

Whatever may have been the origin of cells, both plant and animal cells show such striking similarity in structure as to suggest that either there is a common ancestral type or, with life as we know it, only one basic type of structure (Figs. 2-1 and 2-2) compatible with function. The chemical composition of cells is relatively easy to determine, and many physical properties of cells and their constituents can be measured. However, it is in the organization of these chemicals that the unique property of life and the cellular state is achieved. The chief structural units of cells are large molecules of proteins, carbohydrates, and lipids. Interspersed with the physical framework that results from the ag-

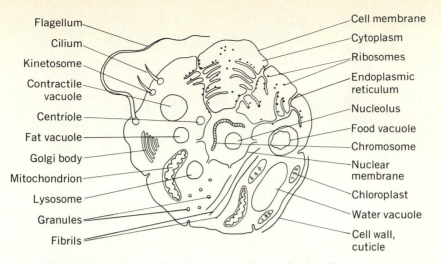

Flagellum
Cilium
Kinetosome
Contractile vacuole
Centriole
Fat vacuole
Golgi body
Mitochondrion
Lysosome
Granules
Fibrils

Cell membrane
Cytoplasm
Ribosomes
Endoplasmic reticulum
Nucleolus
Food vacuole
Chromosome
Nuclear membrane
Chloroplast
Water vacuole
Cell wall, cuticle

Figure 2-1 A generalized cell with parts as seen under the electron microscope (some components enlarged or simplified). *(From P.B. Weisz, The Science of Biology, copyright McGraw-Hill Book Company, 1971; used by permission.)*

gregation of these molecules, and sometimes actually associated with it, are the myriad types of smaller molecules: soluble proteins, amino acids, vitamins, inorganic constituents, etc.

STRUCTURE OF CELLS

Both plant and animal cells are bounded by a membrane, called the **plasma membrane,** which has the important property of being differentially permeable. Physical, chemical, and biological studies of this membrane indicate that it is a complex structure composed of protein and lipid molecules. Formerly most biologists believed that there was a basic unit membrane, interpreted as having two layers of protein enclosing an inner layer of lipid. Recently, however, doubt has been raised about the universality of this unit-membrane concept. It is now clear that plasma membranes vary in both structure and physiological properties.

    Free cells and cells in tissues seem always to have additional layers outside the plasma membrane. In the animal cell, these layers are composed largely of molecules of proteins and sugars, and their integrity depends upon the calcium balance of the cell environment. Plant cells are rather different in two respects. First, the outer layers (or **cell wall**) of plants are composed mainly of carbohydrate molecules. Glucose residues arranged in long chains form the most

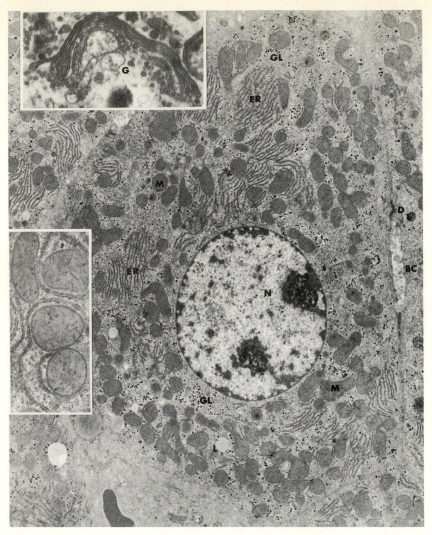

Figure 2-2 A liver parenchymal cell demonstrates many of the specialized organelles found in animal tissue including a nucleus *(N)*, mitochondria *(M)*, rough endoplasmic reticulum *(ER)*, and lysosomes *(L)*. Cytoplasmic regions for glycogen storage *(GL)* can be identified by the abundant electron-dense particles. Bile canaliculi *(BC)* are lined with microvilli and represent an intercellular canal where cellular adhesion is interrupted. At other places along the cell surface disk-shaped desmosomes *(D)* are shown in transverse section (11,500×). One inset includes an enlarged view of the mitochondria and endoplasmic reticulum of the liver parenchymal cell (33,000×); the other demonstrates an active Golgi *(G)* complex as best seen in a young spermatogonia cell (46,000×). *(Photograph courtesy of J. Belton.)*

important constituent, cellulose. Other carbohydrates, as well as fatty acid substances (suberin, cutin, etc.), also are associated with this wall. Contiguous cells in tissues are cemented together by a middle lamella, which is pectic in nature and relies upon calcium for rigidity. The second important difference between plant and animal cells is that cells in plant tissues are in organic connection through strands of cytoplasm called **plasmodesmata.** Electron micrographs have showed that most instances of so-called plasmodesmata in animals do not involve continuity of protoplasm. That is, across the strands that look like plasmodesmata, there is a pair of plasma membranes. It is as if the tissues of Metazoa were composed of cells stuck together, whereas those of plants result from the more or less incomplete partition of "protoplasm." The boundary between animal cells may be exceedingly complex. Structural and chemical properties of the intercellular region suggest that specialization of the periphery of the cell may play an important role in cellular differentiation.

The most conspicuous structure within most cells is, of course, the nucleus. Recent electron micrographs show that the nucleus is bounded by a pair of membranes, and the term **nuclear envelope** is used to refer to both. The inner membrane appears to surround the nuclear contents like a sack. The outer membrane, however, is clearly continuous with a membrane system that permeates the cytoplasm to a greater or lesser extent. This membrane system of tubes, canals, and vesicles (the amount and type depending upon the nature of the cell and its state of activity) is called the **endoplasmic reticulum.** In some cells, there is evidence that the membranes of the endoplasmic reticulum are continuous with the plasma membrane and the outer membrane of the nuclear envelope. Should this prove to be so, the nuclear contents would be topologically outside of the cell, a situation which would have bearing upon the origin of cell organelles (see below). The nuclear envelope is perforated by pores, of rather complex organization, through which nuclear-cytoplasmic interchange presumably takes place.

Electron micrographs of the cytoplasm around the endoplasmic reticulum show it to be far from homogeneous. The tubes of which the endoplasmic reticulum is composed often are associated with small dense granules in synthetically active cells. These are absent where the tubes are continuous with the plasma membrane. Occasionally these granules are found in the intervening cytoplasm. They are known as **ribosomes** and, as mentioned in Chap. 1, are concerned with protein synthesis. (Particles known as microsomes which have been studied by physiologists appear to be artifacts: aggregations of ribosomes and endoplasmic reticulum that appear when the cell is fractionated.) Ribosomes do not occur where the endoplasmic reticulum is continuous, with more or less flattened, concentrically arranged vesicles making up the **Golgi bodies** of both plant and animal cells. It is thought that Golgi bodies arise from the endoplasmic reticulum, with which they appear to be associated. The Golgi bodies apparently function as collecting and packaging centers for the cell. Enzymes formed in the endoplasmic reticulum collect in the Golgi bodies and are packaged into vesicles that bud off from them. In addition, **lysosomes,** membrane-bound spheres which are found in some animal cells, contain digestive enzymes and fuse with vacuoles, emptying their contents into them. **Spherosomes** in plant cells are structurally similar and appear to perform the same digestive function.

Scattered among the tubes and vesicles constituting the endoplasmic reticulum are the **mitochondria** of plant and animal cells. These spherical or tube-shaped structures also have a double-membrane boundary, the inner membrane being thrown into a series of convolutions forming lamellae or villi. The mitochondria are the site of most of the reactions involved in cellular respiration, including the formation of adenosine triphosphate (ATP).

Closely related to the mitochondria structurally, but found only in plant cells, are the **chloroplasts.** Like the mitochondria, they are the site of important reactions providing energy for the cell (indeed for nearly all life). Chloroplasts have a lamellar structure, and upon the alternating layers of lipid and protein molecules are found layers of special pigments such as the various types of chlorophyll, carotenoids, and others (depending upon the plant group studied). Light energy absorbed by the plastid is converted into chemical energy; in a series of steps, energy from oxidation is utilized to phosphorylate ADP to ATP. This photophosphorylation obviously is related to the phosphorylation carried out by the mitochondria. Functional subunits of chloroplasts in higher plants are membrane layers called **grana.**

The cytoplasm of most animal and many plant cells has a structure, known as the centrosome, adjacent to the nucleus. Within this relatively clear area of cytoplasm there are one or two granules, the **centrioles.** These organelles are important in the origin and function of flagella and cilia and in nuclear and cell division. The nine strands making up the outer portion of a flagellum or cilium are continuous with nine tube-like or filament-like structural components of the centriole and basal granule. This remarkable similarity of centrioles, basal granules, and cilia (which is preserved even in such highly specialized cells as retinal photoreceptors) is found throughout the animal kingdom. The centriole and obviously related structures (such as basal granules) apparently have properties that lead to the organization of fibrous protein molecules in special ways, e.g., in the formation of the spindle tubules (fibers) during mitosis. Centrioles are not found in cells of the flowering plants.

Currently an interesting hypothesis concerning the origin of mitochondria and chloroplasts states that prokaryotic (without well-defined organelles) cells invaded other primitive cells and became established as **endosymbionts.** The invaders eventually lost their cell walls and became dependent on the host cell for nutrients. Gradually their division came under the control of the host cell, so that they and the host cell replicate in synchrony. Under this hypothesis, the invading prokaryotes evolved to become the mitochondria and chloroplasts of eukaryotic (with well-defined organelles) cells. Suggestive evidence is the similarity between prokaryotic cells and eukaryotic organelles. For example, neither has a nuclear membrane, both contain only a single DNA molecule, neither possesses membrane-enclosed organelles, and neither divides by a regular mitotic cycle. Further, it has been established that chromosomal DNA cannot direct the synthesis of new organelles without the interaction of the mitochondrial and chloroplast DNA. This suggests that perhaps at one time these organelles had greater autonomy.

Most of the genetic information in the cell is located in the **chromosomes** within the nucleus. The precise state of the genetic material is not known. In the so-called resting or metabolic stage, chromosomes in the nucleus are usually difficult to view. In some instances, portions of chromosomes that have not undergone the usual transformations accompanying mitosis may be seen. Often one or more of these chromosome regions are associated with the **nucleolus,** a usually conspicuous feature of the metabolically active nucleus involved in the manufacture of ribosomes. Presumably, during the metabolic stage, material is exchanged between the cytoplasm and the nucleus, and the role of *m*RNA and *t*RNA in translating the DNA code into protein structure is carried on. The mechanism of this exchange is not clearly understood, however. Evidence clearly indicates movement of nucleic acid from nucleus to cytoplasm. Actual particles usually have not been found in the pores of the nuclear envelope, as seen in the electron microscope.

CELL DIVISION

Mitosis

When cells divide (Fig. 2-3), the first conspicuous change usually is in the appearance of the nucleus. The chromosomes become visible within the nuclear envelope in living or stained cells, and, usually

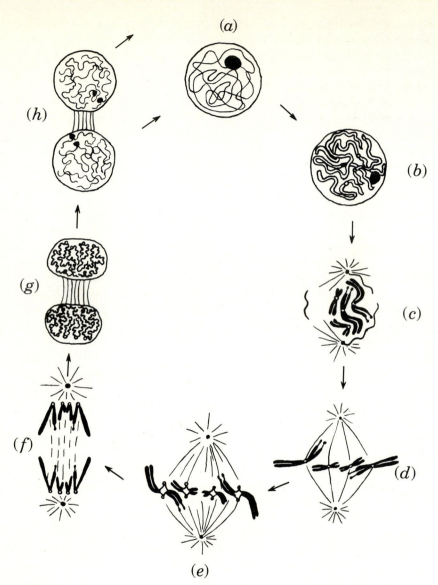

Figure 2-3 Mitosis: *(a)* early prophase; *(b)* late prophase; *(c)*prometaphase; *(d)* metaphase; *(e)* early anaphase; *(f)* early telophase; *(g)* late telophase; *(h)* daughter nuclei. *(Adapted from E.D.P. De Robertis, W.W. Nowinski, and F.A. Saez, General Cytology 3rd ed., W. B. Saunders Company, 1960.)*

concomitantly, the nucleolus decreases in size. This first of the arbitrarily designated stages of mitosis or nuclear division is **prophase.** Toward the end of this stage, it can be seen in the cells of many organisms that the chromosomes are double, each consisting of two half chromosomes **(sister chromatids),** and their ends may be associated with the nuclear envelope. In most organisms the disappearance of the nuclear envelope marks the beginning of **prometaphase.** During this period, or somewhat before, a spindle-shaped bundle of fibers (now known to be microtubules) is organized in the cytoplasm. Toward the end of prophase the chromosomes become arranged in a group at the equator of this structure, called the **spindle.**

The spindle in those organisms in which it can be isolated and studied chemically is composed of fibrous protein molecules containing many sulfhydryl linkages and apparently oriented by the centrioles at either end. In animals (and in some plants) the centrioles also are surrounded by a pompon of fibers, the aster. In the somatic cells of most plants, no asters or centrioles are visible, but they may be conspicuous in the reproductive cells. During the brief stage called **metaphase,** the chromosomes are arranged across the equator of the spindle with at least their spindle attachment points or **centromeres** in essentially a plane at right angles to the long axis of the spindle. Very shortly thereafter, the centromeres appear to divide (they may actually have split at an earlier period), and the sister chromatids—now daughter chromosomes—move to the poles.

The phase of chromosome movement is called **anaphase,** and its mechanism is still not understood. None of the current theories satisfactorily explains the behavior of chromosomes and cells in all organisms. In some animals, for example, certain chromosomes behave with remarkable autonomy. Sex chromosomes may appear precocious or tardy in comparison with the autosomes. Specialized chromosomes may be confined to the germ line and become eliminated in later divisions. In the fungus gnat *Sciara* a monopolar spindle is formed at one division, and one group of chromosomes moves to the "nonpolar" end. In some animals and plants the nuclear envelope does not disappear, and chromosome division takes place within the membrane, which eventually is pinched in two. As more work is carried out on the little-known invertebrates, algae, and fungi, other examples of unusual behavior undoubtedly will be found. Indeed, when proper perspective is reached as a result of systematic study, the higher plant-vertebrate mechanisms may seem unusual.

When the chromosomes have reached the poles of the spindle, a new nuclear envelope, which arises from membrane systems of the cytoplasm, is formed about each group of daughters. During this stage, **telophase,** animal cells usually divide by furrowing and plant cells by cell-plate formation; a new cell wall partitions the old cell. With this formation of two daughter cells, the process of cell division ends. Thus two cells, each with the same genotype, have been produced as a result of **equational division** of the chromosomes (mitosis) and division of the cytoplasm, during which the cytoplasmic organelles are apportioned roughly equally.

During the course of mitosis, the chromosomes (Fig. 2-4) go through an interesting and important series of changes. If a prophase chromosome is compared with an anaphase chromosome, striking differences are seen. The anaphase chromosome is not only easily visible and stainable but fatter and much shorter. By use of appropriate treatments, it can be shown that the anaphase chromosome is in the form of a tight spiral with gyres (turns of the coil) which behave as if they were invested with a stainable substance usually called matrix. The basic thread, or chromonema, of an anaphase chromosome often can be seen to be coiled in a fine series of gyres called minor coils, but the details of chromosome structure and the arrangement of DNA and protein are not agreed upon. The easily seen major coils are imposed upon the minor coils by whatever process causes chromosome shortening. When there is more than one chromatid, as in metaphase, they may be twisted about one another.

Mitosis provides for the equational division of the chromosomes so that, barring a mutational event, each daughter cell receives the *same genetic information.* In many instances, as cells become specialized in form and function, division of the chromosomes may occur without division of the nucleus or of the cell. The result is a cell with more than one nucleus or a nucleus with more than the zygotic number of chromosomes. The latter, which appears to be the more common, is referred to as **endopolyploidy.** In some organisms, e.g., insects, each tissue has its own characteristic degree of polyploidy. In the salivary glands of water striders (*Gerris*) the highly specialized cells may be 2,048-ploid. Tissues of other organisms contain a population of cells of varying degrees of ploidy (usually with a norm at a level above diploidy, e.g., at the tetraploid or octaploid level). This is the situation in the human liver, for example, or in various tissues in the roots and stems of some flowering plants.

Insects in the order Diptera combine endopolyploidy with **somatic pairing** of the chromosomes. Homologous chromosomes are paired in somatic cells as well as in meiosis (see below). In many cells of various tissues endopolyploidy results in the formation of numerous chromosome strands which become somatically paired. The chromosome strands are uncoiled and together form the so-called "giant" or **polytene** chromosome. The best known of these are the salivary-gland chromosomes of fruit flies (*Drosophila*). Each giant chromosome consists of two many-stranded homologous chromosomes closely paired, point for point, along their length. The longi-

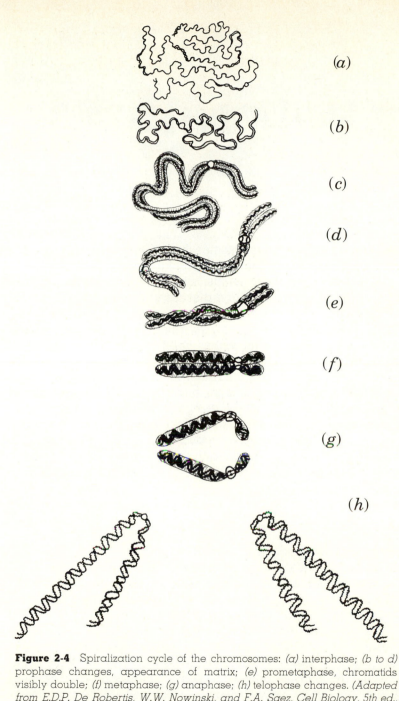

(a)

(b)

(c)

(d)

(e)

(f)

(g)

(h)

Figure 2-4 Spiralization cycle of the chromosomes: *(a)* interphase; *(b to d)* prophase changes, appearance of matrix; *(e)* prometaphase, chromatids visibly double; *(f)* metaphase; *(g)* anaphase; *(h)* telophase changes. *(Adapted from E.D.P. De Robertis, W.W. Nowinski, and F.A. Saez, Cell Biology, 5th ed., W.B. Saunders Company, 1970.)*

Figure 2-5 Polytene chromosome from salivary-gland cell of a *Rhynchosciara angelae* larva at maximum growth. *(Courtesy of C. Pavan.)*

tudinal differentiation of chromosomes is visible as a series of bands of varying size and shape characteristic of a particular chromosome (Fig. 2-5). By studying these bands, structural changes in chromosomes can be analyzed, as will be discussed in Chap. 3.

The significance of endopolyploidy is not well understood. Certainly it correlates with secretory activities of the cells, and it may play a role in development as well as in differentiation. It should be emphasized, however, that no qualitative changes in the genetic material have been demonstrated. Germ-line cells in animals and cells producing spores in plants usually do not become endopolyploid. Reproductive cells therefore retain the zygotic or gametic number of chromosomes, while somatic cells may experience successive increases. If vegetative reproduction in plants takes place by budding or suckering from somatic tissue, offspring with increased chromosome number may arise. Such situations are discussed in Chap. 9.

Meiosis

In addition to these mechanisms for preserving the existing chromosome number or increasing it, organisms obviously must have a mechanism for reducing it. Mechanisms for reducing high endopolyploid numbers are poorly understood, but they have been reported in mosquitos (*Culex*) and onions (*Allium*). The great majority of organisms share the mechanism for reducing the zygotic number of chromosomes to the gametic number; this mechanism is called **meiosis** (Fig. 2-6). Meiosis occurs in tissues that have not undergone endopolyploidy, such as germ-line cells in animals or sporogenous tissue in plants. Again, as with mitosis, its outlines are the same in all organisms, although in animals it results in gametes and in most

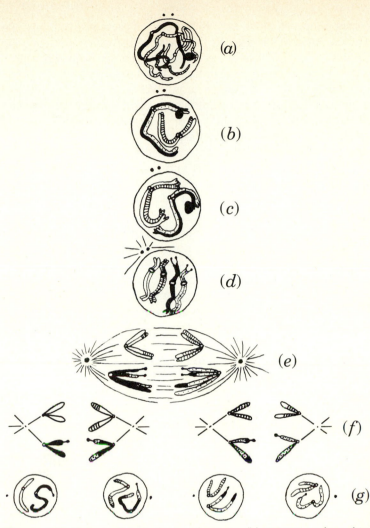

(a)

(b)

(c)

(d)

(e)

(f)

(g)

Figure 2-6 Meiosis. *(a)* single-strand stage; *(b)* pairing to form bivalents; *(c)* four-strand stage; *(d)* opening of bivalents to show chiasmata; *(e)* first anaphase, disjunction of bivalents without centromere division; *(f)* daughter nuclei with crossover chromatids; *(g)* second anaphase, centromeres divide; *(h)* four haploid daughter cells. *(Adapted from E.D.P. De Robertis, W.W. Nowinski, and F.A. Saez, Cell Biology, 5th ed., W. B. Saunders Company, 1970.)*

plants it results in spores. A cell about to undergo meiosis may be called a **meiocyte.** The results of meiosis are nearly always four daughter cells each with half the number of chromosomes of the meiocyte. In the formation of eggs in animals and in the production of megaspores in plants, three of these cells may be much smaller and eventually disappear.

Meiosis achieves these results by two cell divisions but only one division of the chromosomes. The following description refers to the generally observed behavior of the chromosomes as seen with a light microscope. (Variations or exceptions have been noted in some organisms or with special techniques.) When a cell becomes a meiocyte, it usually enlarges somewhat. When the chromosomes become visible in prophase, they often look as if they were single strands instead of appearing double-stranded, as in mitosis. Their subsequent behavior is so complicated that the first meiotic prophase is prolonged and has been divided into a series of substages, the names of which need not concern us here. The first occurrence is synapsis of the chromosomes (present, of course, in pairs) with their homologs, precisely point for point along their length. After pairs, called **bivalents,** have been formed, the chromosomes appear double-stranded. Each bivalent thus comprises four chromatids, two sisters from the chromosome that arrived in the maternal gamete and two paternal sister chromatids. Any chromosome that lacks a homolog, e.g., a sex chromosome in some species, remains as a univalent but undergoes doubling at about the same time as the others.

Apparently at about the time the chromosomes double, the slender chromatids break and rejoin in the bivalents. Very closely associated, coiled, and twisted, they often reunite in nonsister combinations; i.e., instead of sister chromatids rejoining, maternal and paternal chromatids may be connected following a break. This is the phenomenon of cytological **crossing-over.** When the chromosomes have become double, they behave as if they now repelled one another. Bivalents become widely spaced in the nucleus, and members of bivalents are held together only where crossing-over has occurred. If crossing-over has not occurred, the chromosomes in a bivalent frequently separate at this stage. In some organisms crossing-over does not occur in one sex, e.g., male *Drosophila* and *Callimantis*, and special mechanisms hold chromosomes together. As a result of the repulsion (this term is used only descriptively) of the chromosome arms, the bivalents assume forms that depend upon the number and position of crossovers; the latter now become visible as **chiasmata,** or cross-shaped configurations (Fig. 2-7).

At this stage of meiosis, a **chiasma** indicates a crossover. Subsequently, as the chromosomes coil and shorten and become more stainable, the chiasmata (but not the points of crossing-over) are pushed to the ends of the chromosomes. This process, known as **terminalization** (Fig. 2-7), also produces characteristic configurations of bivalents, adjoining loops lying at right angles.

At the end of the first meiotic prophase, the nucleus contains the gametic number of bivalents. At the first metaphase, a spindle is formed, presumably precisely as in mitosis, and the bivalents be-

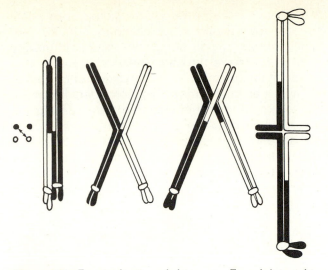

Figure 2-7 Terminalization of chiasmata. From left to right, chiasmata move to ends of chromosomes. (Note that the point of crossing-over does not change.) Far left, cross section of bivalent. Far right, rotation of chromosomes has occurred. *(Adapted from E.D.P. De Robertis, W.W. Nowinski, and F.A. Saez, Cell Biology, 5th ed., W. B. Saunders Company, 1970.)*

come arranged on its equator. During first anaphase, instead of the centromeres dividing as in mitosis, the two centromeres of a bivalent move to opposite poles. Thus the chromosomes do not divide (for a chromosome is defined by its centromere) but disjoin. **Disjunction** results in two daughter nuclei that undergo the usual telophase transformation (or the latter may be much abbreviated). The centromeres of univalents similarly do not divide, a univalent going to one pole or the other (usually according to chance). The distribution of maternal and paternal chromosomes is completely at random.

During the second division of meiosis, the behavior of the chromosomes is like that in mitosis, the difference being that *crossing-over has taken place so that the chromatids attached to a centromere are not identical*. This second division, in which the centromeres divide, results in the formation of the four daughters of the meiocyte. Each has the gametic chromosome number, but the chances of one daughter being genetically like any other are extremely small. Maternal and paternal chromosomes have been segregated at random, chromatids have been segregated at random, and finally, as the result of crossing-over, the genetic material in the parental genomes has been partially exchanged.

This then is the tremendous significance of meiosis and crossing-over. *It provides a continual reshuffling of the genetic material in reproduction.* New gene combinations are continually being produced, and essentially random union of gametes makes it unlikely that any two individuals will have the same genetic makeup. This cytological mechanism of the organism is part of its **genetic system,** the system determining the amount of recombination a population will produce that will be available for the operation of selective agents. The organisms most familiar to us are diploid, sexual, outcrossing organisms such as cats and dogs, oaks and pines. Later (Chaps. 8 and 9) other genetic systems will be discussed as examples of the ways in which the amount of recombination produced by this familiar genetic system may be modified (usually decreased, perhaps to zero).

SUMMARY

The cell is a metabolic unit composed of large and small molecules associated in specific ways, commonly as membrane systems, to form subunits, or organelles, of specialized function. The nucleus of the cell initiates and controls protein synthesis through the functioning of its chromosomes. When somatic cells divide, the cytoplasmic organelles are apportioned between the daughters in roughly equal quantity. By means of mitosis, the chromosomes are divided equationally between the daughters. Meiosis reduces the number of chromosomes in cells that will produce gametes or spores. In the first division of this two-stage process, homologous chromosomes first synapse and then disjoin without division of their centromeres. Cytological crossing-over takes place during the first division, and when the centromeres divide in the second division, four daughter cells with recombined chromosomes and chromosome segments result. Meiosis allows for the recombination that results in much of the variation which is the basis of selection.

REFERENCES

Cohen, S. S.: Are/Were Mitochondria and Chloroplasts Microorganisms?, *Am. Sci.*, **58:**281–289 (1970). An excellent review article on the origin of cellular organelles.

Margulis, L.: The Origin of Plant and Animal Cells, *Am. Sci.*, **59:**230–235 (1971). A concise explanation of the symbiotic theory of the origin of eukaryotic cells.

Novikoff, A. B., and E. Holtzman: *Cells and Organelles*, Holt, New York, 1970. A current survey of cytology with particular emphasis on recent developments in the field.

It seems reasonable to assume that the selective forces involved in the evolution of the genetic mechanisms of early organisms must have been concerned with stabilization of what was at first an almost infinitely variable system. The mechanisms for replicating genetic material generally ensure that it will be exactly duplicated and that, in the offspring, proteins similar to those of the parents will develop. In more complex multicellular organisms, self-regulatory developmental mechanisms are combined with the nuclear and extranuclear genetic material; together they provide a system that usually results in what is thought of as a normal, functioning, wild-type organism. As stated above, the basic phenomenon of genetics is that "like begets like."

CHAPTER
THREE
GENETICS

VARIATION AND MENDELIAN GENETICS

Nevertheless, errors in replication occur; they result in the variability that permits selection. In general, analysis of the nature and transmission of variability from generation to generation is the only means of studying the mechanism of inheritance. If the patterns of variation in organisms are examined, it will be seen that some organisms appear to be more variable than others. Furthermore, the type of variation pattern differs with respect to organisms and the traits studied. In some instances, variation occurs in discrete steps and may be termed **discontinuous.** In other cases, variation appears to be **continuous,** individual organisms not falling into easily characterized discrete classes. Galton attempted to study continuous variation when he made his classic investigations of the inheritance of intelligence and other traits in human beings. Other workers, even before and including Linnaeus, had studied continuous variation by making crosses between varying plants and animals. The science of genetics was not really born, however, until the inheritance of characteristics that varied discontinuously was studied. Organisms having these characteristics could be classified as one or another of a very few distinct types. By observing the distribution in these classes of offspring of an experimental cross of parents with different characteristics, Mendel was able to describe the basic rules of behavior of nuclear hereditary units.

The importance of Mendel's studies was not generally appreciated; indeed, Mendel was urged to suppress his results by other scientists who felt that he was considerably off the beaten path of scientific research. In 1900 Mendel's papers were discovered by three biologists (Correns, de Vries, and Tschermak) who recognized their significance. Almost overnight, genetics became an important and rapidly developing field of biology. However, many scientists felt that Mendel's work had little application to evolution in natural populations or to the prevailing type of continuous variation found in wild and domesticated populations of both plants and animals. It was only after many years of work that the evolutionary significance of Mendel's laws was established.

THE UNITS OF HEREDITY

The units of heredity postulated by Mendel and subsequently termed **genes** were identified as specific regions of the chromosome; they are the segment between two closest points of crossing-over. The specific region of a chromosome at which a gene is located is called the **locus** of that gene. At any one locus, a gene may exist as one of several to many alternative states. The alternative states of a gene are known as **alleles.** More recent work, especially in the biochemical genetics of microorganisms, has led to a functional definition of the gene. Specific genes control the formation of specific polypeptides.

By growing microorganisms on media of known composition, it is possible to show that gene **mutation** may result in the loss of ability to carry out some cellular reaction. For example, a mutant bacterium may lose the ability to synthesize a particular substance, such as tryptophan. Studies of many different genes have shown that the number of alleles of a gene is usually quite large. It is necessary to think of the gene, as revealed by these studies, as a region of the chromosome that is mutationally complex. Benzer has referred to the possible mutational sites within a gene as mutons. Evidence suggests that a single muton may consist of only one or very few nucleotides.

At this level of study, experiments have shown that recombination may occur *within* the limits of a single gene, i.e., within a functional unit. Recombination studies show that the smallest unit that is interchangeable also is about the size of a nucleotide and that the mutons in the functional unit are arranged in linear fashion. Since it appears to make no difference phenotypically how the genes are arranged in an organism heterozygous for two factors, i.e., whether the genes are arranged

$$\frac{++}{ab} \quad \text{or} \quad \frac{a+}{+b}$$

it is interesting to ask the same question about parts of genes. Will protein synthesis take place just as well when the mutants within one gene are distributed between the two chromosomes (the so-called trans state) as when the mutants are on one chromosome (the cis state)? Numerous genetic studies in microorganisms have shown that protein synthesis does not occur when the mutants are in the trans state. Two chromosomes are said to **complement** one another when the mutations under study are in different genes, and, in fact, this is the basis of the **complementation test** to determine whether a group of mutations lies in one or two genes. By the use of the complementation test, functional genes, units called **cistrons,** can be identified. A cistron is an operational unit controlling the synthesis of a specific protein.

In the study of populations, the unit of heredity must be given a strictly operational meaning within the context of the study, as, in fact, it must in all biology. In evolutionary studies, this unit usually cannot be the same as that in biochemical genetics. In studying inheritance in populations in nature, the unit of heredity in most cases becomes a statistical one, for the factors controlling the expression of continuously varying traits are numerous and complexly interrelated. At present, only the methods of the statistician can sort out the interactions of the heredity units (which are assumed to be similar to those affecting discontinuous variation), the developmental systems through which they are expressed, and the effects of the environment on both these systems. The environment of an organism at any particular time or place is unique and not repeated or repeatable. This means that in experimental studies it is important to make replicate experiments in space and time—an unfortunately expensive and time-consuming process. In the following pages, the basic facts of the inheritance of discontinuous and continuous variation will be summarized, together with a discussion of those aspects of gene behavior that are particularly important to evolutionary studies.

MENDEL'S LAWS

The basic rules of heredity deduced by Mendel are familiar to anyone who has had an elementary course in biology. In crossing peas,

Mendel found that when differing parents are mated, the first generation offspring (F_1) resemble one or the other parent. The trait expressed in the F_1 he referred to as **dominant** to that which did not appear (the **recessive**). Crossing the essentially uniform F_1 plants to produce a variable F_2 generation, in which individuals with the recessive trait appeared, showed that the factors responsible for the appearance of the traits are not lost but merely hidden. By a study of the types of progeny in the F_1 and F_2, Mendel deduced that each offspring contains two homologous factors, one received from each parent, affecting the expression of the traits studied. An F_1 offspring from a cross between differing parents must contain two different but homologous factors, one for the dominant trait and one for the recessive. In other words, the F_1 is **heterozygous.** In the formation of the F_2, these factors segregate, and the offspring are produced in the approximate ratio of three with the dominant trait to one with the recessive. The individual showing the recessive condition is **homozygous** for the factors. Further crossing (including backcrosses to the parental types) shows that, of the three with the dominant trait, one will have like factors and the other two different factors, as in the F_1 individuals.

When parents differing in, and homozygous for, several characters were crossed, Mendel found that the factors for the different traits he was studying behave independently. In the F_1 both dominant traits were observed, and in the F_2 each trait segregated by a 3:1 ratio. If the homozygous parents differ in two traits, for example, the proportion of types in the F_2 is the square of a 3:1 ratio, or 9:3:3:1. By backcrossing offspring to the parental types, verification of the number of factors and their independence can be obtained.

It is clear that the behavior of these factors parallels the behavior of the chromosomes now known to bear them. The factors affecting the traits in peas studied by Mendel were on nonhomologous different chromosomes. Later studies showed that factors on the same chromosome were **linked,** i.e., tending to occur together more frequently than would be expected if assortment were independent. When numerous traits are studied, their factors are found to fall into as many linkage groups as the haploid number of chromosomes. Within a linkage group the amount of recombination varies from a very low percentage for genes close together to 50 percent for genes far apart, which is genetically indistinguishable from independence, i.e., occurrence on different chromosomes.

RECOMBINATION

The cause of genetic recombination of linked genes is cytological crossing-over in meiosis (diagrammed in the previous chapter). In studies of inheritance at a gross level, the factor to be defined operationally as a gene and presumably affecting a particular characteristic is the minimum distance between two points of crossing-over. Crossing-over occurs in all organisms in which meiosis and sexual recombination have been found. The basic mechanism may be the same wherever it occurs, and some workers have postulated that meiosis cannot properly take place in the absence of crossing-over (or a specialized substitute).

The precise mechanism of cytological crossing-over is not known. Presumably when the chromosomes are synapsed and twisted together in the first meiotic prophase, chromatids break and the broken ends subsequently rejoin. If nonsister chromatids are joined, crossing-over has occurred. Crossing-over does not take place with equal frequency along the length of the chromosome. It is rare or absent near the centromere and at the very ends of the chromosomes. Near large blocks of **heterochromatin** (differentially staining regions of chromosomes, largely genetically inert at this stage) crossing-over also is reduced. In some organisms, crossing-over does not occur in certain regions of the chromosomes, e.g., in the vicinity of the centromeres; in others, it seems to occur rather evenly throughout the length of the chromosome arm. Perhaps the difference lies, in part at least, in the amount and distribution of heterochromatin. The occurrence of a crossover also interferes with the formation of an immediately adjacent crossover. Interference can be measured by studying linkage and can be shown to vary along the chromosome and between different chromosomes. Probably interference is also a structural phenomenon. Occasionally crossing-over is suppressed entirely, as in male *Drosophila* and female silkworms (*Bombyx mori*).

If the loci of two genes are some distance apart on the chromosome, more than one crossover may occur in that region. Should double crossing-over take place, the effects on recombination depend upon which chromatids are involved (Fig. 3-1). If the same two chromatids of the four associated in the bivalent are involved in both crossovers, the occurrence will not be detected unless a third locus between the original two is also observed. Should the other two chromatids experience the second crossing-over, each chromatid will have one crossover. This crossing-over is referred to as two-strand and four-strand exchange, respectively. Three-strand exchange results in the formation of a noncrossover chromatid and three chromatids with a single crossover. In general with multiple crossing-over between two factors, the resulting chromosomes with an even number of crossovers and those with no crossovers will appear as parental types. Chromosomes with an odd number of crossovers between the genes in question will be recombinant types. Since the number of chromosomes with recombinations equals the number with

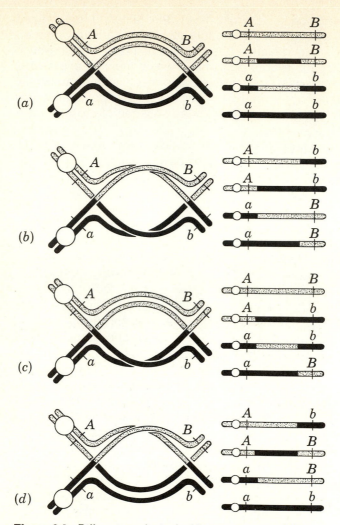

Figure 3-1 Different results in double crossing-over. *(a)* two-strand exchange; *(b)* four-strand exchange; *(c), (d)* three-strand exchange. *(Adapted from E.D.P. De Robertis, W.W. Nowinski, and F.A. Saez, General Cytology, 3rd ed., W. B. Saunders Company, 1954, and M.J.D. White, Animal Cytology and Evolution, 2d ed., Cambridge University Press, 1960.)*

no or an even number of crossovers, recombination cannot exceed 50 percent. There has been an increasing accumulation of evidence that genetic factors also influence crossing-over and play a role in the control of recombination, which is therefore subject to evolution. Recombination is reduced in some organisms by mechanisms such as chromosomal inversions, translocations, or elimination of meiosis altogether (in apomictic organisms; see Chap. 9).

THE EXPRESSION OF GENES

The action of genes in an individual organism has proved quite variable. Genes in a population are rarely found in only two alternative states, or alleles. **Multiple alleles** appear to be the common system governing most characteristics, e.g., blood groups in animals, incompatibility systems in plants, coat color in mammals, and flower color in plants. Interaction of the genes in the genotype or of the developmental pathways resulting in the phenotype produces complex genetic ratios. For example, where pigment systems leading to the formation of a particular color are involved, several genes may control different steps in the elaboration of the pigment. If the functional allele of any of these genes is missing, color is lacking. Such genes are called **complementary genes.**

On the other hand, in many situations the expression of the gene at one locus masks the expression of another gene at another locus. This is known as **epistasis;** the epistatic gene masks or prevents the expression of a hypostatic gene. In chickens, for example, the Leghorn white color is epistatic to many genes affecting color and pattern. If it is present, no color but white will be expressed. It is obvious that complementary effects and epistasis, as well as other sorts of modified expression, are related developmentally. A similar phenomenon occurs where there appears to be one gene or a group of genes, each with relatively small effect, that operate to alter the action of a gene with major effect. These "minor" genes are known as **modifying factors.**

It often becomes necessary to specify the type of expression of genes because, for many factors, it is variable. For certain factors, all or almost all individuals with the same genotype develop a characteristic phenotype that distinguishes them from individuals with other genotypes in a certain range of environments. The genes in such cases are said to have high **penetrance,** since most individuals carrying the gene possess the trait. Other genes do not always produce a detectable phenotypic effect in the individuals that carry them in a given environment. These are genes of low penetrance. The phenotypic expression of a gene also may be variable even though it is completely penetrant. If it is relatively uniform in an essentially "normal" environment, the gene has constant **expressivity,** but if there is interindividual variation in the trait, expressivity is variable.

Studies of the manifold, or pleiotropic, action of genes and systems of genes controlling the expression of particular characteristics have suggested that probably no character of an organism is controlled by only one gene and, conversely, every gene in the genotype of an organism affects a great many (if not all) characters in the complex process of producing the phenotype. But it has been established that the amino acid sequence of a specific polypeptide is determined by one and only one gene. Further, Yanofsky and his associates have conclusively shown that the sequence of codons along a portion of a DNA molecule is correlated, or collinear, with the sequence of amino acids of the protein for which it is coded.

MUTATION

In controlled crosses it is often possible to identify specific genes affecting particular characters that show discontinuous variation. The most obvious characteristic of such identifiable genes is that they change or mutate; indeed, that is how their existence is detected. Mutations occur spontaneously at varying rates; they can also be induced by treating organisms with ionizing radiation, ultraviolet light, or various chemicals. These **mutagens** cause a change in the kind, number, sequence, or structure of nucleotides in the DNA molecule.

There are currently two theories concerning the effects of ionizing radiation. The first states that a single nucleotide in a DNA strand is hit directly by the ionizing radiation. This causes an alteration of the molecule and possible deletion of the nucleotide. The second theory states that free radicals are liberated when some general molecules in the cell, for example, H_2O, are hit and that these highly reactive free radicals initiate chains of destructive oxidation-reduction reactions.

One effect of nonionizing ultraviolet radiations (absorbed by bases of nucleic acids) is to join adjacent thymine groups in DNA (thymine dimers). Although the formation of these dimers may not affect transcription substantially, they do interfere with or inhibit the normal replicative process of the DNA molecule.

Numerous chemical agents are known to induce mutations. Among them are base analogs, molecules closely resembling the pyrimidine or purine bases in DNA. Because of this similarity they sometimes replace the correct base in the nucleotide sequence of the DNA, thereby interfering with the normal coding of that region of the DNA molecule. Acridine derivatives have been shown to cause a transposition in the reading of the bases of the nucleic acid. In the case of acridine, these "frame-shift" mutations are caused by the addition or deletion of a base to one of the strands of the DNA molecule. Other chemicals such as hydroxylamine, aminopurines, and nitrous acid have mutagenic effects.

The discussion of spontaneous mutation rate is difficult because there are several ways in which the rate may be expressed. As far as is known, the mutational event is random; it is not possible to specify what gene will be affected or to assign the cause of a given mutation to a specific mutagenic agent. It would be desirable to know the chance of occurrence of a mutation per cell per division, which expresses change with respect to time. This is very difficult to determine except in microorganisms. Even with bacteria, what is measured is phenotypic change which may involve more than one gene or mutable unit. Rates of from 10^{-6} to 10^{-9} have been measured in microorganisms.

In multicellular organisms, rates must generally be measured differently (except where tissue culture is possible) since the criterion available is the number of gametes producing mutant individuals per generation of the organism. Thus individuals, not gametes, are counted. In the gonads a mutation occurring in a gamete-producing cell may have many or few mutant daughters, depending upon when in gametogenesis the mutation took place. The mutation rate per generation varies with the gene studied but averages about 10^{-5}. When rates per cell division are measured by tissue-culture studies of bone marrow cells, they average about 10^{-6}.

Mutation rate appears to be under genetic control and is therefore subject to change in the course of evolution. Genes whose major effect seems to be affecting the mutation rate of other genes are known. One would expect that on the level of mutation in the population, selection would occur and would result in the maintenance of an optimum level of mutations in the population. This is difficult to study, and few data are available. Heterozygosity is known to affect developmental buffering, or homeostasis, as well as genetic homeostasis (discussed in later chapters).

EVOLUTION OF DOMINANCE

There are several other interesting aspects of mutation about which little is known. Most mutant alleles occurring in the organisms that have been studied in detail are recessive to the wild-type gene. This raises the problem of how dominance-recessiveness arose. It is clear that most mutations will be deleterious to the organism, since they alter a functional system. If they have a major effect, almost certainly the complexly interrelated developmental pathways will be grossly upset and the organism will die. Even if the gene has only a relatively minor effect, however, its incorporation into an integrated genotype will cause the mutant organism to be less fit than its parents, provided that there is no environmental change. (By analogy, the chances of improving the operation of a radio receiver by making a small random change in its circuits are slim indeed.) When deleterious recessive mutations occur in a diploid organism, they are stored in the organism's reservoir of variability. When they are combined (rarely) in the homozygous state, they will be eliminated unless the environment (in the broadest sense) has changed sufficiently to give them positive selective value.

How then does recessiveness arise? There are several hypotheses between which it is difficult to discriminate at present, although all probably have elements of truth. Fisher has suggested that mutations are not necessarily completely recessive on their first occurrence. In the heterozygote form, they will therefore exhibit some selective disadvantage. Fisher then suggested that other loci exist producing genotypes which reduce the disadvantage of the heterozygote. These "modifying" genotypes will be selected for until the original disadvantage disappears, at which stage complete recessivity will have been achieved.

Wright and Haldane have discussed the problem of the origin of dominance in terms of the relation between the gene-produced enzyme and its substrate. In their view, recessive genes are those which are less active than the wild type in the production of a particular enzyme. Selection presumably will have built in a safety factor so that there is an excess of enzyme over levels required by available substrate, and a mutation reducing the amount of enzyme will have little effect in the heterozygote. Biochemical genetics is supplying answers to the questions concerning the quantitative aspects of gene function in enzyme synthesis.

Wright and Haldane have also suggested that the selective value for the modifying factors might be so low that dominance would arise too slowly to have a large chance of appearance when other factors are taken into consideration. In some organisms, there is evidence that the selective coefficients for modifying genes may be considerably higher than those postulated by Wright or Fisher. In either event, selection has played an important role in the evolution of the behavior of genes. Hybridization experiments in organisms as diverse as cotton plants and butterflies have clearly shown that the functioning of a gene may change when it is moved from one genetic background to another. Thus selection altering the background (e.g., "modifiers") can affect the expression of a given gene.

There is considerable evidence that this is exactly what has happened during the development of industrial melanism in the moth *Biston betularia* (see Chap. 7). Early samples of heterozygotes for the melanic allele were quite distinct from the homozygous melanics. By the middle of the twentieth century, the heterozygotes were almost identical to these homozygotes. Clearly, dominance has evolved in this case.

CHROMOSOMAL MECHANISMS

The existence of means of artificially inducing mutations in easily grown organisms with relatively short life cycles, e.g., *Drosophila*, maize (*Zea*), and the mold fungus *Neurospora*, has led to careful studies of linkage and the linkage groups or chromosomes. If it is assumed that the amount of crossing-over between two factors is proportional to the distance between them, then the spacing and arrangement of the genes along the chromosome can be determined. A genetic-linkage chromosome map, based upon recombination data, can be made. In such work it must be kept in mind that there may be interference between adjacent crossovers, that with factors that are far apart, multiple crossing-over may occur between them, and that only parts of the chromosome with easily studied major phenotypic effect can be mapped. There are other means of producing chromosomal maps, however, and these give a check on the method. For example, it is possible to map chromosomes by studying the effects of induced deletions of small portions of the chromosome and of chromosomal changes such as inversions and translocations, as well as other techniques. These have confirmed the linear order of the genes mapped by crossover studies, but these maps vary from genetic maps, often strikingly, in spacing and other details. Regions of the chromosome that are heterochromatic and that seem to lack genes with major effect are not easily studied by the recombination analysis; in the main these are responsible for the differences. Progress has been made in localizing specific genetic effects at visible regions of the chromosomes in *Drosophila* (with its giant polytene salivary-gland chromosomes) and in *Zea* (where the chromosomes have characteristic chromomeres visible under the microscope).

Sex Chromosomes

A specialized sort of linkage occurs in animals and plants with differentiation of sexes. In these organisms, where there are special **sex chromosomes,** as opposed to the other chromosomes (known as

autosomes), one sex usually has two homologous sex chromosomes. The other sex has only one chromosome homologous with these and either lacks the second or has another only partially homologous chromosome. In *Drosophila* and man, the female is the **homogametic** sex with two X chromosomes (every gamete contains an X), while the male is **heterogametic** with an X chromosome and its partial homolog Y (there are two kinds of gametes). Therefore, the transmission of genes located on the sex chromosomes will be different from those on the autosomes. Furthermore, the characteristics affected by these genes will show genetic linkage with sex. Sex chromosomes in these organisms differ from the autosomes in that they are specialized into two different regions. A portion of the two different sex chromosomes in the heterogametic sex will synapse, and crossing-over may occur. In contrast to these *pairing segments*, there are the *differential segments* of the X and Y that do not pair. The differential segment of the Y usually contains few if any genes with detectable effect, and it is heterochromatic and smaller than the differential segment of the X. When it does carry genes, they are passed from father to son and females never show the traits involved. The heterochromatic portion of the Y is necessary for fertility; therefore it cannot be completely without effect.

In some groups, e.g., birds and butterflies, the female is the heterogametic sex, while the male is homogametic. In many species the Y chromosome is lacking, and the male is then designated as XO. More complicated sex-chromosome mechanisms are discussed in Chap. 9. For example, in some organisms there are many sex chromosomes. It should be emphasized that the precise mechanism of sex determination varies from taxonomic group to group, even though the chromosome condition may appear the same. The evolution of sexuality as an aspect of the storage of variability and its release through genetic recombination is discussed in Chap. 8.

Alterations of the Chromosomes

In addition to the mutations discussed above (called gene or point mutations), changes in the structure of the chromosomes take place spontaneously and may be induced by the same agents that cause gene mutation. Although these **chromosomal alterations** are sometimes termed mutations, it is perhaps less confusing to restrict the term mutation to gene changes. Chromosomal alterations also are frequently referred to as **aberrations** or **abnormalities.** This is because they are compared with an arbitrarily selected standard chromosomal phenotype (usually the wild type); they should *not* be taken to represent some unusual or deleterious phenomenon that is inevitably disadvantageous to the organism. As with gene mutation, chromosomal alterations usually have low or negative selective value when they appear but may become stabilized in the population or replace the standard type if their selective value increases as a result of environmental change (if the nucleus and the genes are included as part of the environment). For example, by bringing favorable gene combinations together and holding them they may become selectively advantageous.

Practically any accident that can be imagined as happening to the chromosomes during the course of cellular life and division has been found in laboratory organisms and organisms from the field. Often such modification of behavior is established as a regular feature of particular organisms. A simple classification of chromosomal alterations would include **deletions** (loss of a segment), **duplications** (repeat of a segment contiguously or in another chromosome), **inversions** (reversal of a segment), and **translocations** (transfer of a portion of a chromosome to a nonhomologous chromosome, usually reciprocally). Changes of chromosome number, often thought of as "chromosomal mutation," are discussed in Chap. 9.

Deletions and duplications.   The role in evolution of loss or addition of chromosome material is very poorly understood. When homozygous, deletions usually are lethal. They are a useful tool for mapping chromosomes, however. Other than representing the loss of a possible source of genetic variability, their significance in populations is unclear. Duplications are of importance because they represent a possible cytogenetic mechanism for an increase in the total amount of genetic material. For example, Carlson has shown that duplicated regions observed in the polytene chromosomes of *Drosophila melanogaster* represent a series of genes (pseudoalleles) which are closely linked although superficially they appear to be allelic. (Fig. 3-2). How the total amount of genetic material has changed in the course of the evolution of life is not known. Nor is it possible at present to decide whether the amount of genetic material must increase concomitantly with increase in complexity.

Inversions and translocations.   Inversions and reciprocal translocations are more conspicuous and better-understood chromosomal alterations and their effects are well known. Organisms in which these changes have become a regular feature of the genetic system are discussed in Chap. 9. Here only the cytogenetic aspects of such changes are considered. Unless pairing behavior is studied, an inverted region of a chromosome ordinarily cannot be detected visually except in organisms with polytene chromosomes or con-

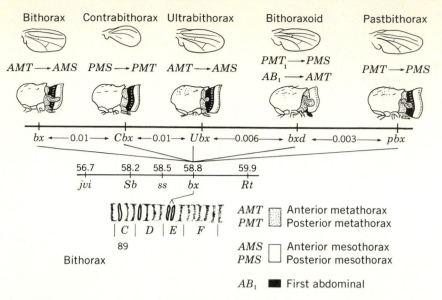

Figure 3-2 Phenotypes and linkage map of the closely linked genes constituting the bithorax locus on the right arm of chromosome 3 of *Drosphila melanogaster*. Note that segment 89E, which contains five loci, consists of two identical segments (duplications). *[From E. Carlson, Q. Rev. Biol., vol. 34 (1959).]*

spicuous chromomeres. During the first meiotic prophase, when homologous chromosomes pair, heterozygosity for an inversion is revealed by the fact that one chromosome must twist in order for synapsis to be accomplished. This characteristic **inversion loop** is also seen in the salivary-gland chromosomes of *Drosophila* and other Diptera where the polytene chromosomes are somatically paired. There are two kinds of inversions: **paracentric inversions** are those in which the centromere is not included in the inverted segment (Fig. 3-3); **pericentric inversions** have the centromere in the inverted segment (Fig. 3-4). If the inversion is a short one, a crossover may never occur with the inversion loop at meiosis. Should the inversion be sufficiently long, however, crossing-over is likely to take place. Crossing-over in the loop of a paracentric inversion will lead to the formation of a dicentric chromosome and an acentric fragment. At anaphase, the dicentric chromosome will be broken, and the acentric fragment will not move properly on the spindle. The meiotic products, in addition to balanced cells, will include unbalanced cells in which whole chromosome regions are lacking. Such cells are nonfunctional. Crossover products in a pericentric inversion will include two chromosomes with genetic duplications and deficiencies, and the result will be lethal or sublethal gametes or zygotes.

From the genetic results, it will appear in both instances as though crossing-over had been suppressed; however, only the *products* of crossing-over have been lost. The number of crossovers within the inversion loop and their distribution among the chromatids of the bivalent determine what effects there will be. Because the genes within an inversion loop are effectively linked in a heterozygote under certain conditions, inversions are discussed in more detail in relation to recombination in Chap. 9. Organisms that are homozygous for a chromosomal inversion can be recognized as such only by

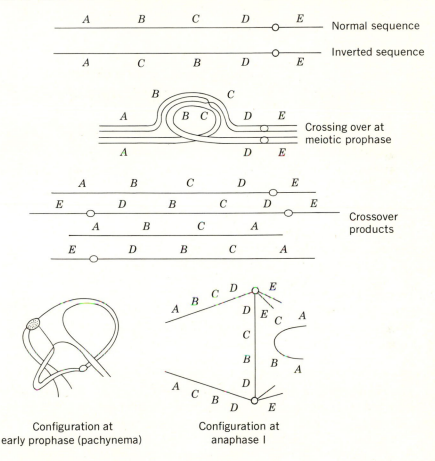

Figure 3-3 Paracentric inversion. The inverted segment of the chromosome does not include the centromere. The inversion loop is formed by the pairing of homologous chromosomes in an individual heterozygous for the inversion. If a single crossing-over occurs within the inversion loop, the chromatid products will be a dicentric bridge (*EDCBDE*), an acentric fragment (*ACBA*), and two noncrossover chromatids (*ABCDE* and *ACBDE*). (From *C.P. Swanson, T. Merz, and W.J. Young, Cytogenetics,* © 1967, p. 107; reprinted by permission of Prentice-Hall, Inc.)

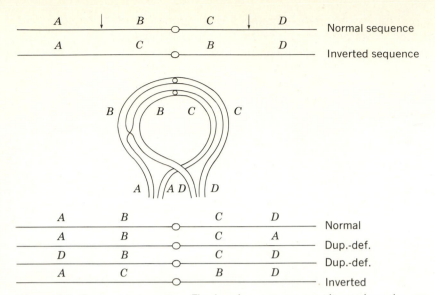

Figure 3-4 Pericentric inversion. The break points are equidistant from the centromere. The results of a single crossover within the inversion loop are four chromatids, two of which have duplications and deficiencies and two of which are noncrossover chromatids. *(From C.P. Swanson, T. Merz, and W.J. Young, Cytogenetics, © 1967, p. 110; reprinted by permission of Prentice-Hall, Inc.)*

mating them with a different chromosomal constitution and then observing the pairing behavior of the chromosomes in the hybrid (except in organisms with polytene chromosomes, where they can be detected by careful examination of the banding patterns). Since synapsis will be unaffected in the homozygote, recombination will not be reduced, but the linear arrangement and linkage relations of the genes will, of course, be changed.

Reciprocal translocations involve *nonhomologous* chromosomes. Here a portion of one chromosome is transferred to another, and vice versa, so that linkage groups are changed. The size of the segments exchanged may be the same or different, large or small. Sometimes one chromosome will exchange a heterochromatic region (with little specific genetic activity) for a euchromatic region (with typical genetic behavior). Heterozygosity for large reciprocal translocations is visible in meiosis, as well as in organisms with somatically paired polytene chromosomes. Synapsis will result in the association of four chromosomes, two standard and two with translocated regions. At metaphase the appearance of this **quadrivalent** will depend, among other things, upon the distribution of chiasmata. Usually the chromosomes separate to form a ring. Chromosome ends are held together by terminalized chiasmata. Disjunction may occur so that adjacent centromeres go to the same pole. Figure 3-5 shows that there are two different possibilities for this sort of disjunction but that either will lead to the production of unbalanced gametes, i.e., those with duplications and deficiencies. Only if alternate centromeres go to the same pole (**alternate disjunction**) can balanced gametes result. The fusion of such gametes randomly will produce standard homozygotes, translocation heterozygotes, and translocation homozygotes in the ratio of 1:2:1. Alternate disjunction effectively inhibits recombination of the genes located around the centromere, since crossing-over occurs in very low frequencies in this region and independent assortment of the nonhomologous chromosomes involved in the translocation is not possible. Organisms that are homozygous for a reciprocal translocation exhibit no meiotic peculiarities since synapsis is undisturbed. The linkage groups are changed, however; this can be detected by genetic analysis.

It is possible for more than one inversion to occur in a nucleus or in a chromosome. If there are several, they may be independent or overlapping, or one or more may be included within another. Detailed studies of these conditions have been made in *Drosophila* and are discussed in Chap. 9. With more than one translocation, the result depends upon which chromosomes are involved. If a different pair exchanges segments after the first translocation, the result will be the formation of two quadrivalents at metaphase of meiosis. If one of the chromosomes experiencing the first translocation exchanges a segment with a third, a ring of six chromosomes is found. Finally, in some organisms, all the chromosomes exchange arms and all are attached in a ring at meiosis. Examples of this are considered in Chap. 9.

CONTINUOUS VARIATION

When one comes to study the genetics of continuously varying characteristics, the problems become much more difficult. Usually it is not possible to identify specific genes controlling specific traits. As has been mentioned above, there is thought to be a continuous spectrum of characters ranging from those which vary qualitatively to those which vary quantitatively. Presumably there is no basic difference between genes with easily detected major effect (often called oligogenes or switch genes) and those with only minor effect individually but which operate as part of a system of an indefinite number of factors (often called **polygenes**, although this term has been used in a more restricted sense). Operationally, it may be said that the difference between the two types of characters depends more on the

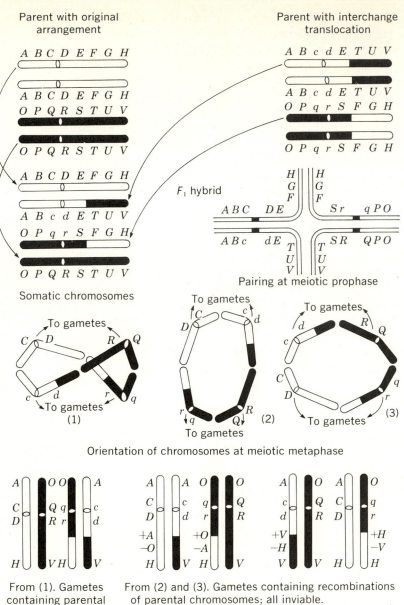

Parent with original arrangement

Parent with interchange translocation

F_1 hybrid

Pairing at meiotic prophase

Somatic chromosomes

To gametes
(1)

To gametes (2)
To gametes

To gametes (3)
To gametes

Orientation of chromosomes at meiotic metaphase

From (1). Gametes containing parental chromosomes; viable.

From (2) and (3). Gametes containing recombinations of parental chromosomes; all inviable. Gene combinations *CDqr* and *cdQR* cannot be obtained.

Figure 3-5 How heterozygosity for an interchange translocation can inhibit recombination between genes located close to the spindle-fiber attachment, or centromere, of the chromosomes concerned. Top row, two pairs of homologous chromosomes as they exist in related genotypes, one of which contains a reciprocal translocation, or interchange. Second row, left, the somatic chromosomes of the F_1 hybrid between the

relative importance of the genetic material and the environment in determining the phenotype than upon the size of individual gene effects. However, there are polygenic characters in which the role of the environment is relatively unimportant, such as number of abdominal bristles in *Drosophila*.

In studying variation in a quantitative character, as large a sample as possible of differing individuals is measured. All the individuals are unique, but their measurements may be grouped into size classes. When these measurements are plotted as a frequency distribution, the nature of the variation can be studied. For instance, the arithmetic mean or average can be calculated for all the individuals in the sample. The amount of variation in the sample can be estimated by the standard deviation S or its square, the variance S^2.

It is usually difficult to separate genetic and environmental components of variation and to study the genetic portion independently. Various laboratory techniques and rather complex mathematical formulations have been developed to study and separate these components. For example, one way to estimate the size of these two components involves the reduction of the genetic component until it is negligible. The variance in the character measured is observed in a population in a "normal" environment. Then the variance of the same character is measured in individuals of a highly inbred line raised in the same environment. Since these individuals may be considered to be essentially identical genetically, the variance observed may be attributed entirely to the effects of environment. The difference between the two variances is then an estimate of the genetic variance, since total variance (phenotypic variance S_P^2) is, in this case, equal to genetic variance S_G^2 plus environmental variance S_E^2. Thus

$$S_P^2 - S_E^2 = S_G^2$$

---

chromosomal types represented in the first row. Right, chromosome pairing at prophase of meiosis, showing positions of centromeres and of the interchange (center of diagram). Third row, the three possible ways in which the ring formed by pairing of the four chromosomes involved in the translocation can become oriented on the spindle at metaphase of meiosis. Left, zigzag arrangement which gives rise to gametes of the parental type. Center and right, open-ring arrangements, both of which produce inviable gametes containing large duplications and deficiencies. Bottom row, the chromosome constitution of the gametes which would be produced by the three types of orientation shown above. Note that all gametes containing the combinations *CD qr* and *cd QR*, shown at right, are inviable. Hence, this translocation inhibits recombination between alleles at gene loci near the centromeres of non-homologous chromosomes. *(From G. Ledyard Stebbins, Process of Organic Evolution, © 1966, 2d ed., p. 51, reprinted by permission of Prentice-Hall, Inc.)*

This procedure for estimating the variance components rests on the assumption that the environmental variance is independent of genotype, an assumption that is often incorrect.

Even if the genetic variance can be determined, further complexities exist. It cannot be assumed for all characters that the effects of genes are additive in a simple cumulative fashion. The genetic variance itself must be broken down into components. There is the **additive component** representing the differences between the homozygotes and heterozygote(s) for each locus. Also to be taken into account are a component resulting from interactions of alleles, i.e., a **dominance component,** and a component resulting from interactions of nonalleles, an **epistatic component.** In many situations the additive genetic component is the only one that can be estimated conveniently. Then the phenotypic variance is partitioned into additive genetic and remainder variances. The latter is a catchall term for the nonadditive components of the genetic variance plus the environmental variance and gene-environment interactions. The proportion of the phenotypic variance attributable to genetic effects is known as the **heritability** ($h^2 = S_G^2/S_P^2$). Sometimes heritability is defined in a narrower sense as S_A^2/S_P^2, where S_A^2 is the additive genetic variance. Heritability is a good estimator of the degree of resemblance between offspring and parent and as such is of great value to the plant and animal breeder.

It is important to note that the heritability of a trait depends on the environment in which it is measured. Thus, for instance, height might have a large heritability in a well-nourished population (tall parents having tall children, medium-sized parents having medium-sized children, and so on). A small heritability for height might occur in a population subjected to a different environment in which children are malnourished and tend to be short regardless of the height of their parents.

If a trait has a high heritability in two populations which differ in the average value of the trait, *this does not mean that the difference between the two populations is genetic*. Lewontin has provided an excellent example explaining this seeming contradiction. Suppose seeds from each of two genetically different highly inbred lines of corn are planted in soil in flower pots, one plant per pot. After they have grown for a time, the plants are measured and the variation in height in each line determined. Virtually all the within-line variation must be environmental, due to differences in conditions from pot to pot. The genetic variability of each line should have been reduced nearly to zero by the inbreeding. Thus the heritability of height in each line would be essentially zero. The average height difference which would be found between the two lines, however, would be due to genetic differences. This assumes, of course, that the pots are randomized so that, say, one line does not get more water than the other. The plants in one line all have the same genotype; the plants in the other line all have a different genotype.

In a contrasting case, suppose a random assortment of seeds from a highly outbred, genetically variable line of corn is planted in two sets of containers containing vermiculite, a porous, inert material. Suppose each set is watered with a carefully prepared nutrient solution used by botanists to grow plants hydroponically. Suppose one set receives a complete solution, containing all the required nutrients, and the other is watered with solution containing only half the needed nitrates and lacking a small amount of zinc, a trace element required for healthy plant growth.

In this case, each plant within both sets is grown in essentially identical environments, and all the observed variation in height would be genetical in each set. The heritability would be 1.0. The plants in the set receiving the incomplete solution, however, would be much shorter than the plants receiving the complete solution. The difference *between the two sets* is thus entirely environmental.

Therefore, we can see that a genetic difference in a trait between populations can exist even though the heritability for the trait is zero in both populations. Conversely there can be a large difference between populations that is entirely environmental, even though the trait has a heritability of 1 in both populations.

Misunderstanding of this property of heritabilities has led to the publication of a considerable amount of nonsense about racial differences in scores on intelligence tests. Because such scores show a rather high heritability in white populations, and because the mean scores in white populations are higher than those of black populations, some people have drawn the conclusion that the difference between white and black on this trait is genetic. The above discussion shows that conclusion to be entirely unwarranted.

Some social scientists have attempted to equalize environments by "compensatory education" and claim failure when the test scores of the disadvantaged (largely blacks) fail to equal those of the advantaged. As Lewontin has pointed out, this is roughly like asking a chemist to correct the difference between the two strains in the experiment with corn plants. The chemist would quickly discover the lack of nitrates in the solution given to one set and might decide that it is the cause of the shorter height of the plants. Addition of the nitrates, however, produces only a small improvement in height; the missing trace of zinc is the crucial factor. Indeed, it took plant physiologists a long time to discover the role of trace elements because some elements could leach out of laboratory glassware in sufficient quantities to permit plants to grow normally.

The lesson here is that the problem of providing "equal" environments is very difficult in controlled laboratory circumstances and virtually impossible in a complex social situation. In summary, heritability is a useful measurement for predicting how much evolutionary change might occur in a given line under given environmental conditions, but it tells us nothing about the differences between lines.

The evolutionist must deal with these complexities since the great majority of traits found to be variable in organisms vary quantitatively and are under the control of multiple factors. Where crosses can be made between races, species, or even genera, the F_1 offspring generally prove to be more or less intermediate and the F_2 show the continuous variation characteristic of polygenic inheritance.[1]

Control of a characteristic by many genes provides a stability of phenotypic expression that may not occur when only single genes are involved. For instance, a single mutational event is unlikely to disturb seriously the expression of a character dependent upon, say, the additive effects of 35 loci. However, a single mutational event $i \rightarrow I$ in the color-inhibitor gene of an onion will result in a white onion rather than a red or yellow one. In view of the possible drastic effects of changes in "major" genes, it is not surprising that most characteristics of organisms are controlled multigenically. Selection would have favored the development of such systems since they tend to reduce the possible deleterious effects of minor events such as a single base-pair substitution.

Polygenic systems that express relatively little variability may store tremendous potential variability simply because they have the ability to respond to selection by producing genotypes which, in the absence of selection, would never be produced. Let us suppose that a character is controlled by 40 loci, at each of which there are $+$ and $-$ alleles, and that the effects of the genes are additive; e.g., the most extreme phenotypes have all loci homozygous $++$ or homozygous $--$. If the gene frequency at each locus were $+.50$ and $-.50$, then in the absence of selection, the probability of a single diploid individual having the extreme $+$ phenotype (being homozygous $++$ at each locus) would be $(1/2)^{80}$, a number infinitesimally smaller than 1 divided by the number of electrons in the universe—for all practical purposes, zero. However, this potential could be realized in perhaps

---

[1] This is an overgeneralization of a complex situation; those wishing further information should consult Falconer, especially Chap. 14, the selections on inbreeding depression and heterosis.

8 or 10 generations by selection favoring individuals with a maximum of + alleles. Multiple-factor systems of inheritance provide, then, an important mechanism for maintaining balance between fitness for the immediate environmental situation and flexibility for response to long-range change in the environment.

SUMMARY

In the majority of organisms, genetic material, DNA, is associated with long protein strands forming chromosomes. The chromosomes are linearly differentiated into functional units, called genes, existing in numerous allelic states, which control the formation of specific polypeptides. Mutation of genes to different allelic states occurs spontaneously with a frequency of from 10^{-5} to 10^{-9} per generation. Meiosis and crossing-over result in recombinational units, usually equivalent to the functional genes. Except for chromosome linkage, genes segregate and recombine independently in the zygotes. Intraallelic interaction or dominance and interallelic interaction or epistasis occur. Some characters are affected by genes with conspicuous major effect, although modifying factors also may be found. Most characters are controlled by a very large number of nonhomologous genes, each with relatively small effect. Study of the resulting quantitative variation is complicated by the difficulty of separating the various fractions of the genetic component of variation from each other and from the environmental component. The basic source of variation is gene mutation. In populations of sexual higher organisms, recombination may be more important than mutation as a source of immediate variability in the short-term analysis.

REFERENCES

Falconer, D. S.: *Introduction to Quantitative Genetics*, Ronald, New York, 1960. A clearly written modern text dealing with quantitative and population genetics.

Levine, P. R.: *Genetics*, 2d ed., Holt, New York, 1968. A well-written text which emphasizes the role that both classical and molecular genetics play in all areas of biology.

Swanson, C. P., T. Merz, and W. J. Young: *Cytogenetics*, Prentice-Hall, Englewood Cliffs, N.J., 1967. An excellent introductory discussion of cytology and its bearing on genetics and evolution.

CHAPTER
FOUR
DEVELOPMENT

The genetic mechanisms described in Chap. 3 presumably evolved because they preserved successful combinations of genetic material. Some protoorganisms may merely have continued growth until accidents led to their disintegration. Many may have died because changing surface-volume relationships disrupted their inefficient internal organization. Some may have fragmented into smaller entities, chance alone determining whether the offspring fragments would have the organization to continue growth. Any mechanisms arising by chance that would tend to ensure that subsequent fragmentation products retained the capacity for growth (and further successful fragmentation) would automatically be perpetuated. Thus evolved the mechanisms that led to a stabilization of the marked variation which must have occurred in early division and development. The origin of these mechanisms is, in a sense, the basic problem in the origin of living systems, as has been discussed in Chap. 1. It has been suggested facetiously that human beings are merely one means that DNA has evolved for making more DNA; it may also be said that DNA is merely one device used by human beings to keep from having nonhuman offspring. Genetic material does not replicate without other components of living systems. The course of evolution has involved increasingly complex systems, including the genetic one.

GROWTH AND HOMEOSTASIS

Presumably the earliest organisms were unicellular (or noncellular). In such organisms only one or two cell divisions (and possibly one fusion of cells) produce separate functioning entities. Here the distinction between heredity and development or differentiation that we are accustomed to drawing for multicellular organisms is often difficult to make. Each cellular component is a hereditary unit that is replicated with greater or lesser accuracy during cell multiplication.

With the development of multicellularity and increased complexity, other problems arise. There is eventually a separation of germ-line cells and somatic cells. Nuclear and nonnuclear replicable components of cells that were present in unicellular organisms now appear to diverge somewhat in function. Greater stabilization and control are characteristic of the nuclear material, which we think of as *the* genetic information (genotype). The nonnuclear material plays a major role in development and differentiation, changing its properties through time and interacting with the nuclear material and the environment (including other cells). A unicellular organism is, in a way, immortal. The "end" of a cell usually is the result of accidental destruction (including the result of predation) or the division of one cell into two. Death comes eventually to the somatic cells of multicellular creatures and may be considered a part of the genetic system. One of the functions of reproduction, of course, is the replacement of individuals that have died. Another equally important function of reproduction, as we have seen above, is to bring about recombination of the genetic material and expression of variability which is the raw material of evolution. Death rate thus is in a real sense a measure of turnover rate in a species and affects the long-range flexibility and variability of that species in a heterogeneous and changing environment. As is discussed in Chap. 9, recombination rates in annual plants and long-lived perennial plants differ.

Increase of size, or growth, is inherent in the idea of continuing reproduction. The term **development** refers to the changes that take place during the life of an organism. Simple changes in surface-volume relationships, which may have constituted development in a protoorganism, seem a far cry from the life cycle of a monarch butterfly (egg, larva, pupa, adult), but the difference is one of degree, not of kind. Organisms change size in growth, and what is a working design at one size may be completely nonfunctional at another. Given the physical limitations of the size of mammalian cells (imposed by such factors, among many others, as the size of protein molecules and rates of diffusion), it is easy to see that a perfect miniature human being the size of an ovum or a sperm cell 6 ft long would be impossible. A genetic mechanism thus ensures not the production of duplicates of the parental multicellular organism but the production of entities that, within certain limits of variation, will *develop* into replicates of the parental type.

Regulatory systems which lead to the production of offspring resembling the parents under a variety of environmental conditions result in what may be termed **developmental homeostasis.** Most kinds of organisms seem to vary greatly only in rather superficial characteristics. In species studied genetically in which numerous mutants have been found, one can recognize a characteristic wild type. This wild-type phenotype is the result of developmental and genetic homeostasis, as well as the action of selective forces of the environment. Genetic homeostasis and selection are discussed in greater detail in Chap. 6.

The *Drosophila* wild type has been extensively described genetically. The human wild type does not have each eye of a different color or six digits on each hand. Redwood trees may vary in height and branch number, but they have characteristic green leaves of distinctive arrangement and rough red bark. Variation is more striking in large organisms than in small organisms, but this does not necessarily mean that small organisms are less variable. In all organisms, it seems to be true that critical developmental systems are relatively immune to genetic alteration. It is advantageous for an organism to avoid reproductive waste by producing optimum phenotypes from a number of minor variant genotypes. In many organisms the processes of development of a specific form have become canalized, leading to a uniform phenotypic expression of individuals in a given population in spite of the genetic variability among them. With this mechanism, genetic variability (of long-range importance) can be present with a minimum of reduction of fitness.

How is this buffering accomplished? Using the model of Lerner, let us assume that gene A produces substance a, which is modified by the action of b (a product of gene B) into r, and that r interacts with c (produced by gene C) to give substance t.

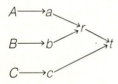

(*After I. M. Lerner*, Genetic Homeostasis, *Wiley, New York, 1954.*)

Now if mutation removes B or if the environment lacks the substrate from which b can be manufactured, then a will accumulate. A system in which high levels of a interact with d (product of gene D) to make s, which in turn can be transformed by c into t, is a buffered system.

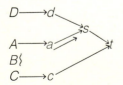

(*After I. M. Lerner*, Genetic Homeostasis, *Wiley, New York, 1954.*)

ORGANISMS: ORIGIN AND FUNCTION

This is a true feedback system, since the exact course leading to normal character expression is determined by the "information" the organism has with respect to the level of *a*. This does not mean that all buffering is genic, however.

It is common practice to draw a sharp line between genotype and phenotype. As a pedagogic device, this is useful for emphasizing the relative permanence and continuity of the genetic information, and although an oversimplification, it has led to considerable progress in the science of genetics. However, it has also led to the impression that the genotype is somehow the basic entity and that the phenotype is merely a crude reflection of the genotype (the image of which has been distorted by the environment). One might well wonder why selection has not done away with the phenotype altogether, permitting the genotype to evolve unsullied. The answer is, of course, that what can be separated in textbooks or in theory cannot be separated in living organisms. If the genetic material were dissected from a fruit fly, one would obtain a long meaningless string of nucleotides, itself an aspect of the "phenotype." It is clear that at this level of study the distinction between genotype and phenotype is meaningless. The genetic information becomes meaningful biologically only when it is translated through contact with the environment. Indeed, the value of the information is judged only by the translation, not the original. Agents of natural selection operate on the phenotype, not directly on the genotype, which merely determines the responses of the developing organism to the environment.

Only in recent years have evolutionists given proper attention to the processes of development that result in the production of an adult functioning organism from a fertilized egg or zygote. These processes are interrelated to form a system which Waddington has termed the **epigenotype.** This may be visualized as a branching system of developmental pathways, each of which leads to one of the components of the adult form (Fig. 4-1). Because the biochemical reactions determining each path are so interlocked with one another (as discussed above), there is a strong tendency for the normal end result to be produced even when there is considerable disturbance at early stages. Thus the paths are canalized or buffered as a result of feedback or cybernetic mechanisms interconnecting the paths.

This epigenetic system must have been the result of natural selection involving the genes that affect more or less directly the expression of particular characteristics of organisms. However, selection also must have involved the many genes that have as their only obvious phenotypic effect the modification of the expression of other genes. Waddington has pointed out that in a population of organisms in a given environment, each individual will have its own genotype, and therefore its own epigenotype, which will eventually

Figure 4-1 The epigenetic landscape. A diagrammatic representation of the epigenotype. The various regions of a developing embryo have before them a number of possible pathways of development, and any particular part will be switched into one or other of these potential paths. (*From C. H. Waddington, Principles of Development and Differentiation, The Macmillan Company, 1966.*)

result in the adult phenotype. Selection to preserve fitness in this particular environment may eliminate genotypes that produce deviant phenotypes. It may also eliminate individuals that are imperfectly buffered against environmental effects. There would thus be selection for a well-canalized epigenetic system.

Should the environment change, some well-buffered individuals would be likely to respond by producing fit phenotypes without the necessity for immediate genotypic change. After a period of time in the new environment, however, genotypic change is inevitable, and it is to be expected that selection would lead to the stabilization of the new developmental paths. When the organisms are returned to their original environment, it would be found that, as a result of this change in the genotype, they no longer produce their original phenotype. Thus what was originally a phenotypic (actually epigenetic) response to environmental change becomes incorporated into the genotype, as a result of selection for a well-buffered developmental system in the new environment. What appeared to be an "acquired characteristic" becomes hereditary through the effects of natural selection. This process, known as **genetic assimilation,** is discussed in Chap. 7.

LIFE CYCLES

Cyclic growth is characteristic of all organisms. Yeast cells go through sequences of fusion and fission, including both a **haplophase** and a **diplophase.** In most higher plants, development occurs both in the haplophase and in the diplophase, although the haplophase (male and female gametophytes) is usually much less conspicuous and of shorter duration than the diplophase (the sporophyte) (Fig. 4-2). In most animals there is virtually no development in the haplophase, which usually is restricted to the gametes (Fig. 4-3). Male hymenopterans, which are haploid, constitute a conspicuous exception; see Chap. 9.

The simple growth-fragmentation-growth cycle hypothesized as the most primitive form of development has been altered by selection in diverse ways. As an example of a complex developmental system, consider the protozoans that cause malaria. The sporozoites of *Plasmodium*, which in a mosquito environment migrate to the salivary glands, will, when injected into the bloodstream of *Homo sapiens*, invade specific cell types. Here they may reproduce asexually, producing merozoites which infect other cells or invade the erythrocytes. Those in the erythrocytes may reproduce asexually, producing merozoites which will infect other erythrocytes, or they may develop into gametocytes and eventually produce gametes which will fuse in

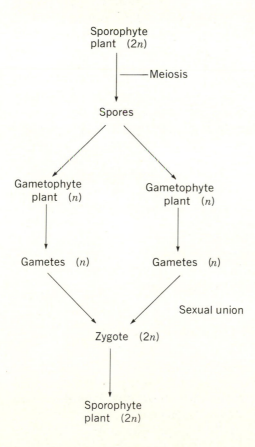

Figure 4-2 Sexual life cycle commonly found in diploid plants. Meiosis produces spores. (In certain plant groups male and female gametes may develop within a single plant.)

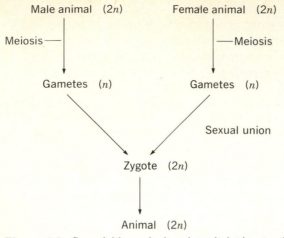

Figure 4-3 Sexual life cycle found in diploid animals. Meiosis produces gametes.

the gut of another mosquito. The motile zygote thus formed migrates to the gut wall and develops into a sporocyst. Sporozoites are formed in the sporocyst by cell division.

The same genotype responsible for the efficient feeding machine that we call a caterpillar also contains the information needed for the manufacture of the highly dissimilar reproducing-dispersing machine called a butterfly. The zygote that develops into a giant sequoia also contains the information necessary for the manufacture of its tiny pollen grains (few-celled male gametophytes). The single cell of the human zygote, through division, gives rise eventually to such diverse descendants as erythrocytes, muscle cells, and nerve cells. These deviations from the simplest cycle of development have been in response to selection operating on the entire life cycle of the organism from zygote formation till death.

The details of how selective agents operate and have operated to produce these systems will become clear only when the mechanics of the systems themselves are elucidated. Therefore it will be necessary to consider developmental systems briefly.

DIFFERENTIATION AND MORPHOGENESIS

Mitosis has been described as a means of ensuring the equal allocation of genetic information to the daughter cells in the course of cell division. That mitosis can accomplish this is easily demonstrated in a number of ways, as previously discussed. For example, if the zygotic nucleus of the dragonfly *Platycnemis* is permitted to divide seven times (to the 128-cell stage) and then all but one daughter cell are killed with a narrow beam of ultraviolet light, a complete embryo still will develop. Obviously all the necessary genetic information has been passed on from the original nucleus to its descendants. In view of the complex mechanism that seems to exist for the purpose of ensuring this successful transfer of necessary genetic information (and considering the demonstrable success of this system), it is pertinent to ask how cells and tissues become differentiated and arranged into a functional organism. Why is a nerve cell so different from an erythrocyte when both are descended from the same zygote?

One answer might be that the two cells were exposed to different environments during development. Even in very early cleavage stages, when few cells are present, the differences in cellular environment may be striking. Differentiation of animal cells may be influenced by such things as their positions relative to the animal and vegetal poles, the outside or inside of the blastula, and proximity to the blastopore in the gastrula. Position may affect the amounts of vital nutrients reaching the cell, the amount of oxygen available, the rate of accumulation of excess metabolites, etc. Once differentiation has begun, the effects multiply exponentially. Various combinations of differentiated elements add to the heterogeneity of the cellular environment, and complex interactions could provide the basis for the development of the entire organism. It can thus be said that development of the organism is controlled entirely by interactions within the cluster of dividing and growing cells. Each cell possesses the same information but uses it differently because it is operating in a different physiological environment.

This picture of development is supported by a vast array of data from experimental embryology. Interactions of cells may be seen in cultures of microorganisms in which density of the culture may affect rate of growth or determine whether growth is possible at all. The literature on induction (by contact or at a distance) and organizers testifies to the potency of effects of cellular environment and to the complexity of the systems that have evolved. None of these data, however, demonstrate that the genetic information plays merely a passive role in development. That mitosis does not parcel out portions of genetic information to the proper parts of the developing organism seems certain. Steward has demonstrated that certain differentiated plant cells retain their totipotency for an indefinite period. Through careful culturing techniques, he obtained complete carrot plants from fully differentiated phloem cells of carrot roots. These phloem cells first produce embryo-like structures, called embryoids, which when properly handled undergo all of the stages in "normal" embryology that eventually lead to a mature plant.

Further, Gurdon and his associates, through a series of ingenious transplant experiments, have shown that in the South African clawed toad (*Xenopus*) the nuclei of the cells remain totipotent throughout development. Fully differentiated nuclei of intestinal cells were transplanted into enucleated eggs from which a number of fertile adult frogs have been obtained. It must be noted, however, that Briggs and King did not obtain similar results when performing transplant experiments with nuclei from differentiated frog embryos of the genus *Rana*. When the nucleus of a frog egg was removed and replaced with one from a frog blastula, normal development ensued. When it was replaced with the nucleus removed from a gastrula or neurula, deformed embryos resulted, in which the only normal tissues were those derived from the germ layer from which the donated nucleus was taken. Whether this discrepancy can be attributed to differences in experimental technique, nuclear stamina, or times of nuclear inhibition has yet to be answered.

Further supporting the assumption that all cells are genetically complete, cells taken from various parts of the body of an organism and examined microscopically do not seem to be deficient in their chromosome content, as would be expected if gross partitioning took place. In those insects having polytene chromosomes with distinct banding in more than one body tissue, it has been reported that the banding does not change from tissue to tissue.

The study of these giant chromosomes has provided other critical data for the interpretation of development. During the course of ontogeny, certain bands become enlarged tremendously and are known as **puffs** (Fig. 4-4). This puffing process is reversible since it is an indication of actively transcribing genes. At the same stage in other tissues, different bands are in the puff condition. Recently, nucleolar genes from amphibian oocytes in the puffed or unwound condition were identified as coding for *r*RNA. The number of puffs that may develop on a chromosome is much lower than the number of bands; therefore at any one time the activity of parts of the genetic code must be restricted.

Perhaps the greatest challenge facing embryologists today is the elucidation of the mechanisms controlling differentiation. The question is of more than casual interest to the evolutionist. If portions of the genotype are turned on and off (as is indicated by the puffing process), the operation of selective agents is quite different from that in a situation in which the entire genotype is always operant. If a portion of the information that controls, say, the color pattern of a caterpillar is inactivated when the adult tissues are differentiating, it might be possible for the larval color to be changed in the course of selection without affecting the adult in any way. Equally, if genes affecting hair were inactivated in endoderm tissue, the form and color of the hair could be changed without any effect on the gut.

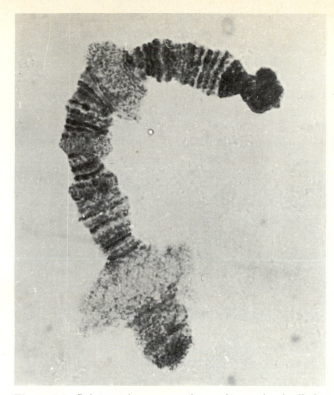

Figure 4-4 Polytene chromosome from salivary-gland cell of a *Rhynchosciara angelae* larva shortly before pupation, showing two puffs. (*Courtesy of E. Pavan.*)

It is known that the cells in different tissues of the same organism (containing identical genetic information) do not have the same complexes of protein. For protein synthesis, then, there must be specific control mechanisms that regulate the quantities of various gene products.

An interesting mechanism has been demonstrated in bacteria by Jacob and Monod, who found that bacterial genes may often be classed as either **structural genes** or **regulator genes.** The primary product of structural genes is *m*RNA, the synthesis of which is a sequentially oriented process initiated at certain regions of the DNA strands. These regions of initiation are called **operator regions.** An operator region may control the transcription of more than one structural gene. The adjacent genes controlled by one operator form a unit of transcription, the **operon.** Regulator genes produce cytoplasmic **repressor** proteins. Such a repressor protein associates reversibly with a particular operator, and the combination of operator and repressor prevents the transcription of an entire operon. Protein synthesis is thus blocked. The repressor protein is also viewed as reacting reversibly with small molecules, called **inducers,** in the cytoplasm as well as with the operator. In certain systems, only the unaltered repressor can associate with the operator and block the operon. Presence of the inducer will then eliminate the effect of the repressor and release the operon from repression. In other systems, the repressor molecule is not able to bind with the operator locus and switch off the structural gene unless it is first bound with a small molecule (the **corepressor**). A simplified diagram of this model is given in Fig. 4-5.

The details of the Jacob-Monod model are beyond the scope of this book. It is necessary only to add that microbial genetics provides considerable evidence for the various processes and systems postulated. It seems clear that some feedback mechanism of this sort must operate at the level of transcription of the DNA code and protein synthesis, just as such mechanisms are believed to be responsible for developmental homeostasis at later stages.

MODIFICATION OF THE DEVELOPMENTAL SYSTEM

Of all the phenomena of morphogenesis, none has received more attention from evolutionists than so-called recapitulation. It was soon observed by embryologists that early developmental stages of vertebrates resembled one another (at least superficially) to a much greater degree than the adults did. This has been interpreted by some workers to mean that in the course of development each organism goes through a condensed version of its phylogenetic history—that man, for instance, goes through a one-celled stage (zygote), a fish stage (when gill pouches appear), and a mammal stage. This generalization was originally called the biogenetic law by Haeckel and is often stated as "ontogeny recapitulates phylogeny." Such a crude interpretation of embryological sequences will not stand close examination, however. Its shortcomings have been almost universally pointed out by modern authors, but the idea still has a prominent place in biological mythology.

The resemblance of early vertebrate embryos is readily explained without resort to mysterious forces compelling each individual to reclimb its phylogenetic tree. It first should be emphasized that an early mammalian embryo resembles a fish embryo, not an adult fish. Virtually all organisms begin development as a single cell. The great diversity of life forms is the result of different courses of development determined in large part by the sets of genetic information that cause alterations of the course of development. However, each change does not mean transformation of the developmental

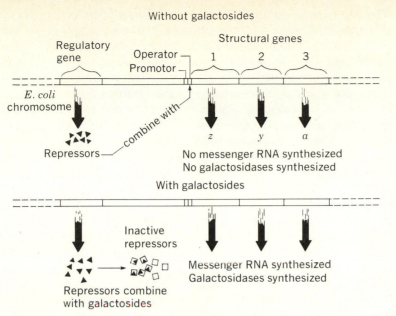

Figure 4-5 A model of regulation of protein synthesis based on the galactosidase (an enzyme) operon in *Escherichia coli*. The presence or absence of the substrate determines whether transcription will occur. (*Adapted from Paul R. Ehrlich, Richard W. Holm, and Michael E. Soulé, Introductory Biology, copyright McGraw-Hill Book Company, 1973; used by permission.*)

system. These tremendously complex integrated systems may be successfully modified only through accumulation of minor changes, with concomitant readjustments of balanced interactions of the various factors. By careful examinations of living and fossil organisms, we can infer these gradual changes of developmental pattern. A good example is the slow change in the vertebrate jaw structure, with the articular and quadrate, which were parts of the jaw in mammal-like reptiles, having been utilized as the ossicles of the hearing apparatus in mammals. All such changes have involved the modification of a preexisting developmental sequence and were possible only when this sequence could be modified without throwing it lethally out of balance. For example, gill pouches (embryonic precursors of gill slits in fishes) became altered into other structures, such as eustachian tubes and thymus glands, in higher vertebrates. This course of evolution avoided the possible complications which might have resulted from altering the entire set of processes producing the pouches themselves. Such alteration might well have caused great disturbance in the inductive systems responsible for, say, the development of the aortic arches.

The idea of recapitulation involves resemblance of developmental stages to ancestral forms. However, there are some cases in which adult forms appear to be similar to embryonic stages of their putative ancestors. For example, the females of some moths and beetles are larviform; certain salamanders do not metamorphose into adults but reproduce as larvae. Many characteristics of adult human beings (relative hairlessness, large head, etc.) are reminiscent of those of young anthropoid apes. The milk teeth of *Australopithecus*, the earliest known fossil man, resemble the adult teeth of *Homo sapiens*, while the permanent teeth of *Australopithecus* are like those of apes. In these and in a great many other similar cases, the developmental system has been altered in evolution so that an intermediate ancestral growth stage becomes the terminal form in the descendant. This phenomenon is known as **neoteny.**

The sequential stages in the development of an individual organism often are thought of as merely steps toward a final goal: the adult. It is surely more realistic biologically to think of ontogeny as the continually changing response of a given body of genetic material to a given environment. The various processes of the epigenotype regulate in varying degree the expression of the initiating genotype. Evolutionary change may involve any of the arbitrarily delimited stages of development.

SUMMARY

A line of descent does not consist of a straight-line sequence of individuals but of a series of cyclic phases. Each complete cycle is a developmental sequence, traditionally thought of as extending from the beginning of one diplophase (zygote) to the beginning of the next. Changes in the genetic information cause a variation in developmental sequence, and the accumulation of these genetically initiated changes constitutes evolution. It is important to remember that the entire life cycle evolves and that all stages of any given cycle are essential to survival and thus equally important from the standpoint of evolution. Many students of evolution, viewing the process from the end of a diplophase, have tended to ignore this fact of life.

REFERENCES

Cellular Regulatory Mechanisms, *Cold Spring Harbor Symp. Quant. Biol. 1961*, vol. 26. See especially the articles by Jacob and Monod.
Ebert, J. D., and I. M. Sussex: *Interacting Systems in Development*, 2d ed.,

Holt, New York, 1970. A treatment of the experimental foundations of modern developmental biology of plants and animals.

Gurdon, J. B.: Transplanted Nuclei and Cell Differentiation, *Sci. Am.,* **219**(6)**:**24–35 (1968). Presents a series of critical experiments which are of fundamental importance in understanding cellular differentiation.

Waddington, C. H.: *The Strategy of the Genes*, Allen, London, 1957. The steps (epigenotype) between genotype and phenotype are discussed in one of the few synthetic works in the field.

PART
TWO

POPULATIONS: PROPERTIES

Biologists working at the population level of organization have been oriented in large degree by the characteristics of the organisms studied. For instance, cytological features of genetic systems are more readily studied in Drosophila and Oenothera than in Papilio or Sequoia. Unusual combinations of circumstances have presented opportunities for studying the operation of natural selection in certain organisms, organisms about which there may be little or no cytogenetic information. Much of our knowledge is gleaned from work on organisms of economic importance, such as crops, domestic animals, and pests. Thus circumstances have made it impractical to produce a unified description of all aspects of evolution within populations.

The theory of population genetics has been created largely to treat diploid, outcrossing organisms. It is therefore convenient to present this body of theory and related examples from nature before discussing the complexities of systems controlling recombination in various kinds of organisms. It is hoped that eventually a theory may be constructed which will consider the interactions of the genetic system of an organism and the evolutionary forces acting upon the organism. In the meantime, the warning of Norbert Wiener must be kept in mind: It is very difficult to study the interactions of two systems with very different rates of time course. This is true when we attempt to understand history on the basis of day-to-day human behavior or when we try to understand phylogenetic history on the basis of individual gene changes in contemporary organisms.

CHAPTER
FIVE
POPULATIONS

In a sense, every phenomenon is unique. No two objects can occupy the same space and time. Sets of energy relations, if recurring with exact precision, at least differ in time. However, the perceptual universe is one of ordered uniqueness. The human mind is an apparatus that functions by imposing relationships upon unique events. Objects having characteristics in common are grouped into a class, e.g., table, race, butterfly, because such group concepts are useful for communication. Indeed, the existence of virtually all organisms depends upon their ability to generalize in some sense from collections of unique events. A "completely unique event," one for which there could be perceived no relationship with any other event, would be totally without meaning.

All scientific understanding is based upon populations of things and events and the patterns of interrelationship thought to exist among them. In order to understand the workings of cells, a biochemist studies the populations of chemical constituents and processes within the cell. For insight into the organization of organisms, physiologists and embryologists study populations of cells and tissues and the interactions among them. At the highest level of biological organization, the population biologist investigates populations of organisms and the relationships within and among them. In this book the term **population** will be restricted to aggregations of individual organisms, the sense in which it customarily is used by evolutionary biologists. *Population biology deals, then, with the properties of aggregations of organisms, particularly those emergent properties not possessed by the individual constituents of the populations.* Populations rarely can be studied in their entirety but must be sampled at one or more points in time. Unfortunately, since all organisms and populations change through time, it is not possible to sample the same population twice.

INDIVIDUALS AND COLONIES

A problem immediately arises with the definition of an *individual* organism. At first sight this appears to be easy, since familiar plants and animals exist as discrete units. However, the situation is complicated by the existence of forms such as lichens. These plants consist of a fungus parasitic upon algal cells included in its thallus. Different lichens have different morphological and biochemical characteristics, which fail to appear unless the correct combination of alga and fungus occurs. The alga and fungus reproduce separately, but the lichen reproduces as well, with propagules consisting of both alga and fungus. The alga can be grown without the fungus, and the fungus can also be cultured without its algal host.

Complex colonial organisms also present difficulties. The colonies of social insects present analogies with organisms, but usually such colonies are referred to as *quasi* organisms. The Portuguese man-of-war, a colonial hydrozoon, can be analyzed into its constituent polyps, which exhibit a striking division of labor. Among the algae and protozoa there are less specialized aggregations of individuals, in which what appear to be units may exist separately or as part of the colony. Even such forms as yeast *(Saccharomyces)* may show different behavior, depending upon environmental conditions. In liquid culture, yeast cells (plants?) are small ovoid cells that reproduce most frequently by budding. Short chains of cells may occur. Grown on a solid medium, however, yeast forms a giant "colony." This structure is a flattened object, several centimeters in diameter, with characteristic color and surface texture as well as biochemistry. Cells from the outermost layer, from the center, and from the portion adjacent to the medium are very different in form and presumably in function. Nevertheless, cells from any region can be used to start a new colony or liquid culture.

In the higher organisms there also may be difficulties in defining individuals. Many plants reproduce vegetatively (see Chap. 9), and if the "offspring" remain attached to the parent, the whole is considered an individual. Should they become separated, each plant usually is thought of as an individual even though it is genetically identical with its "parent." The self-sterile triploid day lily *(Hemerocallis fulva)* is one *genetic* individual throughout its range in much of the eastern half of the United States. Populations of hydra derived from a single budding individual likewise genetically constitute an individual, but ecologically and functionally they consist of many individuals. Populations of genetically identical individuals are called **clones.**

Complexes of individuals belonging to what are called different species may also occur. Many scale insects form amazing compound colonies in symbiotic association with a fungus *(Septobasidium)*.

Forest trees commonly become grafted when their roots touch in the course of growth. It has been found that when a root-grafted tree is cut down, the stump may survive for many years. Although it lacks photosynthetic tissue of its own, it can produce new bark from the cambium so that the cut surface becomes completely covered. Individual organisms are genetically different in these situations, but they are united closely into an ecologically meaningful unit. In the same way, a clone of viviparous onions (see Chap. 9) that are genetically identical constitutes an ecologically meaningful assemblage as it forms part of the environment of other organisms.

An individual is a set of operations (or a machine) programmed in advance to do particular things. In organisms, of course, the program is established by the coded genetic information. A group of genetically identical individuals is one individual reproductively. Ecologically they represent a population of individuals with different epigenotypes. If we had, historically, begun to think about biology in ecological terms rather than taxonomic terms, we would probably now deal with biological "facts" very differently.

It is obvious that to make the definition of individual clear, one must specify whether one's concern is taxonomy, genetics, or ecology. In what follows, a genetic definition of individual will be em-

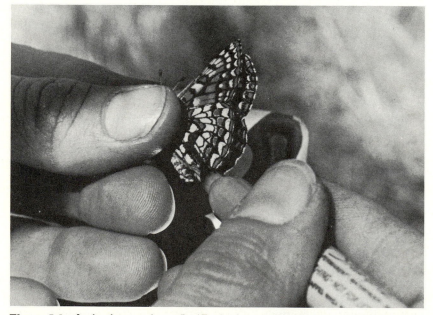

Figure 5-1 A checkerspot butterfly (*Euphydryas editha*) being individually numbered in code by the application of spots on the wings with a felt-tipped pen. (*Photograph by P. R. Ehrlich.*)

ployed. Most of the evolutionary work on populations has been done with organisms among which the discrete individuals are the result of sexual reproduction and thus usually genetically diverse. In sexually reproducing organisms, the most inclusive populations, called **species,** are generally considered to consist of individuals sufficiently alike genetically for successful reproduction to occur if they are given the opportunity to mate. The criteria for just what sort of assemblage may be labeled a species often are hard to establish, and the degree of conformity with these criteria in natural aggregations is usually only guessed.

In this chapter, examples will, in general, replace definitions. A butterfly and a bison obviously do not belong to the same species; a pair of robins raising a brood in the garden obviously do. Near the center of the continuum, problems arise: Could the European brown bear and the American grizzly be part of the same species? Are the oriental and the North American sycamores part of the same species? They have been geographically separated since the Miocene, but their hybrid is a vigorous and much-used street tree. Since our interest is primarily in the process of evolution, rather than in making arbitrary decisions, no answers will be sought to these questions here.

SPATIAL DISTRIBUTION

One property possessed by populations, but not (in the same sense) by their constituent organisms, is **distribution.** At any instant in time, checkerspot butterflies (*Euphydryas editha*) are distributed along an outcrop of serpentine rock on Stanford University's Jasper Ridge Biological Experimental Area (Fig. 5-1). The distribution of adults in successive years is shown in Fig. 5-2. The distribution in each successive year is somewhat different from that in the preceding. Such colonies of *Euphydryas editha* occur throughout the San Francisco Bay area; indeed, they are found along the West Coast from Baja California to British Columbia. It is difficult to specify the limits of the most inclusive population, i.e., of the species, in which the Jasper Ridge individuals could be placed. Most biologists would place in this most inclusive grouping individuals from colonies as far away as Montana.

Similarly, clusters of individuals in various-sized aggregates are found in plants. *Clematis fremontii* var. *riehlii*, a perennial which occurs on limestone glades in the midwestern United States, has been studied in some detail by Erickson. Individuals are grouped into aggregates of several hundred plants, many such aggregates occupying a single glade. The outcroppings of limestone are clustered

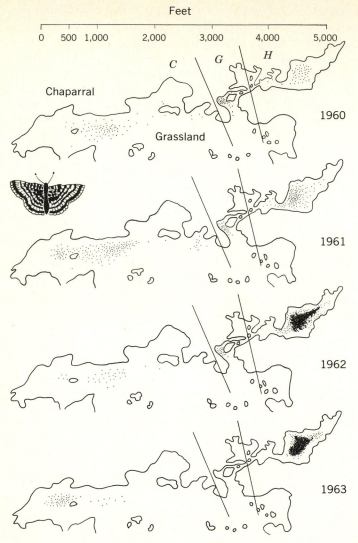

Figure 5-2 Distribution of adult *Euphydryas editha* in 1960 to 1963. Straight lines indicate the borders of areas *C, G,* and *H.* Solid line shows the edge of chaparral and oak woodland which surround the grassland in which the butterfly occurs. Each dot represents the position of first capture of a butterfly. (*Adapted from Paul R. Ehrlich, Richard W. Holm, and Michael E. Soulé, Introductory Biology, copyright McGraw-Hill Book Company, 1973; used by permission.*)

geographically with respect to the mountain systems and rivers. In the Midwest, the plant has a much wider distribution that represents the species population (Fig. 5-3).

In its loosest usage, distribution generally means the smallest geographic area that will enclose all the area normally occupied by the organisms under discussion. On a small world map of the distribution of *Homo sapiens*, the entire United States would be shaded to indicate its occupation by man. Oceanic areas and most of the Greenland ice cap would be left blank. In contrast, if we were mapping the occurrence of man on a large-scale map of Colorado,

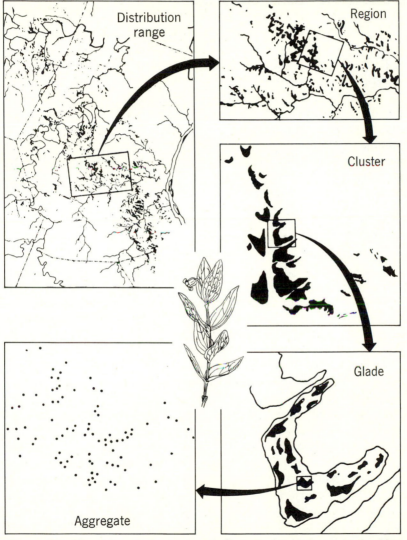

Figure 5-3 Hierarchy of aggregates of *Clematis fremontii* var. *riehlii*. [*After R. Erickson, Ann. Mo. Bot. Gard. vol. 32 (1945).*]

many high mountain peaks and some other areas would be left blank. The problems of such a mapping are obvious. Organisms are mobile at some stage of their life history, and their distributions are constantly changing. Furthermore, no known organisms are uniformly distributed over large areas. Thus the more resolution one strives for in describing a distribution, the more difficult the task becomes.

ECOLOGICAL DISTRIBUTION

The nonuniformity of geographic distributions can usually be explained by the relationships of the organisms with their living and nonliving environments. Gross examples of ecological factors controlling distribution are easily understood; the factors controlling the fine points of the distribution of a given organism virtually are never fully understood. In the San Francisco Bay region *Euphydryas editha* larvae feed on certain plant species of Plantaginaceae and Scrophulariaceae which grow on serpentine outcrops. In this area the butterfly occurs only where both the appropriate plants and serpentine are found together, but the presence of both plant and rock does not guarantee that the butterfly also will be there. On Jasper Ridge, areas of serpentine with abundant larval food plants are often unoccupied although they are immediately adjacent to the colony. They may, for instance, lack suitable nectar sources for the adult butterflies. Not only must *all* environmental conditions be suitable for a habitat to be occupied, but chance must supply access to the suitable area. Thus some areas suitable for *E. editha* may not support colonies simply because no fertilized females have ever reached them. Man has provided many organisms with access to previously uninhabited but suitable regions—as starlings, English sparrows, cabbage butterflies, honeybees, dandelions, and mustard constantly remind us.

Another somewhat different and interesting example of how one organism affects the distribution of a second organism is found in the West Indies and the Bahamas. It has been reported by Carson that on these islands the fly *Drosophila carcinophila* breeds only on the exterior nephric grooves of the black land crab (*Gecarcinus ruricola*). The adult flies depend upon microorganisms which grow on the third maxillipeds and the nephric grooves below for their food source. The larvae are restricted to the nephric grooves and the opposing surfaces of the maxillipeds. That not all *Gecarcinus ruricola* populations have *Drosophila* larvae associated with them rules out any rigid mutualistic relationship between the two organisms. Carson has suggested that the microorganisms the flies feed on may play a role in the conservation and reutilization of water which is reabsorbed at the base of the maxillipeds after passing through and over the filaments of the groove. Further, it is interesting to note that *Drosophila carcinophila* belongs to a group of species generally found in desert or xeric environments. The black land crab, except during the breeding season (when it enters the water), is found almost exclusively in xeric environments and thus, in a sense, provides a highly specialized xeric environment for *D. carcinophila*.

POPULATION STRUCTURE

The **structure** of a population is considered here to be *the totality of all the factors that govern the pattern in which gametes from various individuals unite with each other*. The structure can vary from situations in which combinations might seem to be essentially random, e.g., certain marine animals that release gametes into the sea or some wind-pollinated plants, to those in which the probability of certain combinations is much higher than others. The latter case is certainly the rule, if for no other reason than that close neighbors usually have a higher probability of mating than more distant ones.

Such factors as length of generation and size of individuals also are important. If the variable to be measured is the number of new gene combinations produced in a given area per unit of time, then small organisms will differ from large ones. In any place there are fewer large organisms than small ones and thus less recombination. Organisms with a short life cycle produce more gene combinations than those with long generation time, and their mutation rates also differ.

Especially in higher animals, many behavioral systems have evolved that profoundly affect the structure of a population. Many animals are effectively sedentary in spite of great dispersal potential. Birds often return from long migrations to exactly the same breeding location as in previous years. Twitty has shown that California newts have incredible perseverance and navigating ability, returning precisely to a particular segment of a stream to breed. Indeed, displaced individuals have returned to their home pool over several miles of mountainous country. Specificity within a stream is clearly shown in Fig. 5-4. Butterflies often use their powers of flight merely to patrol a restricted area. In one study of the Jasper Ridge colony of *Euphydryas editha*, 625 out of 647 recaptures of marked adults (98.6 percent) were made in the area (see Fig. 5-2) of previous capture. In an investigation of the population structure of the tropical butterfly *Heliconius ethilla* by Ehrlich and Gilbert, movement was shown also to be sharply restricted and based primarily on the distribution of the pollen

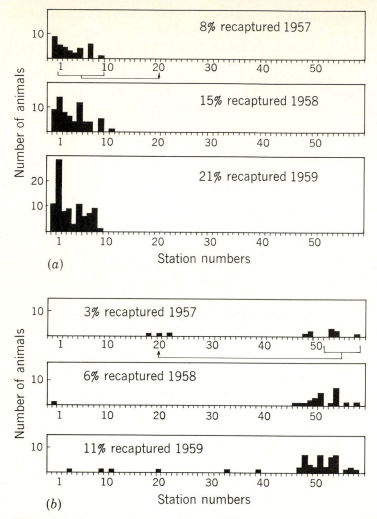

Figure 5-4 Homing behavior of newts. (*a*) Recaptures of individuals displaced in 1956 to a point about ¹/₂ mi downstream from place of original capture. (*b*) Recaptures of individuals displaced in 1956 to a point about 1 mi upstream from place of original capture. Area of original capture and point of release shown by arrows. Station numbers refer to arbitrary subdivisions of stream course. All individuals leave the stream after each breeding season. [*From V. C. Twitty, Science, vol. 130 (1959).*]

sources required by the adults for food. Similar behavioral restriction of physical-dispersal ability seems to be the rule rather than the exception in nonmigratory butterflies.

Few animals seem to be truly nomadic. Most (including most

human beings) stay close to their birthplace, occupying a **home range.** Many animals defend all or part of their home range from intruders of their kind, the well-known phenomenon of **territoriality.** This behavior, common in birds, mammals, reptiles, fishes, and some invertebrates, results in a nonrandom distribution of individuals in the population. They are dispersed more evenly than one would expect in a distribution governed solely by chance. Among other things, this often keeps the population size at a level where the supporting resources of the environment (food, nesting space, etc.) are not strained or entirely consumed. Individuals, often young adults, that do not successfully occupy and defend a territory must find greener pastures or starve; thus a dispersing component is added to the population.

Statistically, the opposite of territoriality is aggregating behavior, which results in more "clumping" than if individuals were randomly distributed. Animals showing this behavior may have little or no known social organization, as in the case of prehibernation aggregations of ladybird beetles or snakes. At the opposite extreme we have the highly social insects, among which there are morphologically differentiated castes, but the reproductives in a colony may consist of a single pair of individuals. At this extreme, selection is largely differential reproduction of colonies, not individuals. In many vertebrates a social hierarchy (peck order) is established in which some individuals dominate others, obtaining perquisites ranging from first choice of mate to first place in line going through the barnyard gate. Dominant males may control large harems and in contrast with their less aggressive brethren may make a large contribution to the pool of genetic information of succeeding generations. Sometimes, social groups, e.g., bands of howler monkeys and gibbons, exhibit territoriality.

NUMBERS OF INDIVIDUALS

One of the most obvious attributes of any population is the number of items, events, or individuals it contains at a given time. The number of individuals in biological populations is of great interest, but unhappily the size of natural populations usually is difficult or impossible to ascertain with accuracy. The most inclusive populations may include billions of individuals, e.g., man, houseflies, some microorganisms, and various algae, or less than 100 individuals (whooping cranes or certain rare plants such as *Pedicularis dudleyi* and *Tetracoccus ilicifolius*).

In the strictest sense, whenever the number of individuals in a population changes, the distribution of the population changes; often a change in distribution also means a change in numbers. Study of Fig. 5-2, on which the position of first capture of *E. editha* adults for four consecutive years is plotted, gives some idea of the problems of dealing separately with distribution and abundance. In area *C* there was little increase in numbers between 1960 and 1961 but some change in the distribution pattern. In area *G* the numbers decreased, and there was a concomitant shrinkage in the area occupied. In area *H* the numbers increased greatly, and the population occupied a region virtually devoid of individuals the year before. It is important to note that the figures, like virtually all representations of distribution, are a conventionalized, static representation of a dynamic situation. The numbers and distribution of individuals in biological populations are constantly changing, the speed of the changes varying greatly from organism to organism.

Population **dynamics,** the study of changes in population size, is of considerable interest to the evolutionist, for, as will be seen, changes in population size affect the evolution of a population in diverse ways. This can be understood intuitively, since each individual in a population is part of the environment of every other individual. Therefore any change in population size is automatically a change in the environment of the population, and populations evolve in response to environmental changes.

The factors that control population sizes are diverse and in many cases poorly understood. Table 5-1 shows estimates of size changes in the three populations of Jasper Ridge *Euphydryas editha* over a 14-year period. Notice that one population, area *G*, declined essentially to extinction in that period and was reestablished. Note also that population size fluctuations in the three populations are quite independent. Recent work by Singer and Gilbert indicates that microclimate is the major factor influencing the size of the population of Jasper Ridge. If the food plant dries out in the spring before the larvae have grown large enough to survive the dry summer in diapause (a resting state), larval mortality is high. The numbers of larvae have not been large enough to cause competition for food. In contrast, the *E. editha* population at Del Puerto Canyon, some 50 mi to the southeast, has a much more limited larval food resource, and competition frequently leads to high mortality. A third population, Snyder Meadow, Nevada, appears to be limited primarily by predation.

Thus in populations of the same species very different factors may limit different populations. In some cases, e.g., Jasper Ridge, the direction of change in population size may not be intimately related to size itself, and in others, e.g., Del Puerto and Snyder Meadow, growth rates may be highly dependent on population size.

TABLE 5-1. POPULATION-SIZE ESTIMATIONS FOR THE THREE JASPER RIDGE
POPULATIONS OF EUPHYDRYAS EDITHA.

Year	Area *C*		Area *G*		Area *H*	
	Min.*	Est.	Min.*	Est.	Min.*	Est.
1960	120	150–210	46	50–60	102	130–150
1961	242	300–400	38	40–50	304	600–800
1962	44	70–90	20	25–30	678	1,200–1,800
1963	54	80–100	8	10–12	708	1,200–1,800
1964	194	300–400	2†	1–2†	1,908	2,400–3,200
1965	156	350–450	2	4–6	1,376	3,000–5,000
1966	244	700–1,000	4	6–10	428	3,000–4,000
1967	320	700–1000	54	70–100	730	3,000–4,000
1968	722	1,000–1,200	90	135–180	648	1,500–2,000
1969	286	700–1,000	108	150–200	906	2,500–4,000
1970	332	700–1,000	132	180–300	650	1,400–2,900
1971	222	300–600	10	15–30	426	600–1,000
1972	948	2,400–3,200	10	15–30	868	2,000–2,500
1973	1,016	1,050–1,250	26	30–60	580	600–800

*   Min. = minimum possible size: twice the number of males captured.
†   The presence of a female was doubtful.

Populations may have their size controlled by different factors
operating in different manners. Indeed the same population may be
subject to different controls at different times. A great deal of
controversy has arisen over what generalizations may be made about
the "control" or "regulation" of population size. We need not be
concerned with this here, except to note that the question of the
efficacy of group selection (page 119) is an important factor in the
argument.

ENVIRONMENT

An individual organism, when such can be recognized, is in a sense
the morphological resultant of the physiological processes of which it
is composed. Each of these processes is interrelated with the other
functions of the organism, and the complex of processes cannot be
separated from the environment, except artificially. The functioning of
an individual is determined by the relationship of its constituent
processes to factors of the environment. Each process has a **range of
tolerance** for the environmental factors, which must fall within the
intensity span of the factors. Organisms do not function unless the
ranges of tolerance of all these processes fall within the intensity
spans of all the environmental factors. These ranges of tolerance of

any organism are, of course, determined by the genotype assembled in the zygote and by the developing epigenotype (see Chap. 4).

There is no satisfactory way, at present, to make sense of the complex interactions of organism and environment. Usually some very rough classification of environmental factors is employed. For example, Andrewartha and Birch, in their classic work *Distribution and Abundance of Animals,* divided the environment of a given organism into four components: **weather, available nutrition, other organisms,** and **a place to live.** These components may be further subdivded as required. All are continually changing in some degree. Just as the range of tolerance of a particular process changes during the course of development (resistance to desiccation, heat or cold sensitivity), so the intensities of environmental factors change cyclically, as well as in complex and little-understood patterns. The soil around the roots of an oak tree may become leached of mineral elements, which are restored when the leaves fall and rot. The food plant of a butterfly dies out in a drought year. The required environmental factors for the establishment of seedlings or for the pollination of flowers may be present for a brief period at only one time of the year, and the behavior of the plant must be closely correlated with the occurrence of these factors.

Interactions of amazing intricacy may be seen in natural populations. For example, other organisms in the environment may be classified also as nutrition (host, prey, food plants of herbivores) or as a place to live (host, trees, etc.). A young muskrat may find all suitable burrow sites (a place to live) occupied by older, stronger individuals (other organisms) and be forced to migrate. During its migration it may freeze (weather), starve before it finds suitable forage (nutrition), or be killed by a coyote (other organism).

Several refinements of the Andrewartha and Birch classification have been suggested. A system of components more broadly applicable might be (1) **medium,** e.g., weather, climate, salinity, turbidity, pH; (2) **resources,** e.g., space, sources of energy, and materials required for synthetic activities; (3) **other nonresource organisms,** e.g., predators and parasites, and (4) **hazards,** e.g., tall buildings as hazards for migrating birds. Under this classification, two caterpillars eating the same leaf would not be part of each other's environment unless they interacted directly. In the absence of such an interaction, each would alter only the resource component of the other's environment.

Attempts at extreme refinement of a classification of environmental components seem relatively fruitless. The essential step was made by Andrewartha and Birch in recognizing that the classification should be made essentially from the point of view of the organism.

This provided a major improvement over the older classification into **biotic** and **abiotic** factors, since the mode of action of the factor in the latter case could never be inferred from the classification. For example, a rock and a bush under which a rabbit hides from a hawk would be abiotic and biotic factors respectively under the old classification. Under the new they would both represent a single resource—space in which to hide. An area lacking a sufficiency of that resource will also lack rabbits.

As part of the environment of an organism, other organisms may serve as vectors of genetic material in infection or reproduction, as well as of propagules. The flowers, fruits, and seeds of the angiosperms show a great diversity of devices effecting successful pollination and dissemination by specific animal vectors. Everyone knows of the instances of pollinating insects carrying the pollen grains (male gamete-producing plants) from flower to flower on their legs or bodies. Less familiar are those orchids in which the flower resembles the abdomen of a female insect so closely that males of the mimicked species attempt to copulate with the flower. The pollen is carried in tiny bags, or pollinia, from one flower to another on the end of the abdomen of the male insects. There is no apparent advantage to the insect.

Similar situations are not unknown among animals. An interesting instance is that of the adult human botfly (*Dermatobia hominis*), which catches mosquitoes and attaches eggs to their bodies before releasing them. The eggs hatch when the mosquito lands on the warm skin of a man, and the larvae burrow in and start to develop. This fly parasitizes a number of mammals other than man.

Besides the interactions among plants and animals commonly observed in the temperate zone, there are less well-known examples of extreme intricacy in the tropical rain forest. On the branches of the giant trees, epiphytic plants accumulate water and soil among their leaves. In this specialized niche, the larvae of mosquitoes and of frogs hatch, grow, and metamorphose. The mosquito fauna is stratified in part by the distribution of the epiphytic plants in which they grow. In the same forests lives the three-toed sloth, the hairs of which are colored greenish by symbiotic algae. The sloth moth (*Bradypodicola hahneli*) spends its entire life on the sloth, its larvae presumably feeding upon the algae. It is clear that different kinds of organisms, e.g., plants and the herbivores which attack them or parasites and hosts, affect each other's evolution reciprocally. Such **coevolutionary interactions,** which are of particular interest to evolutionists concerned with the structure of ecosystems, are discussed in greater detail in Chap. 11.

Changes in climatic patterns strongly affect the kinds of organisms that can exist in an area. Arctic fossils of tropical plants and amphibia testify to warmer times in the past and long-empty desert cities to changes in rainfall pattern or soil fertility. Years of commercial grazing, with no return of essential elements to the soil from decaying plants and animals, have changed the nutritional characteristics of many areas of the Great Plains of the United States, with a resultant change in the flora and fauna. English sparrow populations in our cities have become much smaller since the disappearance of the horse and its seed-laden droppings. Grasslands have diminished in some areas as firebreaks (such as roads) and other sorts of fire control increase.

Man's activities in transplanting organisms have provided many striking instances of change in the influence of other nonresource organisms. *Opuntia* cactus was an introduced pest in Australia until an *Opuntia*-feeding moth (*Cactoblastis cactorum*) was introduced from South America, where *Opuntia* is native. The moth quickly reduced vast areas of dense *Opuntia* to widely scattered clumps. Storksbill, mustard, and wild oats (*Erodium*, *Brassica*, and *Avena*) have replaced native plants in some areas of California. Shipborne rats have virtually destroyed the fauna of small vertebrates on numerous islands.

Changes in the availability of space resources also constantly occur. Silting and slowing down of streams make them untenable for larval and pupal blackflies and other organisms depending on swift-running, oxygen-rich water. Planting trees across the Great Plains has permitted range extensions by tree-nesting woodland birds. The slow accumulation of humus and disintegration of rocks into soil make homes for orbatid mites, centipedes, fungi, and other lovers of dank, dark places. Often the organisms living in areas subject to frequent catastrophic change have specialized genetic systems affecting their genetic behavior (see Chap. 9).

All these examples are of relatively spectacular variation in the environment. Important also are the smaller, more frequent changes: the day-to-day temperature variation that affects the plankton population of a shallow pond or the yearly precipitation changes that determine the condition of a butterfly's food plant. None of these changes can be viewed as an isolated event. Increased moisture may improve the condition of the larval food plant of a butterfly, and a large adult population may result. Among other things, this may mean more food for nestling song birds and, in the long run, more food for hawks. Large numbers of butterflies may mean more caterpillars next year, a year when little moisture means a poor crop of food plant. Thus few adults survive, almost no food plants survive, and topsoil is lost through erosion. Large numbers of song birds from the previous season may find no suitable substitute food, with resultant starvation, emigration, and unsuccessful reproduction. The hawks go hungry.

This somewhat overdrawn example illustrates only a few of the many permutations of effects which might be hypothesized as resulting from a simple change in precipitation. Actually such gross effects are relatively infrequent, for most ecological systems are made up of a great many elements. Complexity leads to less than a one-to-one dependence. Drought which reduces butterfly populations may lead to an increased supply of grasshoppers, and the song birds readily shift their diet. Long-term associations have presumably experienced most of the usual variations in climate, and their members presumably can respond to it. Thus the butterfly food plant will probably survive a drought (perhaps, if it is an annual, as ungerminated seeds) and be ready to return to abundance when moisture reappears. If erosion has not proceeded too far, it may be stopped. Doubtless, however, many organisms would be permanently affected.

COMMUNITIES

Even if a single interbreeding population were found in an environmentally diverse area, it would be expected that, in time, genetic processes would lead to diversification (Chap. 6). Details of the processes involved in the formation of complex communities of organisms are not well known. Denuded areas do become repopulated. During the early stages of repopulation, the aggregations of plants and animals are short-lived. Several different aggregations can be distinguished over a period of time before a relatively stable community develops. Sequential replacement of such stages is referred to ecologically as **succession,** and the terminal stage often is called a **climax community.**

Ecologists do not agree on the best methods of studying succession or terminal communities. Modern workers feel that communities are part of a continuum and that they can be distinguished as units only artificially. Indeed, since there are no genetic connections operating between reproductively isolated populations (although they do influence each other evolutionarily as selective agents), one would not expect communities to be discrete units. Each interbreeding population behaves according to its own cytogenetic processes, producing individuals whose genotypes determine ranges of tolerance that enable them to function. Each entity has had an evolutionary history dependent on, among other things, its own genetic processes. Because of these differences one would not expect any two populations to follow the same historical pattern for even a short time. The community, however defined, results from the overlapping ranges of tolerance of the individual organisms for the various factors of the environment at a particular place (where, of course, other organisms may themselves be environmental factors).

If studied for relatively short periods, the climax communities in a successional series usually appear to be stable or in a steady-state equilibrium. Within such communities, cycling of energy and matter is constant and regulated by feedback mechanisms. Much energy is stored in organic materials: plants, animals, humus, etc. Such a community is disturbed by outside influences only with difficulty. It is thought that organisms from other environments find it difficult to migrate into the community. Energy relations appear to be diverse, and primary producers (green plants), primary consumers (herbivores), secondary consumers (carnivores), and decomposers (usually microorganisms) can be distinguished, as well as omnivores and higher-order consumers. Natural selection has resulted in an ecological unit of great complexity. Organisms have evolved with respect to their position in this complex (or in successional stages leading to it) as the environment of the earth has changed through time.

It is obvious that the different populations of organisms found in a given place are not a random sample of organic diversity. Caribou are found with *Cladonia* (a lichen called reindeer moss) and wolves with caribou. If a butterfly collector seeks the larvae of the pipe-vine swallowtail (*Battus philenor*) in the Arizona desert, he must find the decumbent pipe vine (*Aristolochia watsonii*). He soon learns that this often grows in the shade of other desert plants, especially along the edge of depressions. From observing the searching behavior of adult female butterflies, it seems likely that they use similar associations in their search for a place to lay their eggs. Oaks and hickories are often found together, and oak-hickory forests are a good place to hunt the Virginia deer. Wheat, dogs, houseflies, body lice, and *Treponema pallidum* (the organism which causes syphilis), as well as many other organisms, are often found in association with man.

Since the early 1950s, there has been a continuing trend toward the integration of evolutionary and ecological concepts. In population ecology, the work of Andrewartha and Birch provided a base line by presenting a carefully formulated approach to the question of how the environment influences the probability of an individual's surviving and multiplying. This approach is fundamental to much of the work on evolution within populations described in this book. The problem of how communities evolve is enormously more difficult and considerably more intractable. For a long time, ecologists faced with the complexity of communities tended to throw up their hands and settle for describing that complexity in terms of the more permanent members of each assemblage.

In 1957, however, Hutchinson formalized a new approach to community ecology by defining the **niche** of an individual as a many-dimensional space (hypervolume) the coordinates of which were environmental variables, such as water temperature and salinity. For any individual organism, for example a fish, tolerance limits could theoretically be established for both these factors, and a rectangular **niche area** plotted (if the variables are not independent in their action, the area will not be rectangular). Tolerances to a third environmental variable, such as pH, could be plotted on a coordinate perpendicular to the plane of the first two, and a **niche volume** for the fish specified. Any point within the volumes would have coordinates on all three axes falling within the tolerance range of the fish.

Of course there are many more than three environmental variables affecting all organisms. Thus, to completely define an organism's niche in theory, Hutchinson had to resort to more than three dimensions, since a new dimension must be added for each variable. This is readily done mathematically, and the niche was defined as an n-dimensional **hypervolume.** Although such a hypervolume can in practice be neither pictured nor specified, the general concept has proved to be a seminal one in framing many questions about the characteristics of communities, especially in attempting to explain the diversity of species present and to predict the future patterns of change both in that diversity and in the characteristics of the species.

Hutchinson's modeling of the niche founded a school of community ecology which has flowered in the past decade, led by Slobodkin, MacArthur, Margalef, Cody, Levins, Wilson, Rosenzweig, Roughgarden, and others. The school has been characterized by the assumption that many important features of communities can be mimicked in rather simple models and that those models can then be used to make various kinds of predictive ecological theories. Considerable success has already been achieved in such areas as explaining and predicting the diversity of island faunas and testing various hypotheses about competition. The kinds of interrelated questions which are investigated include the following:

1. Why are there more species in the tropics than in the arctic?
2. What are the constraints upon subdivision of resources by competing organisms?
3. What are the fundamental patterns of predator-prey interrelations?
4. What determines the degree of specialization (or generalization) of a species?

Other approaches to community ecology have gained prominence in recent years, in particular, investigating patterns of energy flow in ecosystems and building complex computer models for the in-depth study of specific systems. The Hutchinson school, however, is one of the greatest interest to evolutionists. It and the related coevolutionary approach seem to hold very great promise of making sense of the extremely complex community events which result from the interactions of numerous populations evolving through the processes described in this book. Some of these complexities will be examined more closely in later chapters.

SUMMARY

Although for some organisms the concept of individual is difficult to define, in most organisms individuals are readily identifiable. Aggregations of individuals are referred to as populations. These, or aggregations of them, may form interbreeding populations; the latter are variously combined to form taxonomic groupings. The functioning of individuals is physiological and is determined by the genotype and epigenotype that set the ranges of tolerance of the organism to intensity spans of complexly varying environmental factors. The factors giving evolutionary coherence to populations are genetic. Communities are aggregations of diverse populations which form part of each other's environment, and they become structured with respect to energy relations. Usually there are no genetic functions between populations in a community, but different kinds of organisms in the community affect each other's evolution. Communities owe their existence only to the mutuality of the tolerance ranges of the constituent organisms at a particular time.

REFERENCES

Andrewartha, H. G., and L. C. Birch: *The Distribution and Abundance of Animals*, University of Chicago Press, Chicago, 1954. A classic work on animal ecology (mostly "autecology").

Ehrlich, P. R., and R. W. Holm: Patterns and Populations, *Science*, **137:**652–657 (1962). Problems of dealing with the properties of populations are discussed and pertinent literature cited.

Odum, Eugene P.: *Fundamentals of Ecology*, 3d ed., Saunders, Philadelphia, 1971. The latest, most comprehensive edition of this standard and important ecological text.

Rickfets, R. E.: *Ecology*, Chiron, Newton, Mass., 1973. A comprehensive synthesis of ecology and evolution which is up to date and beautifully illustrated.

Stebbins, G. L.: *Variation and Evolution in Plants*, Columbia University Press, New York, 1950. See especially the first chapter.

Whittaker, Robert H.: *Communities and Ecosystems*, Macmillan, New York, 1970. A paperback introduction to the concepts of interacting communities and ecosystems.

Wilson, Edward O., and William H. Bossert: *A Primer of Population Biology*, Sinauer Associates, Stamford, Conn., 1971. Many important quantitative aspects of ecology are concisely presented, together with problems and their solutions.

In the first part of this book, the origin of life, the coding and transfer of genetic information, the development of organisms, and some features of populations were discussed. We shall now begin to deal with the very core of evolutionary theory—changes not in individuals but of populations. This chapter will be concerned with the theoretical aspects of the genetics of **mendelian populations.** A knowledge of the basic ideas of population genetics is absolutely essential to an understanding of how the mass of inherited information possessed by a population changes from generation to generation. Familiarity with the simple mathematical ideas presented here will permit the reader to comprehend the more complex situations discussed in ensuing chapters. Although nonmathematical descriptions accompany the various algebraic examples, a firm grasp of the material will be facilitated by working through the algebra.

The examples in this chapter are gross oversimplifications. The integrative aspects of the genotype, multiple alleles, simultaneous operation of different evolutionary forces, and other complicating phenomena are largely ignored. For the moment, it is assumed that a single locus can be torn from its substrate and subjected to conditions of our choice; complex interactions are left for later consideration.

MENDELIAN POPULATIONS

Only sexual organisms constitute mendelian populations, which can be defined loosely as aggregates of **interbreeding individuals.** A more precise definition is neither possible nor desirable, for the word "interbreeding" may refer to any situation from panmixis to almost complete isolation. One might consider the potato beetles on a single potato plant as a mendelian population, or the definition might be broadened to include those in a single potato field, those in a group of adjacent potato fields, or indeed those in a county or larger area. It is therefore important to indicate the scope of a population under discussion and to state what is known of its structure.

Panmixis

A population is **panmictic** if the individuals within it mate at random. Each individual is equally likely to mate with every individual of the opposite sex within the population as defined. The expected frequency of any given kind of mating is the product of the frequency of the type of the male participant and the frequency of the type of his female partner. For example, consider a hypothetical animal which has both black and white forms and exists in a panmictic population consisting of 100 males (90 black, 10 white) and 100 females (70 black, 30 white). The expected frequencies of the various matings are given in Table 6-1.

Statistical study of the frequencies of the various matings might show that observed deviations from these expected frequencies are satisfactorily explained by sampling error, i.e., chance, leading to the conclusion that, *at least with respect to color*, the population is panmictic. To look at panmixis another way, it can be said to occur when the genotypes of the individuals in each mating pair are a random sample of the genotypes present in the population. Complete panmixis seems rare or nonexistent in nature, if for no other reason than that relatives often tend to live close together and thus mate with one another. When this happens, the mating pairs are *not* a random selection from the population and a component of inbreeding is added to the population-genetic picture.

TABLE 6-1.

Mating		Product	Expected frequency
Male	Female		
Black	× black	.90 × .70	.63
Black	× white	.90 × .30	.27
White	× black	.10 × .70	.07
White	× white	.10 × .30	.03
			1.00

Gene pool and gene frequency

The total genetic information possessed by a population may be referred to as the **gene pool** of the population. If the gene pool could be described completely, one would know not only what kinds of information were present but also the frequencies of the different kinds. This chapter is concerned mostly with the distribution within the gene pool of the information at a single locus.

One of the basic ideas of population genetics is that of gene frequency. If it is assumed that there are only two alleles at the locus (A, a) under consideration, there are then N diploid individuals of which D are homozygous for one allele (AA) with respect to the locus studied, H are heterozygous (Aa), and R are homozygous for the other allele (aa). Then $D + H + R = N$, and there are three types of individuals carrying two types of genes. The N individuals have 2N genes at this locus. Since each AA individual has two A genes and each Aa individual has one A gene, the total number of A genes in the population is $2D + H$. The proportion p of A genes in the population is

$$P = \frac{2D + H}{2N} = \frac{D + \frac{1}{2}H}{N}$$

The quantity p, the proportion of A genes in the population, is known as the **gene frequency** of A. By convention, the gene frequency of the other allele (a) is q. Since these are the only two alleles at the locus, $p + q = 1$ and $q = 1 - p$.

Hardy-Weinberg law

If there is random mating in a population, and if the gametes produced by the mates combine at random, there is complete random union of all the gametes produced in the population. As each gamete contains only one of the alleles, the frequency of the two different kinds of gametes (A and a) and the gene frequency are the same. Combining the gametes at random gives

$[p(A \text{ sperms}) + q(a \text{ sperms})] \times [p(A \text{ ova}) + q(a \text{ ova})]$
$p^2(AA \text{ individuals}) + 2pq(Aa \text{ individuals}) + q^2(aa \text{ individuals})$

Populations with this distribution of **genotype frequencies** are in an equilibrium condition.

This equilibrium is described by the Hardy-Weinberg law, which may be stated briefly as follows:

If alternate forms of a single autosomal gene $A_1, A_2, \ldots, A_n$ are

present in a large panmictic population, then in the absence of mutation, selection, or differential migration the original proportions (gene frequencies) of these alleles ($p_1, p_2, p_3, \ldots, p_n$) will be retained from generation to generation, and after one generation the proportion of genotypes will also reach an equilibrium. The genotype equilibrium frequencies are given by the terms of the expansion $(p_1A_1 + p_2A_2 + \ldots + p_nA_n)^2$.

Further discussion of population genetics will center around this law, which is one of the fundamental concepts of biology. An algebraic demonstration of the maintenance of Hardy-Weinberg equilibrium is given in Table 6-2.

If an arbitrary initial population has a gene frequency $p = 0.2$ ($q = 0.8$) and genotype frequencies $AA = .10$, $Aa = .20$, and $aa = .70$, the population reaches equilibrium in one generation and then remains there (Table 6-3).

Note that there is no change whatsoever in the *gene frequency*, which can be determined after the first generation by taking the square root of the frequency of the proper homozygote. Under the conditions described, there is a genetic inertia in mendelian populations. Unless mutation, selection, differential migration, certain

TABLE 6-2. MATINGS AND OFFSPRING IN A POPULATION IN HARDY-WEINBERG EQUILIBRIUM

Type of mating*	Frequency of mating	Proportions of offspring		
		AA	Aa	aa
$AA \times AA$ ($p^2 \times p^2$)	p^4	p^4		
$AA \times Aa$ ($p^2 \times 2pq$)	$2p^3q$	p^3q	p^3q	
$Aa \times AA$ ($2pq \times p^2$)	$2p^3q$	p^3q	p^3q	
$Aa \times Aa$ ($2pq \times 2pq$)	$4p^2q^2$	p^2q^2	$2p^2q^2$	p^2q^2
$AA \times aa$ ($p^2 \times q^2$)	p^2q^2		p^2q^2	
$aa \times AA$ ($p^2 \times q^2$)	p^2q^2		p^2q^2	
$Aa \times aa$ ($2pq \times q^2$)	$2pq^3$		pq^3	pq^3
$aa \times Aa$ ($2pq \times q^2$)	$2pq^3$		pq^3	pq^3
$aa \times aa$ ($q^2 \times q^2$)	q^4			q^4
Totals	1.00†	p^2‡	$2pq$§	q^2¶

* The frequency of types of both males and females is given by the terms of the expression $p^2 + 2pq + q^2$; therefore, with random mating the frequencies of the different matings are $(p^2 + 2pq + q^2)(p^2 + 2pq + q^2) = p^4 + 4p^3q + 6p^2q^2 + 4pq^3 + q^4$.

† The sum of this column is $p^4 + 4p^3q + 6p^2q^2 + 4pq^3 + q^4 = (p^2 + 2pq + q^2)^2 = [(p + q)^2]^2 = (1^2)^2 = 1.00$.

‡ The sum of this column is $p^4 + 2p^3q + p^2q^2 = p^2(p^2 + 2pq + q^2) = p^2(1) = p^2$.

§ The sum of this column is $2p^3q + 4p^2q^2 + 2pq^3 = 2pq(p^2 + 2pq + q^2) = 2pq(1) = 2pq$.

¶ The sum of this column is $p^2q^2 + 2pq^3 + q^4 = q^2(p^2 + 2pq + q^2) = q^2(1) = q^2$.

TABLE 6-3.

Generation	Genotype frequency			Gene frequency	
	AA	Aa	aa		
1	.10	.20	.70	$p = 0.2$	$q = 0.8$
2	.04	.32	.64	$p = 0.2$	$q = 0.8$
3	.04	.32	.64	$p = 0.2$	$q = 0.8$
4	.04	.32	.64	$p = 0.2$	$q = 0.8$
N	p^2	$2pq$	q^2	p	q

changes in the mating pattern, or a drop in population size disturbs the equilibrium, there is no change in the genetic structure of the population. To a very large degree, overcoming this inertia (especially changing the *gene frequency*) is what is described as evolution.

The ideas associated with the Hardy-Weinberg law are basic to any consideration of evolutionary processes, and it is essential that the reader become thoroughly familiar with them. For a discussion of extensions of the Hardy-Weinberg law and a treatment of population genetics in general, the reader is referred to the text on population genetics by Crow and Kimura (1970).

Although populations in equilibrium are rare (or nonexistent) in nature, the law is of great value in describing a situation in which there is *no* evolution at a single locus, as it provides a base line for measuring evolutionary change. Some of the ways in which populations deviate from Hardy-Weinberg equilibrium are considered next.

POPULATION SIZE

In all populations there are some random fluctuations in gene frequency. Because of sampling error, no set of gametes drawn from a parental population will have exactly the same gene frequency as the parental population. In addition, because of chance occurrences in the union of gametes, the population of zygotes formed will not have precisely the same gene frequency as the population of gametes. Finally, even if the deaths among maturing individuals are completely random, sampling error will intervene to produce a filial breeding population in which the gene and genotype frequencies once again deviate from the gene and genotype frequencies of the original population of zygotes. In populations which have large numbers of reproducing individuals these sampling errors tend to balance each other, since they are different in different areas and at different stages in the reproductive process. Changes in gene frequency because of sampling error are therefore usually negligible. However, in populations which have, say, only 20 pairs of adults, these random fluctuations may take on considerable importance.

Thus the size of a population may have a considerable effect on its genetic structure. Biological populations are always finite in size. The gross population size is simply the number of individuals in a population at a given time. Of greater significance is the effective breeding size N, defined as the number of parents responsible for the genetic composition of the next generation. That the breeding size of a population may be considerably smaller than the overall size of a population should be obvious from the situation of man, where individuals under 10 years of age and over 60 years of age are generally excluded from the breeding population.

Effective breeding size

In terms of population genetics, the relevant population size is known as the **effective breeding size** N_e. This is equivalent to the breeding size N only in an "ideal" population of continuing large size in which there are equal numbers of the two sexes, mating is at random, and the gametes are drawn at random from the parents. This ideal population is an abstraction; for practical purposes the effective breeding size is *always* smaller than the breeding size. In some cases, however, the difference may be quite small. Uneven sex ratios, inbreeding, cyclic reduction of breeding size, and nonrandom sampling of the gametes all depress the effective breeding size. As an example, consider a population of N breeding individuals, mating at random, not fluctuating in size, and having the gametes drawn at random from the parents. Table 6-4 shows the effects of different sex ratios on the effective breeding size, where $N_e = 4N\male N\female/(N\male + N\female)$, and $N\male + N\female = 400$. The derivation for this formula is given in Li (1955).

Similarly, if the effective size for four successive generations is 10, 100, 10,000, and 100, the *average* effective size (which is the harmonic mean, the reciprocal of the mean of the reciprocals, of the sizes at each generation) over those generations is approximately 33. Note that this is much closer to the minimum number in the series than to the maximum. Effective breeding size indicates the size of the "ideal" population whose genetic behavior would be the same as that of the population under consideration. This permits valid comparisons of population size. To equate two populations with 400 breeding individuals, one with 200 males and 200 females, and the

TABLE 6-4.   BREEDING SIZE—400

Males	Females	(effective breeding N_e size)
200	200	400.00
100	300	300.00
50	350	175.00
25	375	93.75
5	395	19.75
1	399	3.99

other with 5 males and 395 females, would be to ignore the genetic consequences of the sex-ratio inequality of the latter population. For example, all other factors being equal, it would lose its variability much more rapidly than the former because the males, in essence, would become a bottleneck in the transfer of genetic variability from generation to generation. One sometimes can point out the individuals that are members of a breeding population, but one cannot segregate a group and label it the effective population. Unless otherwise specified, population size will refer in what follows to the effective population size and will be designated simply by N.

Genetic drift

Consider a barrel containing 10,000 black marbles well mixed with 10,000 white marbles, representing the gametes of a population with a gene frequency of $p = q = .50$ at some locus. A random sample of 2,000 marbles from this barrel represents the 1,000 diploid individuals that will make up the breeding population of the next generation. Perhaps the first sample consists of 979 white marbles and 1,021 black marbles ($p = .49$). The gene-pool barrel is then reconstituted with 9,790 white marbles and 10,210 black marbles, and the sampling of 2,000 marbles is repeated. This time assume that a sample of 1,033 white and 967 black marbles represents the gene frequency ($p = .52$) of the breeding population of the next generation. Once again the barrel is reconstituted, with 10,350 white and 9,670 black marbles. A continuation of this process would, under most circumstances, produce a very slight fluctuation of gene frequency around the original figure of $p = .50$. Note that this model is constructed so that the population size remains constant.

Now consider another barrel containing 500 white marbles and 500 black marbles to represent the gametic gene pool of a smaller population with gene frequency of $p = .50$. Suppose 10 marbles representing 5 individuals are withdrawn at random and that 6 are white and 4 are black ($p = .60$). The original gametic population is reconstituted with 600 white and 400 black marbles and the procedure repeated. Now 8 white and 2 black marbles are drawn (gene frequency $p = .80$) and the barrel reconstituted with 800 white and 200 black. On the third sampling one might get 7 white and 3 black (gene frequency $p = .70$), and thus reconstitute the barrel with 700 white and 300 black. It is easy to see that in carrying through this procedure much more violent fluctuations in gene frequency have been caused than with the larger population. This random fluctuation of gene frequency due to chance occurrences in a finite population is known as **genetic drift.** The term drift is quite descriptive, as the frequency of p seems to drift around without approaching any particular value, unlike the directional movements caused by the so-called systematic pressures of mutation, selection, and differential migration.

Figure 6-1 shows the probability distribution of p in 50 and 5,000 offspring from a parental population in which the gene frequency was $p = .50$. In this and succeeding examples it will be assumed that all the conditions for maintaining a Hardy-Weinberg equilibrium are present, except in the factor under study. In other words, the factors will be manipulated one at a time to show how they may alter the equilibrium.

The possible values of p in the two different groups of offspring are given in the two histograms of Fig. 6-1. The ordinate represents the approximate probability of p falling in each interval. As expected, the chances for large fluctuations due to sampling error are much greater for the smaller group of offspring.

Figure 6-1 also shows the fates of large numbers of loci, all of

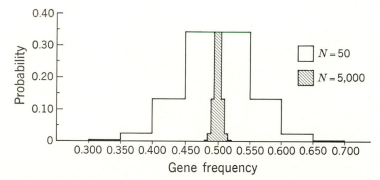

Figure 6-1 Probability distribution of p in 50 and 5,000 offspring from a parental population in which the gene frequency was $p = .50$. For details see text.

which were at gene frequency of .50 in their parental population and all of which were under Hardy-Weinberg equilibrium conditions except for population size. With $N = 50$, one would expect only 3 loci in 100,000 to fluctuate to a value greater than .70, whereas with $N = 5,000$ only 3 loci in 100,000 would fluctuate to a value greater than .52. On the other hand, one may wish to consider the distribution of the gene frequency of a given locus in a large number of populations, all with an initial gene frequency of .50. As shown in Fig. 6-1, with $N = 5,000$, we would expect 99.994 percent of the populations to have a gene frequency between .48 and .52 for the specified locus, whereas with $N = 50$ the same percentage would have a range from .30 to .70.

Decay of variability

The possible consequences of drift in a small population are shown diagrammatically in Fig. 6-2. The gene frequency is analogous to a pinball; moving down the slope (through time), it ricochets from value to value, *as long as it stays on the table*. However, in the absence of mutation and migration, the values of 0 (loss) and 1.0 (fixation) are *dead ends*; once the ball drops into one of these slots it stays there. That is, gene *A* is either fixed (all individuals *AA*) or lost (all individuals *aa*). Thus the gene frequency of *A* can move from any intermediate value to the end points 0 and 1 but not vice versa; the gene frequency will ultimately be 0 or 1 if the population is left undisturbed long enough. Since when the gene frequency reaches 0

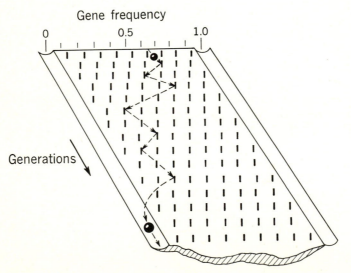

Figure 6-2 Model of genetic drift. See text for explanation.

or 1, heterozygotes no longer can be formed at the locus, the more genes that are lost or fixed, the fewer heterozygotes there are in the population. The process of reduction of heterozygosity through loss and fixation at various loci is known as the **decay of variability.**

The rate of this decay is intimately tied to the population size. Genes are lost at a rate of $1/4N$ per generation and fixed at a rate of $1/4N$ per generation. Thus this decay of variability takes place at the rate of $1/2N$ genes per generation. Consider some extreme examples. If p has a value of .125 in a breeding population of only 4 individuals, 1 individual has the only gene representing p, and only one-eighth of the gametes have that gene. If that single individual fails to reproduce, the gene is lost. If the individual does reproduce, the chance of loss of the gene is $1/2$ when it leaves only 1 offspring, $1/4$ when it leaves 2 offspring, $1/8$ when it leaves 3 offspring, etc. However, in a population of 4,000 breeding individuals, an absolute minimum of 500 individuals carry the gene in question if the gene frequency is .125. All these individuals would then be homozygous for the gene in question, and all must fail to reproduce to cause loss of the gene. Drift (sampling error) is a *mathematical fact*. However, the significance of drift as opposed to selection has been widely debated. It seems certain that drift is of very little importance in large populations (say N greater than 50 to 100), but in small populations drift may be an important evolutionary factor.

Loss of mutations

An additional aspect of loss of variability through sampling error concerns the probability of loss of a mutant gene. Imagine a mendelian population in which N is constant (a pair of adults produce, on an average, sufficient offspring to be "replaced" by two adults in the next generation) and in which a certain locus is at fixation (all individuals homozygous AA). If, in a single individual, A mutates $A \rightarrow a$ producing a single heterozygote Aa, if it breeds, this heterozygote must backcross with an AA individual, no others being available. The offspring of this backcross will consist on an average of 50 percent AA and 50 percent Aa. The probability of each family being of size 0, 1,2,3 . . . , r . . . is assumed to be approximated by a Poisson distribution with a mean of 2 (the average number of offspring). If no offspring are produced, the gene is lost; if 1 offspring is produced, the probability of loss is .50; if 2 offspring are produced, the probability of loss is .25; and if r offspring are produced, the probability of loss equals 2^{-r}. Using the coefficients from the Poisson distribution, one can calculate the limit of the overall probability of loss in that generation to be .3679. Fisher has calculated the probabilities of

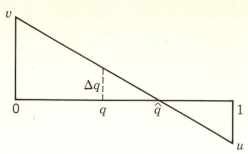

Figure 6-3 Equilibrium with mutation and reverse mutation u = .0004, v = .00002, q = .60. See text for explanation. (*From S. Wright, The New Systematics, Oxford University Press, 1940.*)

extinction for a mutation appearing in a single individual under the condition that the mutation is of no selective value and also under the condition that the mutant has a 1 percent selective advantage over its wild-type allele. These probabilities are reproduced in Table 6-5, which shows that a new mutation has virtually no chance of survival in a population unless selection counteracts the decay of variability.

MUTATION

Mutations are changes in genetic information and as such have been discussed in Chap. 3. Mutation will now be considered as one of the systematic pressures tending to cause deviation from the Hardy-Weinberg equilibrium. If A is the wild-type gene (normally the most frequent allele in the population) and a is the mutated gene, u = mutation rate, $A \rightarrow a$, and v = back-mutation rate, $A \leftarrow a$. Such a system has an equilibrium point, as shown in the following calculations:

$$\Delta q = up \text{ (gain)} - vq \text{ (loss)}$$
$$= 0 \text{ at equilibrium point } (\hat{p},\hat{q})$$

Therefore

$$\Delta q = 0 = u(1 - \hat{q}) - v\hat{q}$$
$$0 = u - u\hat{q} - v\hat{q}$$
$$-u = \hat{q}(-u - v)$$
$$\hat{q} = \frac{u}{u + v}$$

$$\hat{p} = 1 - \hat{q} = 1 - \frac{u}{u + v} = \frac{u + v - u}{u + v} = \frac{v}{u + v}$$

where p = gene frequency of A
q = gene frequency of a
$p + q = 1$

$q = 1 - p$

Δq = change in q

$\hat{q}$ = equilibrium (read "q hat")

A graphic representation of an instance of this equilibrium is shown in Fig. 6-3. Note that Δq is the net change per generation in the frequency of a and that Δq is positive when $q < \hat{q}$ and Δq is negative when $q > \hat{q}$. The change at $q = 1$ (all a) is one-half as great as at $q = 0$ (all A) since $A \rightarrow a = u = .00004$ and $a \rightarrow A = v = .00002$. Thus we can see that in a population meeting all the requirements of Hardy-Weinberg equilibrium except the absence of mutation, the equilibrium value for the frequency of a gene is determined by the mutation rate and back-mutation rate at the locus in question.

SELECTION

The nonrandom (differential) reproduction of genotypes is called **natural selection** if the differential reproductive success cannot reasonably be explained by drift or mutation. One might regard the streams of life of a population as being made up of continually varying, dividing, fusing, and disappearing particles flowing from the past to the present through a series of immensely complex screening sieves. Selection can be said to have occurred when the stream at a lower point differs from the stream at a higher point to such a degree that statistical analysis indicates it is highly improbable that the observed difference is due to sampling error (drift) or mutation. In looking for selection, one must be sure that the stream neither

TABLE 6-5. PROBABILITY OF EXTINCTION OF A MUTATION APPEARING IN A SINGLE INDIVIDUAL*

	Probability of extinction	
Generation	No advantage	1% advantage
1	.3679	.3642
3	.6259	.6197
7	.7905	.7825
15	.8873	.8783
31	.9411	.9313
63	.9698	.9591
127	.9847	.9729
Limit	1.000	.9803

*   From Ronald A. Fisher, *The Genetical Theory of Natural Selection*, 2d ed., Dover Publications, Inc., New York, 1958; reprinted by permission of the publisher.

branches (emigration) nor receives a tributary (immigration) in the stretch observed.

If a population is genetically heterogeneous, the probability of reproductive success of some genotypes will be higher (with possible rare exceptions) than the probability of success of others. Thus certain kinds of genetic information will become more and more common in the gene pool of the population, and other kinds will become less and less common. The gene frequencies p and q will change with time rather than remaining constant, as would be expected under the conditions of Hardy-Weinberg equilibrium.

It is popular to speak of selection as a great "creative force" in evolution, the "cause" of observed trends. In fact, it is a phenomenon observable only *a posteriori*, a description of occurrences. When a nonrandom set of genotypes leaves more offspring than others, selection has occurred. It is therefore *inaccurate* to speak of natural selection "acting" on a population or "operating" in nature. Various factors **(selective agents)** of the environment act or operate in ways leading to differential reproduction of genotypes. The *result* is natural selection. In the broadest view, selection reduces the diversity of living organisms; organisms containing certain types of combinations of genetic information are inviable or do not persist. This process "creates" what we recognize as a certain order in nature, in the same way that the order we see in a team with seven heavy, muscular linemen is "created" by the game of football. It is only in this restrictive sense of combating a trend toward increasing entropy (disorder) that selection is creative.

Critics of the theory of natural selection have claimed that selection can in no way be creative since it functions merely to eliminate certain types. They point out that in the common analogy of natural selection with a sieve, if large and small rocks are screened, the end result is a pile of the smaller rocks from the original pile. Nothing new has been created. This analogy is incorrect, for it ignores the potential mutational and recombinational variability of biological entities. It is as if two small rocks could mate and their offspring included rocks smaller than either parent. If a sufficiently fine sieve were used, screening would create a heap of rocks smaller than any in the original pile.

In the broadest terms one can recognize **natural selection,** in which the environment determines which genotypes are the most "fit," and **artificial selection,** in which man determines which genotypes are the most "fit." It is important to note that while agents leading to natural selection must always operate on the genotype *through the phenotype*, it is sometimes possible for man producing artificial selection to act directly on the genotype. For example, let us suppose that a moth has a melanic form produced by a dominant gene M. If the dominance is complete, only two distinguishable phenotypes will be presented to the environment, melanic and nonmelanic (*M*– and *mm*). As far as agents of natural selection are concerned, the homozygous dominant (*MM*) and heterozygous (*Mm*) genotypes are indistinguishable. In the laboratory the situation would be quite different; appropriate testcrossing would permit the selection of either genotype.

Fitness or adaptive value

In any given environment, the joint effect of the relative survival value and relative reproductive capability of a given genotype constitutes its **fitness,** or **adaptive value.** Fitness has many components, examples of which are the relative fertility, duration of reproductive period, ability to find a mate (in animals), efficiency of pollination mechanism (in plants), and general hardiness of individuals of the genotype in question. We shall adopt the convention that the continuum of adaptive values runs from 0, as for a zygote homozygous for a lethal allele, to 1, for the genotype that donates the largest number of gametes to individuals of the next generation. This maximum number depends, of course, on the organism under consideration. In this book, the term *fitness indicates the success of a genotype in transmitting genetic information to the next generation*, this success being measured relative to all other genotypes along a scale running from 0 (no information transmitted) to 1 (the most information transmitted). Some authors include within the concept of fitness such things as long-range fitness, the ability of a population to meet hypothesized future changes in the environment. There can be no doubt that, as noted under genetic systems and later in this chapter, certain types of populations are better able to adjust to environmental changes than others. However, because of the difficulties of working with this aspect of fitness, it seems best to utilize the concept in a restricted time sense, as above. Further discussion of these and related problems will be found in the last chapter.

It should be noted that these definitions of selection are a long way from the popular picture of "Nature, red in tooth and claw" usually painted by the uninformed. The creative aspect of the process consists almost entirely of the environment, *through selection*, affecting the genetic properties of populations so that they produce the fittest phenotypes.

Types of selection

There are three basic types of selection within populations: **directional, stabilizing** and **disruptive.** Directional selection has occurred when there is a shift in the position of the population mean for the character considered. Stabilizing selection means lowered fitness of extreme individuals and the concomitant reduction of the variance of the character, resulting in a more uniform population. Disruptive selection has occurred when two or more different types have been favored but intermediate types were at a disadvantage. It usually increases the variance and, under some conditions, may lead to fragmentation of the population. Examples of all three types will be found in the next chapter.

Extensive work has been done to describe various types of selection mathematically. Most of this work lies outside the scope of this book, and the interested reader is referred to Crow and Kimura (1970). However, a few examples of this quantitative treatment follow, both to illustrate the methodology and to relieve the reader of the necessity of accepting on faith the results of certain selective processes.

Homozygous recessives completely unsuccessful

Taking an array of genotypes, for example, *AA*, *Aa*, *aa*, one may assign to each one an adaptive value *W*, which indicates the relative differences in their capacity to contribute to the filial gene pool. As stated above, the most successful genotype is given the value 1, while less fit combinations have lower values. Thus we might find the following situation:

Genotype	*AA*	*Aa*	*aa*
Adaptive value *W*	1	1	$1-s$

In this case the homozygous recessives would be adaptively inferior to either the heterozygote or homozygous dominant. The degree of disadvantage is measured by *s*, the selection coefficient, $1-s$ being the fitness or adaptive value. In this situation the coefficient could vary from 0 (no disadvantage, making the comparison pointless) to 1 (homozygous recessives lethal).

When $s = 1$ (homozygous recessives lethal), consequences of complete removal of the recessive individuals from the population are shown in Table 6-6. The relationship between the gene frequencies of any two consecutive generations then is

$$q_{n+1} = \frac{q_n}{1 + q_n}$$

where the subscripts indicate the generation numbers. Thus

Table 6-6. COMPLETE ELIMINATION OF RECESSIVES

Gener-ation	Before or after selection	Genotype frequency			Gene frequency of a
		AA	Aa	aa	
0	Before	p^2	$2pq$	q^2	q
	After*	$\dfrac{p^2}{p^2+2pq}$	$\dfrac{2pq}{p^2+2pq}$	0	$\dfrac{q}{1+q}$
1	Before†	$\dfrac{1}{(1+q)^2}$	$\dfrac{2q}{(1+q)^2}$	$\dfrac{q^2}{(1+q)^2}$	$\dfrac{q}{1+q}$
	After‡	$\dfrac{1}{1+2q}$	$\dfrac{2q}{1+2q}$	0	$\dfrac{q}{1+2q}$
. . .	. . .	. . .	. . .	. . .	. . .

* $p^2 + 2pq$ represents the total after the aa (q^2) genotypes are removed. To find the frequencies of the two remaining genotypes they must be expressed as proportions of the total. These two frequencies are obtained simply as follows:

$$\frac{p^2}{p^2+2pq}=\frac{p(p)}{p(p+2q)}=\frac{1-q}{1-q+2q}=\frac{1-q}{1+q}$$

and similarly $\dfrac{2pq}{p^2+2pq}=\dfrac{2q}{1+q}$

$$\frac{1-q}{1+q}+\frac{2q}{1+q}=1 \qquad q = \text{gene frequency of } a = \frac{1}{2}\times\frac{2q}{1+q}=\frac{q}{1+q}$$

$$p = 1-q = 1-\frac{q}{1+q}=\frac{1}{1+q}$$

† The genotype frequencies before selection are obtained by using the new p and q values and expanding

$$\left(\frac{1}{1+q}+\frac{q}{1+q}\right)^2$$

For example, the zygotic frequency of

$$AA = (\text{new } p)^2 = \left(\frac{1}{1+q}\right)^2 = \frac{1}{(1+q)^2}$$

‡ Frequencies after selection are calculated as in the first footnote, e.g.,

$$\frac{1/(1+q)^2}{1/(1+q)^2+2q/(1+q)^2}=\frac{1}{1+2q}$$

$$q_0 = q_0$$

$$q_{0+1} = q_1 = \frac{q_0}{1 + q_0}$$

$$q_{1+1} = q_2 = \frac{q_0/(1 + q_0)}{1 + q_0/(1 + q_0)} = \frac{q_0/(1 + q_0)}{(1 + q_0)/(1 + q_0) + q_0/(1 + q_0)}$$

$$= \frac{q_0/(1 + q_0)}{(1 + 2q_0)/(1 + q_0)} = \frac{q_0}{1 + 2q_0}$$

$$q_{2+1} = q_3 = \frac{q_0}{1 + 3q_0}$$

These successive q's (gene-frequency values) fall into a harmonic series, i.e., one whose terms are the reciprocals of those in an arithmetic series. When the initial gene frequency is known, the gene frequency for any succeeding generation can be found by substituting in the equation $q_n = q_0/(1 + nq_0)$. The change in gene frequency per generation is again symbolized by Δq and is given by

$$\Delta q = \frac{q}{1 + q} - q = \frac{-q^2}{1 + q}$$

Note that the *rate of change* of gene frequency is itself *a function of the gene frequency*. When the gene frequency is high, the gene is removed from the population rapidly. A few representative values are given in Table 6-7.

The reason for this change in rate of removal is that the proportion of recessive genes in the *heterozygotes* increases rapidly as the gene frequency decreases. Where $q =$ gene frequency of *a*, the percentage of *a* genes in the heterozygotes is as follows:

q	AA	Genotype frequency		Percent in heterozygotes
		$A\alpha$	$\alpha\alpha$	
.9	.01	.18	.81	10
.1	.81	.18	.01	90
.01	.9801	.0198	.0001	99

The recessive genes in the heterozygotes are "hidden" from selection, since only the homozygous recessives are lethal. The lower the gene frequency, the smaller the proportion of recessive genes exposed in homozygotes; progress toward removal of the gene from the population slows down accordingly. This result is of particular interest

TABLE 6-7.

Gene frequency	Decrease per generation
.9	.426
.5	.167
.1	.009
.05	.0024
.01	.000099

to students in eugenics. If a particular undesirable gene (*a*) had a gene frequency $q = .01$ in the human population, so that q^2 (*aa*) individuals made up .0001 of the individuals (1 defect per 10,000 "normals"), it would take 100 generations (roughly 2,500 years) of a program of sterilization of defective individuals to halve the gene frequency and reduce the number of defective individuals to 1 in 40,000. The problem of carrying out such a program without mistakes for such a protracted period makes it highly unlikely that such meager results would justify the effort involved. On the other hand, selection against dominants is relatively highly effective. If a dominant gene became lethal, for instance, it would be removed from a population in one generation.

Homozygous recessives relatively unsuccessful

The situation where the dominants are favored over the recessives but the homozygous recessive individuals make some contribution to the gene pool of the succeeding generation is probably more common than that of complete homozygous recessive lethality.

If fitness values of 1 are assigned to the two dominant genotypes, and $1 - s$ to the homozygous recessives, after one generation of selection there would be p^2 *AA* individuals, $2pq$ *Aa* individuals, and $q^2 - sq^2$ *aa* individuals out of a total of $p^2 + 2pq + q^2 - sq^2 = 1 - sq^2$ individuals. Using the same procedure as in the previous example, we see that the change in gene frequency per generation is

$$\Delta q = \frac{pq + q^2(1 - s)}{1 - sq^2} - q = \frac{-sq^2(1 - q)}{1 - sq^2}$$

Thus the change in gene frequency under these conditions is small when q is very large or very small and is relatively large when the value of q is intermediate. When q is large, progress is slow because of the relatively large reproductive contribution of the

TABLE 6-8.

q	Δq
.99	$-.00961$
.90	$-.06807$
.50	$-.07143$
.30	$-.03298$
.05	$-.00119$
.01	$-.00005$

recessive homozygotes (in contrast with the dominants). As q becomes small, the sheltering effect of the heterozygotes slows progress, as it does when the homozygous recessives are lethal. A few sample values are given in Table 6-8, where the selection coefficient operating against the homozygous recessives is $s = .5$.

Homozygotes inferior to heterozygotes

As a final detailed example of selection models, consider the case in which both homozygous genotypes are inferior to the heterozygotes, i.e., the locus shows overdominance with respect to fitness, so that

Genotype	AA	Aa	aa
Adaptive value	$1 - s_A$	1	$1 - s_a$

The proportions after one generation of selection will then be $p^2(1 - s_A)$ AA, $2pq$ Aa, and $q^2(1 - s_a)$ aa out of a total of $1 - s_A p^2 - s_a q^2$. The latter quantity, obtained by summing the values for the three genotypes and simplifying, is thus

$$p^2(1 - s_A) + 2pq + q^2(1 - s_a) = p^2 + 2pq + q^2 - s_A p^2 - s_a q^2$$
$$= 1 - s_A p^2 - s_a q^2$$

Since the initial gene frequency is $q_0 = pq + q^2$, q_1 (the frequency of a genes after one generation of selection) is

$$\frac{pq + q^2(1 - s_a)}{1 - s_A p^2 - s_a q^2} = \frac{pq + q^2 - s_a q^2}{1 - s_A p^2 - s_a q^2} = \frac{q - s_a q^2}{1 - s_A p^2 - s_a q^2}$$

Then the change in q per generation is

$$\Delta q = \frac{q - s_a q^2}{1 - s_A p^2 - s_a q^2} - q = \frac{q - s_a q^2 - q + s_A p^2 q + s_a q^2 (1 - p)}{1 - s_A p^2 - s_a q^2}$$

$$= \frac{-s_a q^2 + s_A p^2 q + s_a q^2 - s_a p q^2}{1 - s_A p^2 - s_a q^2} = \frac{pq(s_A p - s_a q)}{1 - s_A p^2 - s_a q^2}$$

Therefore, if $s_A p > s_a q$, then Δq is positive and the frequency of a is increasing, and if $s_A p < s_a q$, then Δq is negative and the frequency of a is decreasing. When $s_A p = s_a q$, $\Delta q = 0$ and the frequency of a is at equilibrium. In this situation, then, the equilibrium value of q ($\hat{q}$, read "q hat") is determined solely by the magnitude of the selection coefficients:

$$s_A \hat{p} = s_a \hat{q}$$

Thus, since $\hat{p} = 1 - \hat{q}$

$$s_A(1 - \hat{q}) - s_a \hat{q} = s_A - s_A \hat{q} - s_a \hat{q} = 0$$

$$s_A = s_A \hat{q} + s_a \hat{q} = \hat{q}(s_A + s_a) \quad \hat{q} = \frac{s_A}{s_A + s_a}$$

similarly,

$$\hat{p} = \frac{s_a}{s_A + s_a}$$

If the adaptive values of the three genotypes remain constant, the equilibrium value will also remain constant. If some incident, such as the arrival of a group of migrant individuals, shifts the gene frequency away from the equilibrium value, the selective forces will restore the equilibrium. Therefore, we refer to this as a stable equilibrium.

Balanced polymorphism and the retention of variability

The selective system in which the heterozygotes are superior to either homozygote results in the retention of both alleles in the population rather than a trend toward fixation of one or the other. A situation in which two or more forms of an organism persist in the same population, with the rarest form in a frequency *too high to be accounted for by mutation alone*, is known as **polymorphism.** When heterozygotes are favored over homozygotes, the establishment of a gene-frequency equilibrium creates a **balanced polymorphism.** This type of polymorphism is important in evolution in part because it permits a certain amount of variability to be retained in the population. This means that the population may be able to react very rapidly to an environmental change and thus avoid extinction. For example, a hypothetical population of locusts living in a semiarid environment shows the following array of adaptive values at a locus: *BB*, .50; *Bb*, 1.00; *bb*, .40. The heterozygotes are physiologically superior to either homozygous type, and the *BB* nymphs are slightly more resistant to desiccation than the *bb* nymphs. In such a population the gene frequency would reach a stable equilibrium at

$$\hat{B} = \frac{s_b}{s_b + s_B} = \frac{.60}{.60 + .50} = .545$$

(Note that the *bb* adaptive value is .40 but the selection coefficient is .60, since $1 - s$ equals the adaptive value.) The maintenance of the equilibrium at $B = .545$, $b = .455$ by this selective system is shown in Table 6-9.

Suppose that a climatic change suddenly increases the rainfall in the area occupied by our hypothetical locusts, encouraging the growth of a mold which is fatal to *BB* and *Bb* eggs but to which the *bb* eggs are relatively immune. The adaptive values *W* are now $BB = .00$; $Bb = .00$; $bb = 1.00$. The survival of the population now depends entirely on the presence of *bb* eggs. If a prerain (polymorphic) adult population of 100 pairs and an average egg production of 100 eggs per female are assumed, there would be 10,000 eggs exposed to the mold. Of these, 2,070 (.207×10,000) would be of the resistant kind, presumably giving the population a reasonable chance of survival.

On the other hand, if the prerain population had been monomor-

TABLE 6-9. BALANCED POLYMORPHISM*

Generation	Before or after selection	Genotype frequency				Gene frequency	
		BB	*Bb*	*bb*	Σ	*B*	*b*
0	Before†	.297	.496	.207	1.000	.545	.455
	After‡	.148	.496	.083	.727	.545§	.455
1	Before	.297	.496	.207	1.000	.545	.455
	After	.148	.496	.083	.727	.545	.455

*	Succeeding generations continue this pattern as long as assumptions hold.
†	Random mating is assumed in the calculation of this row, giving the following genotype frequencies: $B^2 = .545^2 = .297$; $2Bb = 2(.545)(.455) = .496$; $b^2 = .455^2 = .207$.
‡	The selection pressure is included by multiplying each genotype frequency by its adaptive value: .297(.50) = .148; .496(1.00) = .496; .207(.40) = .083.
§	*B* gene frequency is given by $(D + \frac{1}{2}H)/N$. Half of .496 = .248, thus (.148 + .248)/.727 = .545. Gene frequency of *b* is $1 - .545 = .455$.

phic (all individuals *BB*, perhaps because of strong selection against *Bb* and *bb* individuals), the outcome would almost certainly be different. If a mutation rate $B \rightarrow b$ of 10^{-5} is assumed, only 1 egg in 10 billion would be of the surviving genotype. (The chance of both members of a pair of alleles being mutant in a single individual is the product of the chances of either one being mutant: $10^{-5} \times 10^{-5} = 10^{-10}$ = 1/10,000,000,000.) The advantage of balanced polymorphism to this hypothetical population is obvious.

How much genetic variation exists in natural populations has long been a question of central importance to population biology. Experimental evidence was difficult to obtain, since most visible mutational changes were in characters whose expression is affected by many genes. *Drosophila* eye color, for instance, represents the final result of the action of multistep pigment pathways mediated by many genes. Nor was there any means of determining the incidence of genetic variants without visible effects. In the absence of direct experimental data, many population geneticists concluded on theoretical grounds that the total amount of genetic variation maintained in natural populations could not be large. They reasoned that different genetic variant types would have differing effects upon overall fitness, one type being optimal in a given environment. If the fitness of the optimal genotype is 1.0, then the more of the other suboptimal types present, the less the average fitness of the population (the frequency of the optimal type being less). They thus concluded that the maintenance of genetic variation in populations imposed a **genetic load** in terms of reduction in average population fitness. The amount of genetic variation maintained in natural populations was not thought to be large because too great a genetic load would so reduce average fitness that the population would be driven to extinction.

Within the last few years, the technique of gel electrophoresis has provided a direct experimental approach to this problem. A tissue homogenate is placed in a supporting gel and subjected to an electric field. The soluble proteins of the homogenate migrate in this field, the rate of migration depending upon their charge (and to some degree upon their size). Most proteins have differing charge distributions and thus migrate at different rates. The value of this approach is that it allows investigation of discrete gene characters. A locus whose enzyme's substrate is known can be studied independently of other loci by simply placing a gel, after electrophoresis, in an assay mix containing the enzyme's substrate and reagents for rendering the reaction product visible. Only at that position on the gel to which the enzyme has migrated will the reaction occur, producing a band of visible reaction product. Because this technique is experimentally simple to carry out, it has rapidly become widespread. It should be noted, however, that electrophoresis characterizes variant types in terms of their mobilities in an electric field, and such measures of rate of movement depend not only upon protein charge but also upon time, temperature, current, and other variables. Comparability thus requires careful standardization.

Allelic alternatives characterized by differences in electrophoretic mobility are called **allozymes.** They are not to be confused with **isozymes,** which are different forms of an enzyme reflecting differences in subunit composition; for example, α and β subunits may combine as $\alpha\alpha$, $\alpha\beta$, or $\beta\beta$.

The first estimate of the amount of genetic variation in natural populations obtained by electrophoretic analysis was reported by Lewontin and Hubby in 1966 for natural populations of *Drosophila pseudoobscura.* From their studies they concluded that if all populations are considered, the average individual is heterozygous for 12.3 percent of his genes. Similar studies with other species of *Drosophila* and with butterflies, certain grasses, mice, and human beings, have yielded essentially the same result. This is much more genetic variation than was predicted by the arguments of genetic load. Later treatments of genetic load, notably by Sved, Reed, and Bodmer, have taken account of possible threshold effects in fitness as well as demonstrating that the interaction of epistasis and linkage between different loci also reduces load.

Characterizing the amounts of genetic variation by electrophoresis requires independent demonstration that the observed differences indeed reflect genetic differences, rather than other factors (separate bands may also occur because of binding of ions or cofactors, enzymatic modification of the protein, etc.). The great strength of the initial report by Lewontin and Hubby lies in its careful documentation of the genetic basis of each variant type examined.

It should be noted that estimates of the amount of genetic variation in natural populations based upon surveys of electrophoretic variation are almost certainly underestimates, as not all structural changes in an enzyme will affect its mobility in an electric field. Such estimates also assume that the sample of the genome is random, i.e., that the level of variation seen in the loci examined is typical of all loci. This assumption may not be valid if the levels of heterozygosity reflect the nature of the loci examined. The degree to which enzyme function and amount of genetic variation are related is not yet clear. Although some workers, e.g., Johnson, argue for a strong correlation, little clear evidence exists.

It has been suggested by Kimura and others that the large amounts of genetic variation detected by electrophoresis do not contradict the predictions of the genetic-load theory if one assumes that the different allozyme types do not produce differences in fitness. They argue that the electrophoretic variation simply reflects genetic "noise," being minor changes in protein structure sufficient to alter electrophoretic mobility but not enzyme activity. Such variants, being functionally equivalent, would respond similarly to selective forces; they are referred to by Kimura as being *selectively neutral*. This theory of selective neutrality of allozyme polymorphisms is highly controversial and has been a hotly debated issue in population genetics. Alternative arguments are being advanced strongly by Ewens, Johnson, and others that the existence of these polymorphisms has a selective basis. For example, Johnson finds that the occurrence of particular enzyme polymorphisms in butterflies is correlated with the pattern of certain environmental factors, transcending species boundaries. Regardless of how genetic polymorphism is maintained in natural populations, there is no question that it is an immediate source of genetic variability, the necessary prerequisite for genetic change in populations.

Heterozygotes inferior to homozygotes

When selection is *against* the heterozygotes, an unstable equilibrium point exists at $p = q = .50$ if the two homozygotes are equally fit. Any deviation from this value leads to extinction of the allele that is made less frequent by the deviation. As shown by the values given in the discussion of selection against recessive homozygotes (page 108), the rarer an allele is, the larger its proportion in the heterozygotes. Therefore when selection is against the heterozygotes, the less frequent allele is at a disadvantage which *increases* as it becomes rarer. This means that there is no tendency to return to the equilibrium point once a deviation has occurred; instead, the situation proceeds to fixation of one allele or the other. If the homozygotes are not equally fit but both are fitter than the heterozygote, an unstable equilibrium point exists at some value other than $p = q = .50$.

Fundamental theorem of natural selection

In 1930 Fisher proposed what he called the fundamental theorem of natural selection. His statement dealt with the speed of evolutionary change in terms of the genetic variance in a population. In terms of our treatment here, the theorem states that the mean fitness $\overline{W}$ of a population in a given generation will be greater than that of the preceding generation unless the population has reached an equilibrium. In that case, $\overline{W}$ ceases to change. This theorem has been validated for any one-locus system with any number of alleles;

however, it does not necessarily hold for systems with more than one locus.

Other aspects of selection

Certain other patterns of selection pressure often discussed in the evolutionary literature deserve brief mention here. These include frequency-dependent selection, density-dependent selection, and group selection.

Frequency-dependent selection. In many circumstances, the fitness of a genotype will depend on its frequency in the population. Often, for instance, a genotype which is at high frequency in a prey population appears to be at a selective disadvantage relative to less common genotypes because a predator feeds primarily on the most common kind. Thus when a polymorphic insect population is fed upon by birds, the birds may tend always to eat the most common form differentially because they develop a "specific search image" for that pattern. Such frequency-dependent selection against the most common form is, theoretically, capable of maintaining a polymorphism.

An interesting case of frequency-dependent selection has been hypothesized by Dolinger and his coworkers to maintain variation in alkaloids in certain populations of lupine plants. Alkaloids are defensive chemicals evolved by the plants which poison herbivorous animals. Specialist herbivores, such as the larvae of small butterflies, which spend their entire lives on a single plant, most rapidly evolve resistance to the commonest alkaloid poison pattern. Thus the plant alkaloid genotype which is most common in the population at a given time also tends to be under the heaviest herbivore attack and an alkaloid-pattern polymorphism is maintained. There is a beneficial side effect of this pattern of selection for the plant population as a whole, since it tends to prevent the insect population from evolving one single resistance mechanism which would permit them to thrive on all plants.

Density-dependent selection. In many, if not most, circumstances the fitness of a genotype will vary with the density of the population in which it occurs. Consider as an oversimplified example, a polymorphic butterfly population in which one brightly colored form mimics the warning coloration of another "model" species of butterfly which is distasteful (see Chap. 7) and the other form does not. If the population size is small, the number of mimics is small relative to the number of models. A bird attacking a butterfly of the pattern shared by the model and mimic will usually get a distasteful model, and its aversion to the pattern will be reinforced. Thus the fitness of the mimetic form will be high. But if the population size increases enough, the mimics will become more common than models. The birds will often have a positive experience when they attack a brightly colored butterfly—they will taste a mimic rather than a model. Birds will then concentrate their attacks on the mimic-model pattern and the fitness of the mimetic form will be greatly decreased.

The most general example of density-dependent selection involves the high fitness at low population densities of genotypes with high reproductive potential, as contrasted with a high fitness at high population densities of genotypes able to survive in crowded conditions. In theory, populations subject to harsh or seasonal environments (which cause frequent "diebacks") should be largely composed of individuals with the former kind of genotype, while those in mild, stable environments should be composed largely of the latter. Obviously there is a continuum between the two extremes of selective regimes. The two extremes are sometimes referred to as r selection and K selection (in ecological symbolism r is a measure of how rapidly a population can increase when freed of environmental restraints, and K is the carrying capacity of the environment, i.e., the maximum number of individuals that can be sustained).

Since mildness vs. harshness and seasonality vs. nonseasonality vary along altitudinal and longitudinal gradients, it has been suggested that altitudinal and longitudinal gradients might be found in r and K selection. What appear to be exactly such gradients have been found in various organisms in nature. One can recognize what have been called r and K strategies. A K strategy is one in which much energy is used, through a long period of development, to produce individuals capable of competing successfully for the resources of the environment. Such a strategy results in relatively few offspring. An r strategy is one in which, early in the life span, energy is invested in the production of relatively large numbers of offspring or propagules.

Roughgarden has investigated mathematically the effects of density-dependent selection on population size. In a mild environment, selective values result in the evolution of phenotypes having a high K and a low r. Conversely, in a harsh, seasonal environment, there is selection for phenotypes having a high r and a low K. Selective values W for the genotypes in this one-locus model are assumed to decrease as a linear function of density. Roughgarden has also shown that there are genetic mechanisms which produce stable polymorphism between high-r and high-K alleles in a moderately harsh, seasonal environment. Finally, r and K strategies are related to energy flow in populations. An r strategist is an individual who makes a larger contribution to the productivity of the population providing crowding effects are negligible. A K strategist is an individual whose contribution to the population's productivity is not easily reduced by crowding effects.

Such mathematical analyses can be tested in the field. For example, Solbrig has studied the reproductive strategies of the common dandelion (*Taraxacum officinale*). Electrophoretic analysis of individuals from three field populations with differing habitats in Ann Arbor, Michigan, revealed that at least four different genotypes (biotypes) were present in these populations. The proportion of each genotype (designated A, B, C, and D) varied from population to population (Table 6-10), A and D, being particularly extreme.

The different proportions of these two genotypes may be explained either as a function of microenvironmental requirements or as examples of r and K strategies. To test which explanation is correct, plants with A and D genotypes were grown experimentally, in varying proportions, under different environmental conditions. Since dandelion seeds are produced by apomixis, it is easy to obtain a relatively large number of virtually genetically identical individuals for such experiments. Under all environmental conditions, plants with the D genotype outcompeted plants with the A genotype (K strategy) both in the number of plants that survived and the biomass produced. On the other hand, plants with the A genotype produced more seeds and produced them earlier under all environmental conditions (r strategy). (Note that, in Table 6-10, percentage of biotypes in a population is not a measure of competition.)

These data suggest that at least the plants from the three field-study populations represent mixtures of genotypes that differ in their ability to compete and in their strategies of survival.

Solbrig has suggested that plants with the A genotype are predominately found in disturbed sites, because survival in this type

TABLE 6-10. PERCENTAGE OF BIOTYPES IN THREE POPULATIONS OF TARAXACUM OFFICINALE

Population and habitat	No.	Biotype, %			
		A	B	C	D
1. Dry, full sun, highly disturbed	94	73	13	14	
2. Dry, shade, medium disturbed	96	53	32	14	1
3. Wet, semishade, undisturbed	94	17	8	11	64

of habitat is dependent upon producing a large number of seeds not all of which would be removed by environmental disturbance. Plants with the *D* genotype are more common in undisturbed habitats, where competition for available resources is an important factor.

Group selection. Evolutionists have often been puzzled by the occurrence of certain traits whose evolution seems difficult to explain by differential reproduction of individuals. Take, for example, a male baboon sacrificing itself to protect the troop. Since the altruistic individual dies, selection appears to act to weed genes for altruism out of the population. As a solution to this dilemma it has been proposed that the actual unit of selection is, in these cases, the group, not the individual. Populations with altruistic genes thrive; populations lacking them die out.

Since population extinction appears to be a very common phenomenon, there would seem to be some basis for such "group selection" to operate. The mechanism suffers from a serious flaw, however. How are populations with "altruistic genes" to preserve them in the face of nonaltruistic mutants and recombinants? How do the self-sacrificing male baboons avoid being replaced over time by cowardly males? An extremely rapid rate of population extinction would seem to be required for group selection to be an effective evolutionary mechanism.

A more likely explanation of the evolution of phenomena such as altruism is *kin selection*. The individuals saved by the male baboon's self-sacrifice are likely to be his close relatives, especially his children. Thus his sacrifice actually *increases* the frequency of his genes (or genes like his) in the next generation, and it is by genetic representation in the next generation that fitness is conventionally defined.

It is important to note that many important mechanisms occur at the population level *but did not arise by group selection*. Social behavior (including altruism), polymorphism in the mimetic butterflies or plant alkaloids, and so forth, are populational phenomena which can arise by selection with *individuals* as the units of selection. There is no need to invoke group selection to explain their evolution.

MIGRATION AND POPULATION STRUCTURE

When a population is not completely isolated from other populations, its gene frequencies are subject to alteration through the incorporation of migrants, which, as a group, have gene frequencies deviating from that of the recipient population. This is called **differential**

migration or **gene flow.** At any locus the change in gene frequency per generation is given by

$$\Delta q = -m(q - q_m) = -mq + mq_m$$

where m = number of migrant individuals divided by population size of recipient population
q = gene frequency in recipient population
q_m = gene frequency in migrant group

Manipulating the right-hand side of this equation by adding and subtracting the quantity $mq_m q$, we get

$$-mq + mq_m q + mq_m - mq_m q = -mq(p_m) + mp(q_m)$$
$$= -mq(1 - q_m) + mq_m(1 - q)$$

This final equation is in the same form as the expression for Δq in the discussion of the action of mutation and back mutation (page 102); indeed, the situations are analogous.

The most interesting problems associated with migration involve its interactions with finite population size. These problems are discussed in detail by Cavalli-Sforza and Bodmer. Figure 6-4 shows how the distribution of gene frequencies changes with changes in m or N, where N is the effective population size and the gene frequency of the migrants is .50. Diagrams such as Fig. 6-4 are known as stationary frequency distributions. Each curve in the figure represents a probability density function of the form

$$y = Cq^{4Nmq_m - 1}(1 - q)^{4Nm(1 - q_m) - 1}$$

where C is a constant making the function integrate to 1 and the other notation is as above. Such a function may represent the manner in which the probability is distributed over the possible events. The area under each curve is unity, and the area between the curve and each section of the abscissa (q axis) is the probability that the gene frequency will lie along that stretch of the q axis. Thus in Fig. 6-4 one can see at a glance that, under the given conditions, there is a much smaller probability that q will lie between .4 and .6 when $m = 1/4N$ than when $m = 4/N$. Stationary frequency distributions are a very convenient way of illustrating the effects of various evolutionary forces on different kinds of populations and are widely used for this purpose. Readers interested in further information on probability density functions and other subjects relating to the mathematical treatment of probabilities are referred to any introductory text on probability theory.

Stationary frequency distributions may be used to represent the distribution of the gene frequency under consideration in a large number of populations under the same evolutionary conditions, the distribution of the gene frequencies at a large number of loci subjected to the same pressures within a single population, or as the probability distribution for the chances of a given gene frequency occurring in any one generation. Thus the $Nm = 4$ curve ($m = 4/N$) in Fig. 6-4 may be interpreted in the following ways. It can be said that *under the same given conditions* the gene frequencies of a large number of populations (or loci within one population) would tend to cluster rather tightly around the value of .50 or that among all loci within a population subjected to the same conditions the probability of any given locus having a gene frequency between .35 and .65 is high. Finally, the curve represents the probability of the gene frequency at one locus having a given value in the generation observed. Thus the chance of observing a value of $q = 1$ in any one generation is vanishingly small.

The type and amount of movement of genetic information

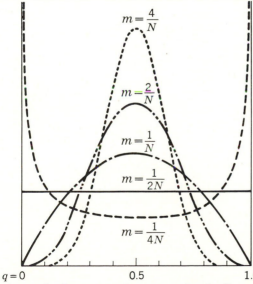

Figure 6-4 Distribution of frequencies of a gene among subdivisions of a population, where the gene frequency of the migrants is $p = q = .50$. The abscissa represents gene frequency. Each curve represents a different set of evolutionary pressures and can be interpreted in three ways: (1) the distribution of gene frequencies at the same locus in a large number of populations subject to the same pressures, (2) the distribution of gene frequencies at a large number of loci in the same population, subject to the same pressures, (3) the probability of the gene frequency at a locus subjected to those pressures having an observed value in any given generation. In all cases the area between the curve and segment of the abscissa represents the probability of q having a value in that segment (since the total area under the curve is unity). For further explanation of this type of diagram see text. [*From S. Wright, Genetics, vol. 16, (1931).*]

(gene flow) found within and among populations may be important factors in determining their evolution. Obviously, a situation in which a group of semi-isolated subpopulations randomly exchange genetic information among themselves (the "island model") is quite different from a situation in which the gene flow is unidirectional along a linear array of subpopulations (the "river model"). In turn, both of these differ from a situation in which a group of organisms is continuously distributed over a large area. In the latter case, although semidiscrete clusters of individuals may not exist, the probability of mating by two widely separated individuals may be very low because of their remoteness alone. The effects of such **isolation by distance** have been dealt with mathematically by Wright, who showed that the amount of local differentiation in a population is largely a function of the size of the panmictic units (neighborhoods) of which it is composed. When a population is divided into semi-isolated subpopulations, or when some degree of inbreeding is found in the population as a whole, the general result is a reduction in the frequency of heterozygotes. An important aspect of this change in genotype frequencies is illustrated by the effects in such a population of selection against the homozygous recessives. This would be much more effective in a subdivided or inbreeding population because the reduced number of heterozygotes would shelter fewer recessive genes.

JOINT PRESSURES

Up to this point, only single evolutionary forces acting on isolated loci have been considered. However, in virtually all cases studied, two or more pressures act jointly to affect the gene frequency at a given locus. In addition, the gene frequencies at different loci are not independent of each other, and the gene frequency at one locus may have a profound effect on the gene frequency at another. To appreciate this, one need only recall the phenomenon of linkage. An evolutionary description of a population solely in terms of gene frequencies will always be incomplete.

Some progress has been made in describing mathematically the results of various types of interactions in mendelian populations. Whether a completely satisfactory mathematical description of the simultaneous action of all evolutionary forces (varying with the environment) on an integrated genotype will ever be possible is an open question. Progress in the development of computers gives reason for hope, but the extreme complexity of the situation to be analyzed would require a computer of undreamed-of sophistication. For the moment we must be satisfied with combining gross oversimplifications. There is solace in the fact that these simple models seem to approximate some natural situations and have proved quite useful in describing them.

As a short excursion into more complex situations, consider Figs. 6-5 to 6-12. Figures 6-5 to 6-8 illustrate the effects of different selection pressures in populations of different sizes. In Figs. 6-5 to 6-7 mutation and back-mutation rates are considered constant and equal ($u = v$). In Fig. 6-5 the population size is $N = 1/40v$; in Fig. 6-6, $N = 10/40v$; in Fig. 6-7, $N = 100/40v$. In all three figures the solid line is the case with the least selection ($s = -v/100$), the broken line the case with selection 10 times as severe (not represented in Fig. 6-5 since it is practically indistinguishable from the preceding), and the dotted line the case with selection 100 times as severe. Note that selection in the very small population (Fig. 6-5) merely alters the symmetry of the distribution slightly, the probability of loss or fixation remaining high. As the population size increases (Figs. 6-6 and 6-7), the selection effects become much more pronounced. Figure 6-8 illustrates the distribution when the heterozygotes are favored and there is no difference between the selective values of the two homozygotes. Again $u = v$, $N = 1/40v$, and $s = 100v$. (Note that in these figures and in Figs. 6-9 to 6-12 the selection coefficient is not used as defined earlier but is given both positive and negative values. Thus $s = 100v$ is an index of the advantage of the heterozygotes, whereas above $s = -v/100$ is an index of the disadvantage of the allele under consideration.)

Figures 6-9 to 6-12 show the distribution of gene frequencies in populations of different sizes and different states of subdivision, under various selection and mutation pressures. Figure 6-9 depicts a small population under virtually no selection or mutation pressure. The majority of alleles are fixed or lost at random. An intermediate-sized population under opposing selection and mutation pressures is shown in Fig. 6-10. There is random variation around modal values established by the opposing pressures. The case of a large population with gene frequencies at equilibrium points determined by the magnitudes of opposing selection and mutation pressures is covered in Fig. 6-11. Finally, Fig. 6-12 gives the gene frequencies in subdivisions of a large population fluctuating around modal values established by opposing forces of migration and selection.

EVOLVING IN A VARIABLE ENVIRONMENT

Many of the problems of evolutionary theory would be far more tractable if populations lived in environments which were homogeneous in space and unchanging through time. With very minor ex-

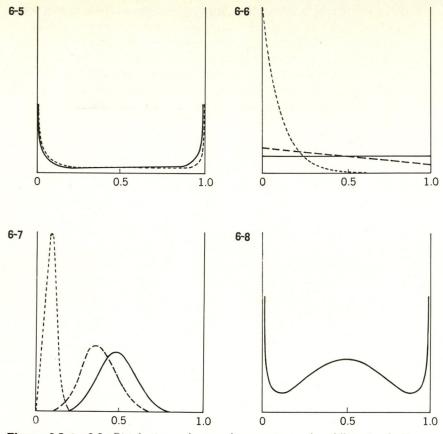

Figure 6-5 to 6-8 Distributions of gene frequencies under different selection pressures and in populations of different sizes. For details see text. [*From S. Wright, Proc. Natl. Acad. Sci., vol. 23 (1937).*]

ceptions, however, such situations are rarely even approximated in nature. This means that many models used in evolutionary biology may be unrealistic because they assume environments uniform in space and time when populations may be evolving specifically in relation to the variability of the environment.

There are, therefore, very important questions to be answered about evolution in variable environments. Suppose, for example, an insect population lives in an area in which two species of plants bear fruits that are suitable as food for the insect. Plant species *A* is abundant, but its fruit is well protected by alkaloid poisons and requires considerable "physiological effort" on the part of the insect for its utilization. The insects must continually produce specific enzymes which can detoxify the poisons. Plant *B* is relatively rare but

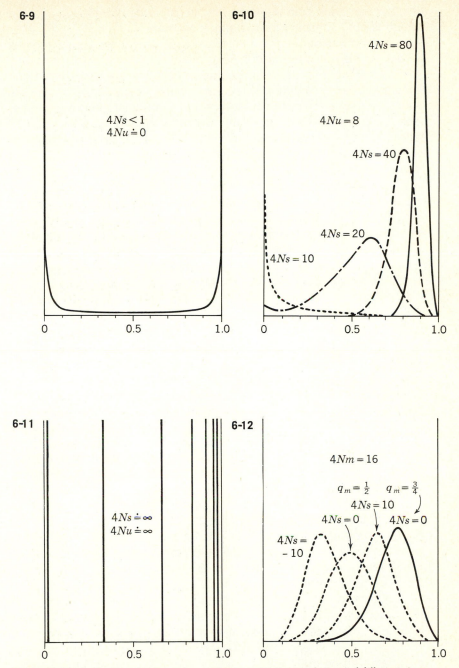

Figure 6-9 to 6-12 Distributions of gene frequencies in populations of different sizes and different states of subdivision under various selection and mutation pressures. See legend to Figure 6-4 and text for details. [*From S. Wright, Genetics, vol. 16 (1931).*]

is more nutritious than *A*—*if* the insects produce a digestive enzyme which is not necessary for the attack on plant *A*.

Levins has begun the analysis of such situations. One of the first questions he would ask is whether, from the point of view of the insect, the environment tends to be *coarse-grained* or *fine-grained*. At one extreme if each individual insect can spend its entire life in a single fruit, the environment is coarse-grained. The environment (or rather one factor of the environment) exists for individuals as a pair of alternatives: they may live either in a fruit of species *A or* in a fruit of species *B*. At the other extreme suppose each individual insect always took just one bite of fruit from an individual plant, moving on to the next plant for the next bite. In that case the fruit environment would be fine-grained, presenting itself to the insect as a mixture of mouthfuls from plants of species *A* and *B*. Coarse-grained and fine-grained can, of course, be thought of as the ends of a continuum; e.g. our hypothetical insect might spend an average of a day feeding on an individual plant, but for the purpose of the discussion which follows we shall consider them to be discrete states.

Having established whether the environment at a given time was coarse- or fine-grained one could then ask the following kind of question about evolutionary strategies. Would it be better for the population to have a single "generalist" genotype which would produce a phenotype possessing both enzymes (and thus capable of feeding on both plants) or to have a mixture of two "specialist" genotypes each producing a phenotype with the enzyme specific to either plant *A* or *B*? The answer to such a question obviously depends, among other things, on the grain of the environment and the relative fitnessess of generalist and specialist genotypes. In a fine-grained environment the generalist strategy would be more success-ful. In a coarse-grained environment the specialist strategy would obviously be optimal, given that the generalist was not as good as either specialist at the specialist's speciality, unless the frequency of plant species *B* was so low that the physiological effort of producing the enzyme was greater than the energy gained by feeding on *B*. If *B* were sufficiently rare, selection would presumably favor individuals which always bypassed species *B*. Then the population would become monomorphic for specialists in eating species *A*.

It is intriguing to imagine possible strategies. Would a mixture of specialists always be best in a coarse-grained environment? Under what circumstances, for instance, might a genotype which could *produce* either phenotype within 10 min of exposure to the ap-propriate fruit be optimal?

Levins has attempted various graphical methods of approaching such problems. His general approach can be illustrated by a simple example. Consider an array of phenotypes which are exposed to two different environments, A and B, either in space or in time. Each phenotype will have a fitness in environment A and in environment B. Two possible situations are shown in Fig. 6-13a and b. When these same data are graphed using the fitness in the two-environments as the axes (Fig. 6-13c and d), each point on the graph represents the fitness of a given phenotype in the two environments and all the points *on the line* (but not those enclosed by the line) make up a *fitness set.* The set in Fig. 6-13c is convex; that in Fig. 6-13d is concave.

Levins then defines functions $F(W_A W_B)$ of W_A and W_B which describe the relationship of fitness in each environment to the overall fitness of a phenotype $\overline{W}$. Such functions he calls adaptive functions. For instance such a function might be the linear average of the fitnesses in the two environments

$$\overline{W} = aW_A + bW_B$$

where a is a constant giving the proportion of the total environment of the population which is environment A, and b (=$1-a$) is the proportion which is B.

The function can be graphed along with the fitness set if it is reorganized thus:

$$W_B = \frac{\overline{W}}{b} - \frac{a}{b}W_A$$

Since a and b are constants, this equation describes a family of straight lines all with one slope a/b. The position of each line is determined solely by the average fitness value $\overline{W}$. Of interest is the line with the highest value of $\overline{W}$ that touches the fitness set, e.g., line $\overline{W}y$ in Fig. 6-13e. The phenotype at that point would be the one most favored by selection.

If the mix of environments (a,b) were different, a set of lines with another slope would be generated by varying $\overline{W}$. Again looking at Fig. 6-13e one can see that as the proportion of a increased, the lines would slope more steeply to the right. The point of contact of the line with the highest $\overline{W}$ with the fitness set would move smoothly downward and to the right, and the characteristics of the optimal phenotype would change gradually.

Contrast this situation with a similar adaptive function but a concave fitness set [Fig. 6-13f]. As the proportion of a increases, the optimum phenotype changes dramatically between a_2 and a_3 from one which has high fitness in environment A and medium fitness in B to one with very high fitness in B and rather low fitness in A.

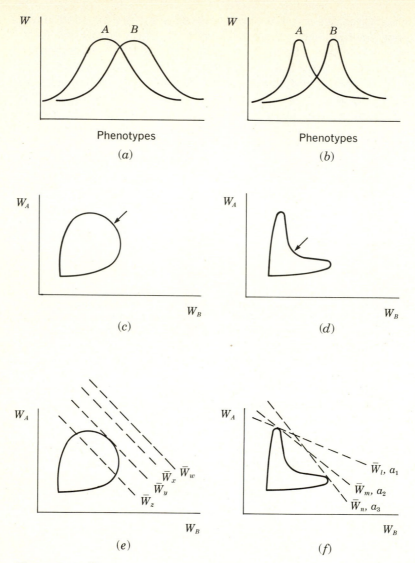

Figure 6-13 Phenotypes in two environments, A and B. See text for complete explanation. Note that in the situations portrayed in (a) and (b) each phenotype, i.e., each position on the abscissa, has a different fitness in each of the environments, the sole exception being the phenotype where the curves cross. That phenotype is represented by the arrows in the fitness sets of (c) and (d)/(e). In each of the straight lines graphs the equation $W_B = \dfrac{\bar{W}}{b} - \dfrac{a}{b} W_A$ with (a)/(b) fixed and $\bar{W}$ varying. In (f) the equations are the same, but both $\bar{W}$ and a are varying.

This discussion will serve to introduce the kinds of analyses which have been attempted. Levins has given us a way of looking at evolution in changing environments which may eventually prove extraordinarily productive, but things are still at a very exploratory stage. We must know much more about actual fitness sets, and in particular about the characteristics of adaptive functions, before many of the ideas generated by this sort of analysis can be properly tested. Advanced students interested in further exploration of this topic should consult Levins' classic Evolution in Changing Environments, with the caveat that it will be necessary to consult other literature (especially papers in recent years of *American Naturalist*) to keep up with this rapidly changing field.

SUMMARY

In this chapter an attempt has been made to give a brief introduction to evolutionary processes from the viewpoint of population-genetic theory. One of the fundamental concepts of biology is the Hardy-Weinberg law, which states that in an idealized population and in the absence of evolutionary forces, the gene frequencies of autosomal alleles in the population will not change and after one generation the proportion of genotypes will reach an equilibrium. The ways in which mutation, migration, drift, and selection may cause deviations from this equilibrium have been formulated mathematically. The effects of these forces depend not only on the interactions among them but also on the structure of the population and the feedback effects of this structure on the forces themselves. Theoretical descriptions of possible responses of populations to various combinations of factors have been developed. Considering populations as shifting and interacting arrays of gene frequencies has given the evolutionist tools for analyzing his observations with some degree of rigor and precision. Population-genetic theory is extremely useful in describing what may happen in natural populations and in interpreting data gathered from such populations.

REFERENCES

Cavalli-Sforza, L. L. and W. F. Bodmer: *The Genetics of Human Populations*, W. H. Freeman and Co., San Francisco, 1971. A superb and thorough compendium of human genetics, as well as an invaluable reference source.

Crow, J. F. and M. Kimura: *An Introduction to Population Genetics Theory*, Harper & Row, New York, 1970. An up-to-date survey of population-genetics theory of which the first two-thirds require only calculus as background.

Dobzhansky, Theodosius: *Genetics of the Evolutionary Process*, Columbia University Press, New York, 1970. This successor to *Genetics and the Origin of Species* is an important source for both theoretical and experimental approaches to population genetics.

Dolinger, P. M., P. R. Ehrlich, W. L. Fitch, and D. E. Breedlove: Alkaloids and Predation Patterns in Colorado Lupine Populations, *Ecologia*, **13:**191–204 (1973). This paper on plant-herbivore coevolution has profound implications for the design of evolutionarily sound agricultural systems.

Fretwell, Stephen D.: *Populations in a Seasonal Environment*, Princeton University Press, Princeton, N.J., 1972.

Haldane, J. B. S., and Jayakar, S. D.: Polymorphism Due to Selection of Varying Direction, *J. Genet.*, **58:**237–242 (1963).

Johnson, G. B.: Allozyme Variation and Systematic Biology, *Annu. Rev. Ecol. Syst.*, **4:** (1973).

Levins, R.: *Evolution in Changing Environments*, Princeton University Press, Princeton, N.J., 1968.

Li, C. C.: *Population Genetics*, University of Chicago Press, Chicago, 1955. Every student should be familiar with this clear, concise, and generally excellent text. The bibliography contains references to most of the theoretical literature, including very important general works by Fisher and Haldane and the classic papers of Hardy and Weinberg.

McKusick, V. A.: *Human Genetics*, 2d ed., Prentice-Hall, Englewood Cliffs, N.J., 1969. An introduction to human population genetics.

Population Genetics: The Nature and Causes of Genetic Variability in Populations, *Cold Spring Harbor Symp. Quant. Biol. 1955*, vol. 20. The papers in this volume are a good sampling of thinking in population genetics. See especially the papers in the section Integration of Genotypes.

Roughgarden, J.: Density-dependent Natural Selection, *Ecology*, **55:**453–468 (1971).

Wilson, Edward O., and William H. Bossert: *A Primer of Population Biology*, Sinauer, Stamford, Conn., 1971. An excellent brief introduction to population genetics with problems and their solutions.

Now that some of the theoretical aspects of population genetics have been discussed, it is appropriate to ask whether it is necessary to rely on inference for the investigation of evolutionary mechanisms or whether it is possible to study them directly. Especially in recent years, examples of evolution in natural populations have been investigated, and some of these will be considered in the first part of this chapter. The remainder of the chapter is devoted to more general discussion of some aspects of evolutionary changes in populations.

CHAPTER
SEVEN

CHANGES IN
POPULATIONS

EXAMPLES FROM NATURE

Differential mortality in sparrows

There have been numerous efforts to demonstrate the results of natural selection by comparing the characteristics of surviving and nonsurviving individuals. A brief summary of one of the earliest of these studies, Bumpus' work on sparrows, will serve to represent them all. In the winter of 1898, after a severe snow, rain, and sleet storm, Bumpus brought 136 stunned English sparrows into his laboratory at Brown University. Of these birds, 72 revived and 64 died. Bumpus measured the total length, wingspread, weight, length of beak and head, length of humerus, length of femur, length of tibiotarsus, width of skull, and length of keel of sternum on all the birds. His measurements showed that these characters generally were closer to the mean in the surviving birds than in birds that died.

The observed differential mortality of sparrows with extreme measurements has been cited as an example of stabilizing selection. Indeed it is if the observed variability has a genetic basis and if statistical analysis of Bumpus' data convinces us that the observed differentials cannot reasonably be explained as a chance occurrence. Thorough reanalysis of the measurements by Johnston and his coworkers indicates that the following hypotheses are adequately generated by the data: under winter stress, (1) large males are more likely to survive than small ones; (2) intermediate-sized females are more likely to survive than large or small ones; and (3) small subadult males are less likely to survive than individuals with the normal proportions of their sex. Thus, the observed differences in size in sparrows may be related to ability to withstand winter stress as well as to characteristics more closely associated with reproduction. If these hypotheses are correct, there were elements of both stabilizing and directional selection resulting from the stress of the storm.

Many other studies also have demonstrated the correctness of the widespread notion that selection often results in the elimination of deviant individuals. As previously mentioned (Chap. 6), this is only one aspect of natural selection. The examples which follow often involve complex interactions of the so-called basic types of selection. It is to be expected that stabilizing selection in the form of failure of extreme deviants occurs in virtually all natural populations.

Industrial Melanism

During the past 120 years, dark forms of numerous cryptically colored (camouflaged) species of moths have appeared in certain areas of northern Europe and North America. In many of these areas, the melanic forms have become predominant, replacing or partially replacing protectively colored "typical" forms. These changes have taken place primarily in heavily industrialized areas and have been especially spectacular in England, where they have been studied extensively. It has been estimated that in 1848 in the area of Manchester the dark form of the moth *Biston betularia* made up a maximum of 1 percent of the population and that in 1898 in the same area it made up more than 99 percent of the population. In most of the known cases, the melanism is produced by a single dominant gene.

The following hypothesis has been developed to account for the phenomenon of **industrial melanism.** The spread of melanic forms seems to be intimately connected with the pollution of woods by soot in industrial areas. Apparently in unpolluted areas the dark forms are removed from the populations by visual predators (those which hunt by sight) because they are conspicuous when resting on lichen-covered tree trunks (Fig. 7-1). This disadvantage outweighs a possible selective advantage of the larvae of the melanics, which may be *physiologically superior.* In polluted areas the situation is reversed; the typical (light, mottled) forms, which are nearly invisible on

Figure 7-1 Two individuals of *Biston betularia*, one typical, one melanic, resting on an unpolluted lichen-covered tree trunk. [*After H. B. D. Kettlewell, Proc. 10th Int. Congr. Entomol., vol. 2 (1958)*.]

unsooted, lichen-covered bark, are conspicuous on sooty trees where pollution has killed the lichens (Fig. 7-2). Thus the melanics are protectively colored in the polluted area; in addition, any physiological advantage they possess is magnified under the stress of eating contaminated food. Selection therefore strongly favors the melanics in industrial areas and the typicals in unpolluted areas. In polluted districts a **directional selection** moved the frequency of melanic individuals toward 100 percent.

What evidence supports this hypothesis? Some of the most compelling is the strong correlation, in time and space, of industrialization and the appearance of the melanic forms. This is so striking as to make some relationship between the two phenomena virtually certain. The composition of various English populations of *Biston betularia* is shown in Fig. 7-3. The populations in the industrial midlands are highly melanic, as are those in eastern England, on the downwind side of the industrial areas where pollution fallout has been most intense.

The most elegant demonstration of one factor responsible for industrial melanism is the observations and experiments on differen-

Figure 7-2 Two individuals of *Biston betularia*, one typical, one melanic, resting on a soot-covered tree trunk. [*After H. B. D. Kettlewell, Proc. 10th Int. Congr. Entomol., vol. 2 (1958).*]

tial predation by Kettlewell and Tinbergen. In an unpolluted wood in Dorset, equal numbers of melanic and typical individuals were released; predation was observed and photographed from a blind. Spotted flycatchers (*Muscicapa striata*), nuthatches (*Sitta europaea*), yellowhammers (*Emberiza citrinella*), robins (*Erithacus rubecula*), and thrushes (*Turdus ericetorum*) ate 164 melanic individuals but only 26 typical individuals ($P<.01$).[1] Very impressive motion pictures were made of these experiments, with repeated sequences of birds searching tree trunks and eating the conspicuous moths without noticing adjacent, protectively colored individuals.

A series of release and recapture experiments supported the visual-predation hypothesis. Known numbers of marked individuals of

[1] Probability much less than 1 in 100. This means that a statistical test has indicated that a deviation from the expected frequencies of this magnitude would, on the basis of chance alone, be expected much less than 1 percent of the time. The notation ($P <.02$) means probability less than 2 in 100. Many biologists conventionally consider "significant" any result where the probability of chance alone being responsible is less than 5 percent. The best discussion of the use of statistics in biology may be found in R. R. Sokal and F. J. Rohlf, "Biometry," Freeman and Company, San Francisco, 1973.

both types were released in an area, and the percentage of each recaptured later at a light was recorded. The percentage of recovered moths contrasting with their background (melanics in unpolluted and typicals in polluted areas) was considerably lower than the percentage of cryptically colored moths recaptured. For instance, near Birmingham (where the population is 85 to 87 percent melanic) 154 melanics and 73 typicals were marked and released. Later 98 marked moths were recaptured, 82 melanics (53 percent of 154 released) and 16 typicals (25 percent of 64 released). If experimental error is ignored and the survival value of the favored melanics is set equal to 1, the survival value $(1-s)$ of the typical phenotype becomes :25/.53 = .47. Hence, s is equal to .53, coincidentally the same value as the

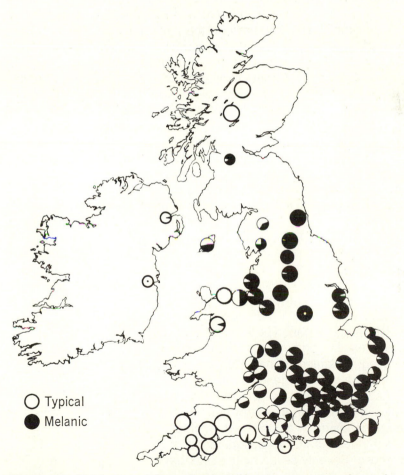

Figure 7-3 Distribution of *Biston betularia*, showing proportion of typicals and melanics. [*From H. B. D. Kettlewell, Heredity, vol. 12 (1958).*]

percent recovery of the melanics. It is highly unlikely that these results were due to chance alone ($P < .01$). Similar experiments in unpolluted areas have given reverse results. The relationship between the release-recapture experiments and the observations of visual predation is this. The release-recapture work shows that selection occurs; the observations show what the selective agents are.

There have been numerous experiments testing the viability of larvae of the melanic form. Ford, working with the moth *Cleora repandata*, produced backcross broods by mating melanic heterozygotes with typical recessive homozygotes. As in all test crosses, a 1:1 ratio of melanics to typicals was expected, and when the broods were well fed, there was no significant deviation from this ratio. However, when the caterpillars were starved every day (put under physiological stress), the ratio found was 51 melanic to 31 typical ($P < .02$), a significant departure from the expected 41.5 of each type. Kettlewell exposed six backcross broods to stress and found 108 melanic survivors as opposed to 65 typical individuals ($P = .01$).

There can be little doubt that, under certain conditions of stress, the larvae of the melanic moths are better able to survive, but the most recent work on the subject indicates that the situation is more complex than was previously thought. In some recent experiments, the expected deficiency of nonmelanic individuals was not found. In other experiments, the results showed interbrood heterogeneity. While the offspring from some matings showed a significantly higher proportion of melanics, the offspring from others did not. Furthermore, a study of backcross broods of *Biston betularia* raised between 1900 and 1905 showed no surplus of melanics but rather a slight (not statistically significant) deficiency of them. It is possible that early in the evolution of industrial melanism, melanic larvae were not physiologically superior and that this superiority, where it now exists, is a rather recent development. Perhaps it was only with the easing of the severe selection against melanic adults that melanic individuals increased sufficiently in populations to permit selective reorganization of the melanic genotype to gain the physiological advantage. Just as selection for modifier genes increased the dominance of the genes producing melanism (Chap. 3), so could selection enhance the effects of the melanic genes on viability.

Kettlewell tried experiments to see whether melanic moths tend to settle on dark surfaces and typical moths on light surfaces. He painted the inside of a barrel with alternate black and white surfaces and then released in it an assortment of moths to see which surfaces they chose. Seventy-seven moths selected the noncontrasting background, while forty-one selected the contrasting background. It has been suggested that this choice is possible because the moth can determine the degree of contrast between the scales around its eyes and the background on which it is resting.

In some areas, melanics are becoming predominant where the countryside appears to be unpolluted. Two reasons may be given for this. First, pollution is often greater than meets the eye. Smog clouds tend to drift a long distance, and, in spite of its overall green appearance, the countryside may actually have a considerable layer of soot and industrial chemicals. A second reason is that man's other activities alter the countryside, and these changes, e.g., decimation of the predators in an area, may be enough to shift the balance to the melanics.

Although recessive melanics are known, the spread of industrial melanism has been due to the spread of dominant genes. The possible reasons for this were discussed in Chap. 3 in relation to the origin of dominance.

It should be pointed out, however, that the spread of industrial melanism may be on the decline. During a recent 3-year period, areas in which only melanic moths previously had been found were sampled. Of the 972 specimens that were collected 25 were nonmelanic. The suggestion has been made that this possible reversal of a trend is the result of Britain's efforts to improve the polluted environment of its industrial areas.

Industrial melanism is an example of **transient polymorphism.** Remember that polymorphism is the occurrence in the same habitat of two or more distinct forms of a species in such proportions that the rarest of them cannot be maintained by recurrent mutation. Transient polymorphism is the situation in which the two forms coexist while one is in process of replacing the other. Another example of what is probably transient polymorphism involving melanic moths has been studied in an old woods in Scotland. In these woods, which are essentially free from pollution, one species of moth (*Cleora repandata*) had a population in which 10 percent of the individuals were melanic. At rest on lichen-covered trunks (light background) the typical forms were very inconspicuous. On dark trunks the melanics were inconspicuous, but their protective coloration was (to the human eye) not as effective as the camouflage of the light form. In flight, in the dark woods, the light forms were much more conspicuous. (Three were observed to be taken on the wing by birds in a period during which no melanics were observed to be eaten.) If the advantages were the same both at rest and in flight, progress toward fixation would be more rapid than in the situation described. Nevertheless fixation will still occur unless conditions change.

Microevolution in British Lepidoptera

British lepidopterists have pioneered in studies of other types of evolutionary changes in populations of butterflies and moths. Long-term studies of the gene frequency at a single locus in the scarlet tiger moth (*Panaxia dominula*) have been made by Fisher, Ford, and Sheppard. In the only colony (near Oxford) where the gene in question has been detected, its frequency has been estimated for a series of years starting in 1939, and population-size estimates have been made for all years since 1941. Three phenotypes occur in the colony: *dominula* (homozygous for the common allele), *medionigra* (heterozygous), and *bimacula* (homozygous for the rare allele). The frequency of the rare gene was .012 before 1928, .092 in 1939, and .111 in 1940. After 1940 it dropped rapidly, leveled off around 1947, and since then has fluctuated between .011 and .037. There seems to be little doubt that these changes are due largely to fluctuating selection pressures. Among other things, the gene is known to affect color pattern, mating behavior, fertility of the males, and larval viability. There has been some controversy over the possible role played by drift in this situation; the degree to which random changes interact with selection has not been determined.

Dowdeswell, Fisher, Ford, and others have made a long series of studies of the frequency distribution of spot number on the underside of the hind wings of the satyrine butterfly *Maniola jurtina*. Over most of southern England the spot distributions are remarkably uniform, in spite of the great diversity of environments. However, the spot distribution in females was found to differ sharply between populations in Devon and Cornwall (Fig. 7-4). Furthermore, the change was found to be extremely abrupt; indeed, in 1956 the border between populations with the two kinds of spot distribution was found to be a hedge which was not a barrier to the passage of individual butterflies. There was no sign of a gradient between the two types; if anything, the difference was greatest at the point of contiguity. In 1957 the border between the two forms was found to be 3 mi east of its 1956 location, and the transition was more gradual. The boundary itself was a strip some 150 yd wide which was occupied by an intermediate population. Spot distributions also have been studied extensively on a small archipelago, the Scilly Isles, off the western tip of Cornwall (Fig. 7-5). Numerous differences were found between the islands.

Each island tended to have a characteristic population which remained the same from year to year. On the larger islands the spot distributions were quite similar to one another. In contrast, the populations on the smaller islands had spot distributions which were significantly different from each other as well as from those of the

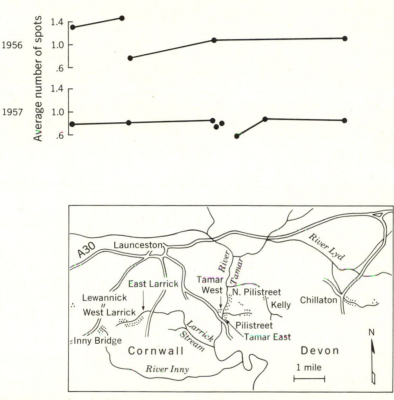

Figure 7-4 Graph of the average number of spots on the hind wing of English *Maniola jurtina*, correlated with localities showing where the specimens were obtained. Collecting sites are shown by small dots on map, and the east-west position at which samples were taken by the large dots on the graph. Thus the cluster of three dots on the 1957 graph indicates the average number of spots from samples taken at Tamar West, Tamar East, and N. Pilistreet. Note that the point of discontinuity in the spotting pattern shifted east between 1966 and 1967. The River Tamar forms the Devon-Cornwall border here. (*Adapted from Paul R. Ehrlich, Richard W. Holm, and Michael E. Soulé, Introductory Biology, copyright McGraw-Hill Book Company, 1973; used by permission.*)

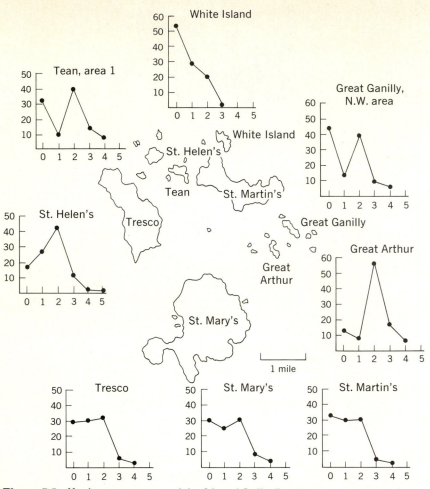

Figure 7-5 Northeastern region of the Isles of Scilly. Distribution of spots on hind wings of female *Maniola jurtina* on three of the large islands as well as on five of the smaller islands (vertical axes of graphs, percent; horizontal axes, number of spots on hind wings). Distribution on the large islands is comparable, but there is significant diversity among the distributions on the small islands. (*Adapted from L. E. Mettler and T. G. Gregg, Population Genetics and Evolution, Prentice-Hall, Inc., © 1969, based on E. B. Ford, Ecological Genetics, Methuen & Co. Ltd., 1954.*)

larger islands. Ford believes these differences are the result of strong selection pressures rather than genetic drift, as suggested by others. He has shown that on the Isle of Tean (a small island) the spot distribution changed and stabilized without there being a reduction in population size. This new spot-distribution frequency was correlated with a vegetation change that occurred after a herd of cattle was removed. Further, Ford has reported that in a drought year an isolated population on one of the larger islands almost became extinct. However, as the vegetation returned to normal, the population increased and still the original spot distribution was maintained. According to Ford, the spot distributions on the larger islands result from a similar selection for the average of many conditions whereas the populations on the small islands are selected for different and more or less homogeneous habitats.

Spot distributions certainly are under strong selective control, although it seems clear that the selective value is manifest in other effects of the genes involved rather than in inconspicuous spot changes. It is difficult to formulate any other explanation for the sharp and mobile border between the Devon and Cornwall types than the shift from one highly integrated gene complex to another. The types seem highly successful, as they extend for a considerable distance in either direction from the border (especially eastward) and are relatively undisturbed by the heterogeneity of the habitats they occupy. It does not seem likely that the environmental factors responsible for the change in selection coefficients actually change as abruptly as the spot frequencies; the reason for the sharp border almost certainly lies in the genetics of *Maniola*.

Polymorphic land snails

The microevolution of polymorphic land snails of the genus *Cepaea* has been studied in detail, mostly with the very variable species *C. nemoralis* (Fig. 7-6). The shell of this snail may be yellow, brown, or any shade from pale fawn through pink and orange to red. The lip of the shell may be black or dark brown (normally) or pink or white (rarely), and up to five black or dark brown (rarely transparent) longitudinal bands may decorate it. The genetic basis of many of these characters is fairly well understood, and fossil evidence shows that this polymorphism has existed since before the Neolithic.

The snails have been studied extensively in Europe, where the frequencies of different forms vary greatly from colony to colony. The roles played by selection and drift in accounting for the intercolony differences have been the subject of controversy. Lamotte in France originally claimed that genetic drift accounted for the observed diversity. However, Cain and Sheppard demonstrated that, at least in some English colonies of the snails, selective forces were at work. Near Oxford they found that the frequencies of the different kinds of shells were correlated with the microhabitat of the snails (Fig. 7-7). Collections were made in six main types of habitats: downland beech woods, oak woods, mixed deciduous woods, hedgerows, open areas

Figure 7-6 The snail *Cepaea hortensis* shows variation in banding and coloration similar to that found in *C. nemoralis*. (*Courtesy of W. Dowdeswell.*)

with long coarse herbage, and open areas with very short turf. Analysis of the frequencies of color and pattern types in these samples showed that they were highly correlated with the color and uniformity of the background. For instance, the percentages of effectively unbanded shells in the five localities with the most and the least uniform backgrounds (uniformity decreases to the right in each series) were as follows:

Most uniform	100	100	100	93	79
Least uniform	39	35	34	25	4

The percentages of yellow shells in the five greenest and in the five least green localities (greenness decreases to the right in each series) were:

Greenest	76	64	45	43	41
Least green	17	8	4	0	0

This association between the frequencies of shell types and the character of the habitat suggested that visual predators eating the most conspicuous snails might be a selective agency causing differences between colonies. One such predator is the song thrush

(*Turdus ericetorum*). In the summer of 1951, Cain and Sheppard studied a colony of snails in a small hillside bog in Wutham Woods, near Oxford. Thrushes remove snails from the colony, crack their shells on stones on a nearby bank, and eat the soft parts (Fig. 7-8). Thus a sample of the predated portion of the population could be obtained by collecting the broken shells from around the thrush anvils and a sample from the entire population by collecting

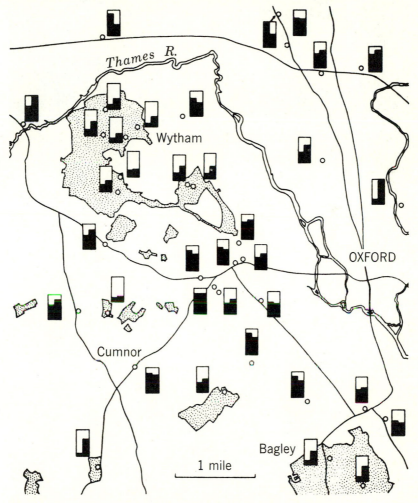

Figure 7-7 Map of Oxford district showing the correlation of *Cepaea* shell type and environment. In each histogram, the left-hand column represents percent of yellow shells and the right-hand column, percent of effectively banded shells. Woodlands are stippled; all colonies outside woodlands are in hedgerows or rough herbage. [*From A. J. Cain and P. M. Sheppard, Genetics, vol. 39 (1954).*]

Figure 7-8 A thrush anvil stone with shells of *Cepaea nemoralis* and *C. hortensis.* (*Courtesy of W. Dowdeswell.*)

individuals from the bog. Of 560 individuals taken from the bog, 296 (52.8 percent) were unbanded, while of 863 shells collected around the rocks only 377 (43.7 percent) were unbanded. This significant difference ($P<.001$) indicates that unbanded individuals in this colony were less likely to be eaten than banded individuals.

Similar methods were used by Sheppard in 1950 to study two other colonies. He found a decrease in the percentage of yellow snails killed as the season progressed, and the rate of decrease appeared to be the same in both localities. There was no evidence that this was due to the thrushes hunting in different areas or to a change in the percentage of yellow shells in the population at large. Apparently the selective value of the yellow phenotype was at least partly a function of the background on which it occurred. Early in the spring, when the woodland floor was predominantly brown, the yellow shells were relatively conspicuous and thus at a selective disadvantage. As the season progressed, the background became greener and this disadvantage lessened. By late April or early May the yellow shells were selectively neutral; by mid-May they were at a selective advantage.

These data indicate a rather strong selective pressure. Because of this, one would expect populations living on uniform backgrounds to be composed only of unbanded individuals and those living in rough tangled habitats to contain only banded individuals. How then is the polymorphism maintained? Shifts of selective values with the seasons would delay, but not prevent, the removal of the less favored varieties. Interchange of individuals among colonies in different habitats would account for some of the variability, but the range of movement of snails is too small to support this hypothesis for more isolated colonies.

The answer is that there are physiological factors genetically correlated with pattern type. Experiments with *Cepaea nemoralis* have shown that unbanded individuals (especially yellow ones) are more heat-resistant than banded individuals. Yellow snails are more resistant to cold than pink snails, and unbanded snails are more cold-resistant than banded snails. These and similar characteristics indicate that color and banding are subject to strong nonvisual selection because they are associated with important physiological advantages.

Arnold has also studied the correlation of banding morphology frequencies and habitat types. Along the river Segre, south of Andorra, snail populations were sampled from two types of habitats, arid hillsides and more temperate sites along the river. Arnold found that the hillside populations possess a higher number of unbanded snails while the sites along the river have a greater proportion of banded snails. These differences could be correlated only with very local environmental differences and support the earlier work suggesting that the polymorphism results, among other things, from selection for certain physiological factors. In some cases, heterozygotes may be expected to be more viable than either homozygote, so that physiological selection would tend to establish a stable polymorphism. These nonvisual selective forces, interacting with the selection pressures created by visual predation, seem to be responsible, in large part, for the observed pattern of variation in *Cepaea*.

Several other factors may be of importance in some situations. One is predation in which the selection pressure is a function of the relative frequency of the type of individual predated. Certain predators may form search patterns that result in selection against the commonest type in the population, without regard for which type is commonest. This sort of predation pattern could lead to a stable polymorphism. Random processes, once considered to be the prime factor in differentiating *Cepaea* populations, may have relatively minor importance. Undoubtedly drift plays some role in the smaller populations, and it may account for some patterns of variation recorded by Lamotte in France (Fig. 7-9). Recently established colonies may not have achieved equilibrium with their environment, and their composition may be strongly influenced by the genetic information possessed by the snail or snails that established them. This influence of the genetic endowment of the individuals involved in

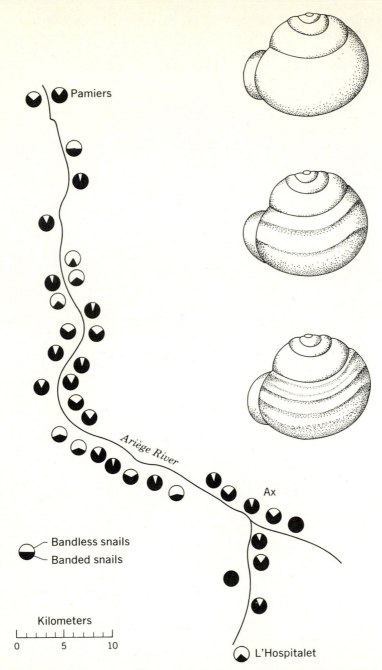

Figure 7-9 Colonies of snails along the Ariege River in the Pyrenees. (*From V. Grant, The Origin of Adaptations, Columbia, 1963.*)

Pamiers

Ariège River

Ax

Bandless snails
Banded snails

Kilometers

0   5   10

L'Hospitalet

starting a new colony, known as the **founder principle,** is discussed further in Chap. 10.

One of the arguments used to demonstrate that visual selection does not play an important role in determining the characteristics of *Cepaea* colonies is that in mixed colonies of the two closely related species *C. nemoralis* and *C. hortensis* the phenotype frequencies of the two snails are uncorrelated. Clarke has satisfactorily explained this by showing that both species respond to visual selection pressure but in different ways. It appears likely that the selective values of various pattern and color genes differ in the different genotypes of the two species. As a hypothetical example, a strongly banded pattern might be at a selective disadvantage in a certain dry habitat. In one species the genes for strong banding may also be involved in producing individuals resistant to desiccation. In the other species there may be no such system. Thus in the same habitat one species may have a population with a large proportion of strongly banded snails because of heavy selection for resistance to desiccation whereas the other species may have no banded snails at all because of the selection against banding. Although the genes controlling the various factors seem to be homologous, they cannot be divorced from their genetic environment. It will be recalled that in Chap. 6 genes were treated as if they were independent entities; results such as these studies of *Cepaea* constantly remind us that this is an oversimplification.

Island water snakes

Camin and Ehrlich have studied microevolution in populations of water snakes (*Natrix sipedon*) on the islands in western Lake Erie. The snakes on these islands have variable banding patterns, divided into four classes, A, B, C, D (type A snakes being unbanded and type D snakes being completely banded; see Fig. 7-10). Except in the area of western Lake Erie (and one locality in Tennessee), all known *N. sipedon* populations are made up of type D individuals. Virtually all the snakes taken from the mainland surrounding Lake Erie are type D. On the islands a large proportion of the snakes are of the other types, including numerous individuals that are totally unbanded. The islands have very little inland water, and the snakes are restricted largely to the lake shore. The marginal flat limestone rocks, limestone cliffs, and pebble beaches are the only suitable habitats.

Large samples of water snakes were taken from the islands, and pregnant females were kept alive until their young were born. The distribution of pattern types in the litters was compared with that of the

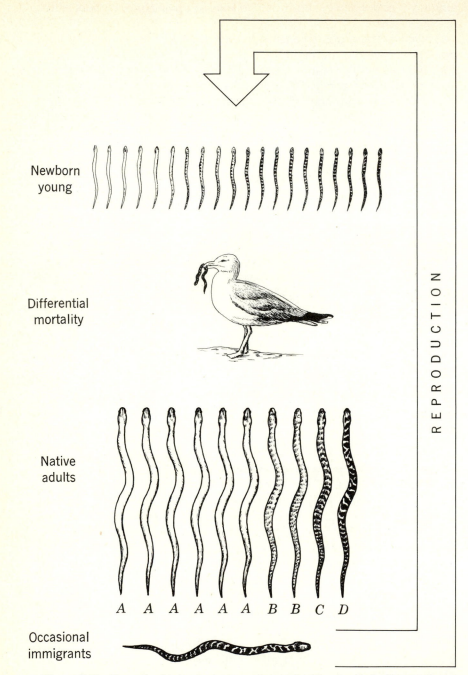

Figure 7-10 Natural selection in water snakes on the islands of Lake Erie, as shown by the differential in frequencies of banding types in young and old snakes.

adults. The distribution of the adult and litter pattern types from one group of islands is shown in Fig. 7-11. In spite of the difficulties of statistical comparison of the cluster-sampled litter population with the random-sampled adult population, it was possible to show a significant difference in banding pattern between the two populations. The percentage of relatively unbanded individuals (A and B) was higher in the adult population than in the litter population.

The observed significant differences between the young and adult populations can be accounted for only by differential elimination of pattern types or by pattern changes in the individual snakes. The evidence is overwhelmingly in favor of the former hypothesis, since snakes kept in the laboratory show no evidence of pattern change in ontogeny. In addition, individuals of all pattern types have been recorded from both adult and litter populations; only the *frequencies* differ.

Evolutionary agencies other than selection are easily disqualified. A high proportion of unbanded pattern types might be maintained by migration from other unbanded populations, but the nearest

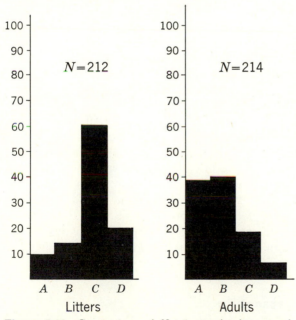

Figure 7-11 Comparison of *Natrix sipedon* litters and adults from the Bass complex of islands in Lake Erie. Ordinate, percent in class; abscissa, banding type. A, least banded; D, completely banded. [*From J. H. Camin and P. R. Ehrlich, Evolution, vol. 14, (1958).*]

such population is in Tennessee. To maintain the unbanded genotype by mutation alone would require a mutation rate far above that known for any locus ever studied in any organism, even if it is assumed that color pattern is a single factor trait, which it is not. Nor will genetic drift account for the observed differences. There is no sign that the populations on the islands ever approach a size at which drift is likely to be a significant factor. Seven collectors once captured 400 snakes on an island in 5 h, and three collectors at another time took 234 snakes in 4 h. The juvenile-adult shift is toward unbanded individuals on all the islands studied, indicating a systematic pressure rather than random drift. Therefore, by a process of elimination, selection seems most likely to be responsible for the change in pattern-type frequencies.

To prove that selection has taken place, one does not have to discover all the selective agents involved. However, it is interesting to speculate on the factors producing the observed situation. To the human eye, unbanded snakes are cryptically colored on the flat limestone rocks of the island peripheries, and banded individuals are very conspicuous. It is likely that banding would help to break up the outline of the snakes in their more typical, less uniform swamp habitat. There are visual predators present which eat snakes (gulls, herons, hawks, etc.), and man kills many with firearms.

The selective force is obviously very strong. If the pattern spectrum is divided arbitrarily into two halves (banded and un-banded), the snakes in the banded half have only about 25 percent of the chance of survival of the unbanded half (s for the banded phenotype equals approximately .75). This raises the question why, with such heavy selection, any banded individuals at all remain in the population. The answer appears to be that differential migration brings a steady influx of genes for banding into the gene pools of the island populations. Snakes have been observed swimming far from land on many occasions, and the distance from the shore to the islands is not too great to be spanned by migrating individuals. Thus the mainland populations form a reservoir of banded individuals, some of which periodically reach the islands. The resultant interaction between selection and migration has produced a situation unusually amenable to analysis.

Polymorphism in Drosophila

Classic examples of microevolution are found in the work pioneered by Dobzhansky on inversion frequencies in some 30 species of fruit flies of the genus *Drosophila* that show polymorphism in chromosome type. This type of investigation is possible in *Drosophila* because of

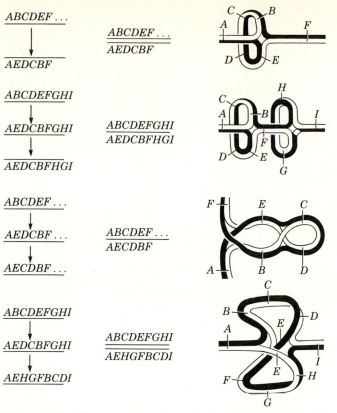

ABCDEF...
↓
AEDCBF

ABCDEF...
AEDCBF

ABCDEFGHI
↓
AEDCBFGHI
↓
AEDCBFHGI

ABCDEFGHI
AEDCBFHGI

ABCDEF...
↓
AEDCBF...
↓
AECDBF...

ABCDEF...
AECDBF

ABCDEFGHI
↓
AEDCBFGHI
↓
AEHGFBCDI

ABCDEFGHI
AEHGFBCDI

Figure 7-12 Chromosome pairing in the salivary glands of individuals which are heterozygous for inversions; upper row, a single inversion; second from top, two included inversions; lower row, overlapping inversions. (*From T. Dobzhansky, Genetics of the Evolutionary Process, Columbia University Press, 1970.*)

the giant polytene chromosomes found in the salivary glands of its larvae. These chromosomes show the close pairing usually associated with the zygotene stage of meiosis. Their size and this somatic pairing make them very useful tools for research. Inversion heterozygotes in *Drosophila* can be detected by examining the salivary chromosomes for the characteristic inversion loops (Fig. 7-12).

The relationships between the different sequences of the banding in the very variable third chromosome of *D. pseudoobscura* have been studied extensively (Fig. 7-13). Each banding sequence (inversion) has its own vernacular name. The three most widely discussed inversions in evolutionary literature are Standard (ST), Arrowhead (AR), and Chiricahua (CH). When two different kinds of chromosomes

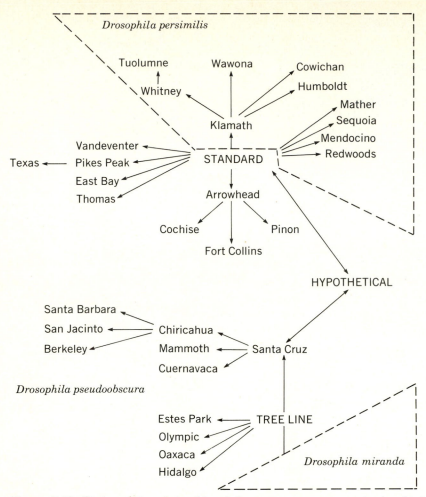

Figure 7-13 Phylogenetic relationship of the gene arrangements in the third chromosome of *Drosophila pseudoobscura*, *D. persimilis*, and *D. miranda*. (*From T. Dobzhansky, The Genetics of the Evolutionary Process, Columbia University Press, 1970.*)

occur in the same population, three types of individuals will be present: two inversion homozygotes and one inversion heterozygote. For example, where only Standard and Chiricahua chromosomes are present, the three different kinds of individuals are the two homozygotes, ST/ST and CH/CH, and the heterozygote, ST/CH. The homozygotes can be distinguished only by progeny testing or careful study of the pattern of banding, while the heterozygote is easily recognized by the inversion loop.

The different kinds of chromosomes in *D. pseudoobscura* have different geographic distributions, apparently because of their varying adaptive values in different habitats. Figure 7-14 shows the geographic variation in frequencies of three different chromosome types in a series of localities transecting the southwestern United States. The chromosomal frequencies also change with altitude, as can be seen in Fig. 7-15, which shows the proportions of three different types at different elevations in the Sierra Nevada of California. Superimposed on this geographic variation is a seasonal variation in frequency. For instance, in the Sierra Nevada, the Standard gene arrangement is commonest at lower elevations, becoming

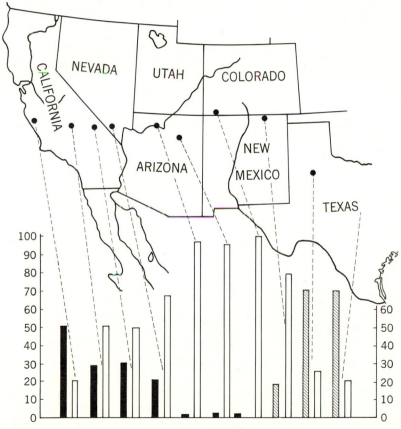

Figure 7-14 Frequencies (in percent) of different kinds of chromosomes in populations of *Drosophila pseudoobscura* in southwestern United States. Black columns, Standard; white columns, Arrowhead; hatched columns, Pikes Peak. [*From T. Dobzhansky, Evolution, vol. 1 (1947).*]

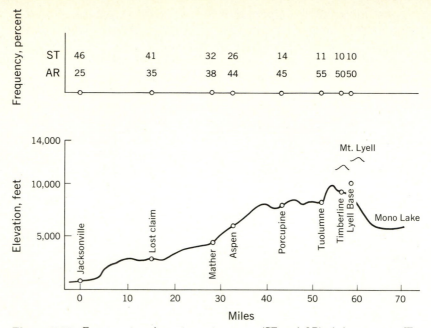

ST	46	41	32	26	14	11	10	10
AR	25	35	38	44	45	55	50	50

Figure 7-15 Frequencies of two inversion types (ST and AR) of chromosome III in populations of *Drosophila pseudoobscura* at different elevations in the Yosemite region of the Sierra Nevada. (*From V. Grant, The Origin of Adaptations, Columbia University Press, 1963.*)

progressively less common with increasing altitude. Arrowhead, on the other hand, has a frequency that is positively correlated with altitude, being greatest in the subalpine zone (about 10,000 ft) and least at the base of the mountains (850 ft). However, in general, Standard chromosomes tend to increase in frequency as the season progresses, whereas Arrowhead chromosomes tend to decrease in frequency. Thus the frequencies of these chromosomes in late-summer populations at high elevations tend to approach the frequencies of spring populations at lower elevations.

Seasonal variation of inversions was intensively studied in a population of *D. pseudoobscura* from San Jacinto Peak, California. This population exhibited marked changes in the frequency of the ST and CH chromosome arrangements from March to October. The frequency of the ST chromosome decreased from March until June, when it increased until October. The frequency of the CH chromosome showed an increase from March until June and a decrease during the latter half of the season (Fig. 7-16).

These changes may best be explained as the result of rather strong selection pressures. Dobzhansky has tested this hypothesis in a series of experiments in which he reared *D. pseudoobscura* in population cages. His experimental populations were started with known frequencies of chromosome types, and then he repeatedly sampled them to determine what changes in frequency, if any, had occurred. He found that in populations maintained at 16.5°C there is no change in the chromosome frequency. However, in populations maintained at temperatures above 20°C the frequencies change, usually arriving at an equilibrium point at which all the original types are still present but in frequencies quite different from the initial ones.

The data from one population-cage experiment are given in Fig. 7-17. The cage colony was constituted on Mar. 1, 1946, with individuals selected so that the population had 10.7 percent ST and 89.3 percent CH chromosomes. Throughout the remainder of the year the percentage of ST chromosomes increased, at first rapidly and then more slowly, until at the end of the year (some 15 generations later) it had leveled off at about 70 percent. This pattern of increase and the establishment of an equilibrium strongly suggested a selective advantage of the structural heterozygotes (ST/CH) over both homozygous types. From the standpoint of population genetics, this would be comparable to overdominance for fitness at a single locus. Similar

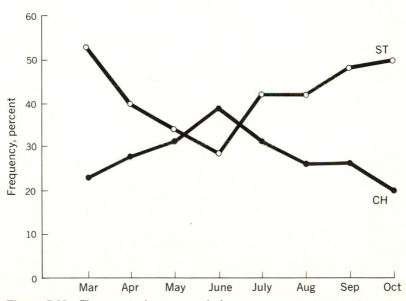

Figure 7-16 Changes in frequency of chromosomes carrying two inversion types in natural populations of *Drosophila pseudoobscura* in the San Jacinto Mountains during the advance of the season from March to October. [*From V. Grant, The Origin of Adaptations, Columbia University Press, 1963.*]

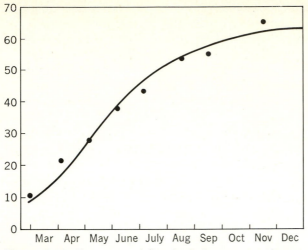

Figure 7-17 Frequency of Standard chromosomes (in percent) in different months in a single population cage. [*From T. Dobzhansky, Evolution, vol. 1 (1947).*]

results could be obtained, however, by negative assortative mating (unlikes mating). This latter possibility was ruled out by taking a sample of eggs from the population cage and raising the larvae under optimum conditions so that almost all survived. The different genotypes proved to be present in the expected Hardy-Weinberg frequency, demonstrating that mating was random and that differential fecundity or fertility was not involved. Samples of adult flies taken from the population, however, showed the following deviations from Hardy-Weinberg frequencies:

	ST/ST	ST/CH	CH/CH
Observed number	57	169	29
Expected number	78.5	126.0	50.5
Deviation	−21.5	+43.0	−21.5

These results, along with the establishment of equilibrium at about 70 percent, clearly indicate that there is differential elimination of the homozygous types between the egg and adult stages. It has been demonstrated that in eggs laid by wild flies there is no significant deviation from the expected Hardy-Weinberg frequencies but that heterozygotes are significantly more common in adult males thán one would expect in a population in Hardy-Weinberg equilibrium.

An extreme case of selective advantage of structural heterozygotes occurs in a population of *Drosophila tropicalis*, where there are two common chromosome sequences. Both types of structural homozygotes *die* early in development, and only the heterozygotes survive to breed. Although only half of the zygotes formed are viable, this population flourishes.

Long-term changes in inversion frequencies have been reported. In California before 1941 only 4 Pikes Peak (PP) chromosomes were found among 20,000 chromosomes studied. In 1957 PP chromosomes were found in all 10 localities sampled in the state, the mean frequency being about .08 (a 400-fold increase). This increase occurred at the expense of CH chromosomes.

Recently Cory and his associates have shown a correlation between this change in inversion frequencies and the increase of pesticide residues in the environment. It has been suggested that pesticides from extensive sprayings in the agriculturally important Central Valley of California have been carried eastward and deposited with precipitation in areas such as the Sierra Nevada and the San Jacinto Mountains (Fig. 7-18). Also, extensive treatments of DDT in certain forested areas have further contributed to the pesticide residues in the Sierra Nevada. Cory has noted that significant changes in inversion frequencies occurred in a population of the San Jacinto Mountains in 1946, a year after DDT was licensed for use in California agriculture. Further, in 1954, and again in 1957, within a year after massive forest sprayings, similar changes were observed in a population from the Sierra Nevada.

This is only a brief outline of some of the highly interesting work done on the genetics of natural populations of *Drosophila*. There can be little doubt that the different gene constellations or supergenes in the inversion chromosomes (remaining together as rather stable blocks because of the effect of the inversions in suppressing the results of crossing-over) have different adaptive values under different conditions. Many examples of chromosome frequency changes correlated with environmental changes have been elucidated. Less success has attended attempts to determine when advantages and disadvantages occur in the life cycle. Describing the exact nature of these adaptive changes also has proved difficult. It is impossible to reproduce natural conditions in the laboratory, and very little is known about the life history and ecology of the various *Drosophila* species. Studies of *D. pseudoobscura* and *D. persimilis* breeding in "slime fluxes" (yeast- and bacteria-infected sap exudations on trees) in the

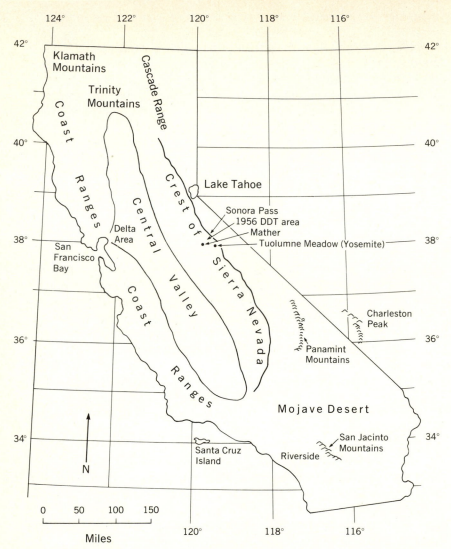

Figure 7-18 Map of California showing the Sierra Nevada and the San Jacinto Mountains and their relationship to the agriculturally important Central Valley. [*From L. Cory et al., Nature, vol. 229 (1971).*]

Sierra Nevada and laboratory studies of many species have yielded valuable information. Much more work is needed, however. The complexity of the problem may be appreciated by considering the factors determining pupation site in *D. melanogaster*, a characteristic known to be of considerable selective importance in laboratory populations. Among other things, temperature, humidity, moisture content of the medium, larval density, and length of larval period all affect the place in which the larva chooses to form the pupa. The ecogenetics of *Drosophila* will supply a fertile field for research for a long time to come.

Examples from man

As one might expect, natural selection plays an active part in shaping the genetics of populations of our own species. An outstanding example of this is the selective control of the frequency of the sickle-cell gene. Individuals who are homozygous for this recessive gene show distortion of their erythrocytes, accompanied by severe anemia and general, serious, and painful disability. The condition is usually fatal. Heterozygous individuals may be detected by the distortion (cells become sickle-shaped) of their red blood cells, which occurs when the oxygen concentration of the blood is reduced. Sickle-cell heterozygotes, however, apparently are protected to some degree against malaria and thus are favored in malarial areas. Allison found that the frequency of sickle-cell heterozygotes is higher in adults than in young children, indicating that this genotype is at a selective advantage. The advantage of the heterozygotes is about of the magnitude theoretically necessary to maintain the frequency of the sickle-cell gene (about .20). This advantage appears to be responsible for the maintenance of a balanced polymorphism at the sickle-cell locus throughout much of Africa, in spite of the very low viability of the homozygous recessives.

Glass and his coworkers studied the frequencies of genes controlling a number of characteristics (blood groups, mid-digital hair types, etc.) in a Dunker religious community in Pennsylvania where the population was less than 100 individuals. The gene frequencies at several loci deviated strongly from those found in the surrounding population and in the population from which the group was originally derived (western Germany). Drift is tentatively considered to be responsible for these deviations. There was little or no deviation at different loci known to be under rather strong selection pressure, for example, Rh.

A study of blood-group frequencies in the Parma Valley of Italy shows that genetic drift has been an important evolutionary factor in the small highland village populations of this valley. Significant variation in frequencies of blood groups existed between the small, rural, isolated populations of the upper valley. The variation between the larger populations of the lower hill country was not as significant, and on the plain the variation between the largest and most mobile populations of the valley was negligible. This pattern was found to be the same for all blood-group loci, thus eliminating the possibility that variation between the villages was due to selection pressures.

The one known example of selection against heterozygotes involves the Rh locus in man. When an egg of an Rh negative mother (double recessive) is fertilized by an Rh positive sperm, the resultant heterozygous fetus runs a relatively high risk of death due to antigenic incompatibilities between it and the mother. This selection is apparently compensated to some degree by a tendency of families with Rh problems to have repeated pregnancies until a number of children are raised successfully. This, however, is probably not sufficient in itself to account for the continued presence of the polymorphism in the human population. The gene-frequency equilibrium point, although no longer at .5 (see Chap. 6, selection against heterozygotes), is still unstable. It is possible that migration has helped to prevent fixation in the human population as a whole. Fixation of the Rh positive genes may be approached in some subpopulations and fixation of Rh negative in others. Intermixing of these populations may then reestablish the polymorphism.

Examples from plants

Unfortunately botanists have found few situations in nature that are as amenable to the sort of analysis discussed in the foregoing cases. Kemp in southern Maryland studied pasture seeded with a grass-legume mixture and subsequently partitioned. One half was protected from livestock, while the other was used for grazing. Three years later, plants of bluegrass (*Poa pratensis*), orchard grass (*Dactylis glomerata*), and white clover (*Trifolium repens*) from each half of the pasture were dug up and transplanted to an experimental garden with uniform conditions. The grazed half yielded a high proportion of genotypes that produced a low prostrate growth. Those from the ungrazed half were erect and showed no tendency to procumbency. This clearly shows that there was a heavy selection in these populations for adaptive growth forms.

Prunella vulgaris, a member of the mint family, is a weedy species some strains of which were introduced into North America from Europe. It occupies a variety of habitats, including gardens and lawns. Nelson has shown that survival, vegetative growth, and flowering of this plant are severely limited when it occurs as a weed in lawns. Different strains of *P. vulgaris* vary in their ability to adapt to life in a lawn environment. A genetically determined variety that is low and spreading in growth habit in both vegetative and flowering phases has been identified. In addition to the genetically based, low-growing lawn forms, there are other types which are low and spreading in lawns, which have tall, ascending shoots in garden culture. Thus *Prunella vulgaris* has produced a lawn ecotype (genetic race) and a phenocopy (see P. 170) of this lawn ecotype. The genetic and evolutionary relationships between the ecotype and the phenocopy are not yet known.

An extremely complex system of mimicry has been found in the genus *Camelina*, plants of the family Cruciferae. Various types of *Camelina* occur as weeds in fields of cultivated flax (*Linum*, family Linaceae). It has been hypothesized that as the cultivation of flax became more efficient, *Camelina* was subjected to a series of increasingly severe selection pressures. For instance, there was a selective advantage for the *Camelina* seeds to remain with the flaxseeds during the winnowing process so that they would be sown along with the flax. Those plants that grew were from seeds with the correct aerodynamic properties; the other seeds were never planted. The more thorough the winnowing, the stronger the selection pressure. This selection produced *Camelina* seeds that mimicked flaxseeds, not in appearance but in distance blown by a given amount of wind. Similarly, selection favored the production of tall spindly *Camelina* plants that would not be shaded out of existance in the dense stands of cultivated flax. Such selection has produced flax mimics not only in the genus *Camelina* but also in other plants, including *Spergula* and *Silene* (Caryophyllaceae). It should be noted that the evidence for selection here is more inferential than in the preceding examples.

Disruptive selection in mimetic butterflies

When selection favors two or more phenotypic modes, it is said to be **disruptive.** Experimental work with bristle numbers in *Drosophila* has shown that a pattern of selection in which extremes are favored over intermediates can produce a bimodal population with increased variance. Apparent examples of the operation of disruptive selection in natural populations are found in arrays of mimetic butterflies. For instance, the widespread and much-studied African swallowtail butterfly (*Papilio dardanus*) (Fig. 7-19) has a wide variety of mimetic females, although the males never show mimicry. Presumably this is because "normal" color pattern and wing shape of males are important in making them sexually acceptable to the females. Several different forms of the females commonly occur in the same locality, each one accurately mimicking a different distasteful species of butterfly.

There is evidence (summarized by Sheppard, 1961) that certain combinations of characteristics give the best mimicry of different

Figure 7-19 Mimicry involving an African swallowtail butterfly (*Papilio dardanus*). Upper butterfly, male *P. dardanus*; right-hand column, three danaine butterflies (*Danais chrysippus*, *Amauris niavius*, and *A. echeria*); left-hand column, forms of female *P. dardanus* mimicking the danaines. (*After R. C. Punnett, Mimicry in Butterflies, Cambridge University Press, 1915.*)

models and are at a selective advantage. Others do not look like any model and are at a disadvantage. In at least some cases, selection seems to have reduced the possibility of the production of poorly protected combinations by increasing the linkage between the loci concerned in producing the pattern. This permits linked groups of loci (supergenes) to be selected as a unit and the superior combinations to be preserved. (The phenomenon is quite comparable to holding together gene constellations in *Drosophila* populations by inversions.) In mimetic butterflies disruptive selection may also operate through the accumulation of modifier genes which further perfect the resemblance. This is supported by hybridization experiments in which supergenes are transferred to a new genetic environment. There they do not produce phenotypes that mimic the model as precisely as before.

Resistance to antibiotics and insecticides

No discussion of evolution would be complete without mention of the response of some organisms to man's attempts to reduce their population size or eradicate them. Striking and important examples of the response of natural populations to human endeavors center around the phenomenon of resistance. Indiscriminate application of insecticides to large areas of the earth's surface has constituted a very potent selective force. In the vast majority of cases, the large population sizes of pest insects have contained sufficient residual variability to allow them to develop strains resistant to virtually all the compounds that the ingenuity of the organic chemists can produce. The chemists are, of course, severely limited by the survival requirements of nontarget organisms such as man. Even so, there is considerable evidence that man and his domesticated plants and animals have not escaped from powerful synthetic pesticides. For example, the concentration of DDT in the fat deposits of Americans averages about 12 parts per million. The people of India and Israel have much higher concentrations.

It is interesting to note that insects have met the challenge in diverse ways. There have been examples of behavioral resistance in which insects no longer alight on sprayed surfaces and of many kinds of physiological resistance in which the penetration or action of the insecticide is prevented by various mechanisms. Parallel to insecticide resistance has been the appearance of strains of microorganisms that are highly resistant to antibiotics. This has been caused by the overuse of these antibiotics by well-meaning doctors when other treatments might suffice or be better. So far, the increase of these organisms has been countered largely by a scramble to find new chemical weapons to use against them rather than by the application of methods that are biologically more sophisticated and infinitely more beneficial in the long run.

Such alternative approaches are well known. Some resistant

insects prove to be less viable than their nonresistant relatives when they exist in an environment free of insecticide. In these cases, moratoria on insecticide applications would give time for the agents of natural selection to return the populations to their previously susceptible condition. By intelligent use of insecticides at critical moments, a reasonable level of control may be attained with a minimum danger of creating resistant strains (a danger maximized by broadcast application and spraying by the calendar).

Where moratoria are not feasible, multiple applications of many different poisons may make it impossible for the population simultaneously to develop resistance to all. Because it is not possible, however, to affect one member of a community without affecting the entire ecosystem, this approach is fraught with unforeseen dangers. It is well known that accumulation and concentration of chemical poisons take place along food chains. Many bird populations, e.g., those of the peregrine falcon and brown pelican, are in danger of extinction because of the high concentrations of chlorinated hydrocarbons, e.g. DDT, DDE, and PCBs, that accumulate in the visceral fat of these birds. This buildup of chlorinated hydrocarbons, among other things, leads to the formation of thin-shelled eggs which cannot be hatched.

Thus if the target organisms are not eliminated, their predators and the predator's predators may be severely affected. An example of such a situation is the history of attempts to control cotton pests in the coastal Cañete Valley of Peru, as reported by Smith. Against the advice of ecologically sophisticated entomologists, who recommended the use of cultural control methods and inorganic and botanical insecticides, synthetic organic pesticides were widely introduced in the valley in 1949. At first the use of these pesticides, principally the chlorinated hydrocarbons DDT, BHC, and toxaphene, was very successful. Cotton yields increased from 494 kg/hectare (440 lb/acre) in 1950 to 728 kg/hectare (648 lb/acre) in 1954. Insecticides

were applied like a blanket over the entire valley. Trees were cut down to make it easier for the airplanes to treat the fields. The birds that nested in these trees disappeared. Other beneficial animal forms such as insect parasites and predators disappeared. As the years went by, the number of treatments was increased; also, each year the treatments were started earlier because of the earlier attacks of the pests.

Trouble started in 1952, when BHC proved to be no longer effective against aphids. In 1954 toxaphene failed against the tobacco leafworm. Boll weevil infestation reached extremely high levels in 1955–1956, and at least six entirely new pests had appeared which were not found in similar nearby valleys that had not been sprayed with organic pesticides. Furthermore, the numbers of an old pest, larvae of the moth *Heliothis virescens*, exploded to new heights and showed a high level of DDT resistance. Synthetic organic phosphates were substituted for the chlorinated hydrocarbons, and the interval between treatments was shortened from 1 or 2 weeks to 3 days. In 1955–1956 yields dropped 332 kg/hectare in spite of the tremendous amounts of insecticide applied. Economic disaster overtook the valley. In 1957 an ecologically rational integrated control program was initiated in which biological, cultural, and chemical controls were combined. Conditions improved immensely.

Many pestiferous insects are more readily and economically controlled by interfering with their biology in a nonchemical way than by the application of insecticides. For example, draining swamps in which mosquitoes breed, introducing predators to control imported pests, releasing multitudes of sterilized males to compete for mates with wild males, and disseminating laboratory-grown pathogens have all proved effective.

LABORATORY POPULATIONS

The term artificial selection pertains to man's control of the genotypes that contribute to the gene pool of succeeding generations. Artificial selection is carried out by plant and animal breeders and by scientists wishing to study the effects of selection in the laboratory. It may have a purposiveness directed at a single trait, another respect in which it differs from natural selection. In natural selection, selective agents affect all the phenotypic characters controlling fitness and it has no purpose other than reproductive success. Nevertheless, even under the most carefully controlled laboratory conditions, *natural selection still is found in conjunction with artificial selection*. A great deal of work in artificial selection has been concerned with so-called quantitative characters. Quantitative characters are those influenced by many pairs of alleles at many different loci, such as height and intelligence in man, bristle number in *Drosophila*, or color pattern in water snakes.

The work of Mather and Harrison is a rather typical example of artificial selection. Here selection was for changes in the number of abdominal chaetae in *Drosophila melanogaster*. The diagram of Fig. 7-20 summarizes the results of more than 100 generations of selection for higher bristle number. For 20 generations, selection was extremely effective. At generation 20, reduced fertility and fecundity in the population made it necessary to discontinue selection in order to

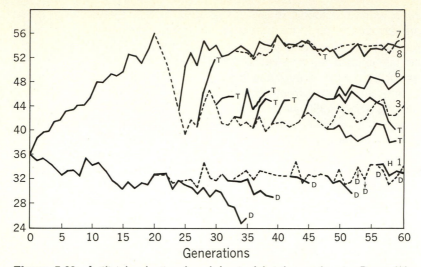

Figure 7-20 Artificial selection for abdominal bristle number in *Drosophila melanogaster*. Ordinate, mean number of bristles; abscissa, number of generations; lines under selection are solid, lines not under selection are dashed; *T* indicates deliberate termination of a line, *D* that it died out through sterility; circled numbers indicate different lines. From line 1, all the selections were for low number of bristles, except that marked *H*, which was for high number. [*From K. Mather and Harrison, Heredity, vol. 3 (1949).*]

keep the strain from dying out. In the absence of selection over several generations, the population reached an equilibrium chaeta number higher than the original one but much lower than that achieved at the peak in generation 20. From the line in which selection was suspended, a new selected line was extracted at generation 24, and progress was rapid to a point near that achieved with the first selected line. When selection in this line was relaxed, there was only a slight regression in bristle number. In addition, after some 85 generations in the continuously selected line, further response was achieved.

Artificial selection often produces rapid results at first. Then a plateau is reached at which further progress is difficult or impossible, or the viability of the line reaches such a low ebb that either selection must be discontinued (relaxed) or the line is lost. Generally, if selection is discontinued before a plateau is achieved, the relaxed lines regress toward the control level. If selection ends after the population has reached a plateau, there may be little or no regression. Continuous selection of a population that has achieved a plateau often will not produce appreciable results for long periods. However, if selection is continued long enough, progress once again may be made.

One reason for these phenomena presumably is the balance between artificial and natural selection. Although the details are neither clear nor uniform, it appears that natural selection must lead to a balanced system in which the best possible relationship of characteristics determining fitness is produced. In other words, *fitness must be maximized*. The available evidence seems to indicate that, especially in animals, a high degree of heterozygosity in the genotype produces a high degree of physiological fitness. It also seems likely that extremes of quantitative characters often are produced by a high degree of homozygosity at the loci concerned. Therefore artificial selection for high or low bristle number may well be countered by natural selection for fitness if the bristle-number extremes are produced by homozygosity at a series of loci.

One might make a crude analogy to an airplane. Trying to improve the airplane by making the motor more powerful will do little good if increased speed would tear off the wings. This problem might be solved by strengthening the fusilage or the structural members of the wings, but this would not help if it made the airplane too heavy to get off the ground. In an organism, as in an airplane, a viable balance of all the various factors that ensure successful functioning must be attained. There is a limit to how much one factor alone can be modified before the "working combination" is seriously disrupted.

Lerner (1954) has produced a mathematical model which might explain the establishment of a plateau below the maximum level of expression of a character under selection. He bases this model on a system in which there is an obligate level of heterozygosity determining fitness. Crossing-over can convert potential genetic variability to free genetic variability, permitting further selection without loss of fitness. The whole situation can then be looked at in terms of shifting states of balance. Strong selection at one or a few loci places a stress on the balanced genotype. After the expression of the character has been shifted a certain distance, this stress will result in loss of fitness, followed by either extinction or the attainment of a new balanced state. If selection is relaxed before either of these events, the line tends to regress to the control level.

A tremendous volume of literature on artificial selection has accumulated, as work on economic problems (improvement of domestic animals and plants, studies of resistance, etc.) has produced information of great value. Much of our understanding of such diverse problems as the origin of dominance, the integrative properties of genotypes, and the efficacy of selection under varying conditions has been the direct or indirect result of investigations of such prosaic matters as egg laying in chickens, the weight of swine, rust resistance in wheat, the yield of corn and cotton plants, and the productivity of bovine mammary glands. The reader wishing a well-organized introduction to this vast and complex subject is referred to Lerner (1958).

GENETIC HOMEOSTASIS

At this point it would be well to mention an important steady-state property of mendelian populations, the often-observed tendency of populations subjected to directional selection to regress toward the original mean. Lerner has called this phenomenon **genetic homeostasis.** A mendelian population has characteristics above and beyond those of its component individuals. For instance, it would be meaningless to say that an individual is in Hardy-Weinberg equilibrium. Populations tend to retain a genetic composition that produces a maximum number of individuals with a high degree of fitness. This is essentially a stabilizing selection operating against deviant individuals. A genotype showing a high degree of fitness is adapted not only to the environment in the classical sense but also to its genetic environment, i.e., the gene pool in which the genotype occurs. In other words, the frequency and distribution of genes in the population help to determine the fitness of any genotype within the population. Well-integrated genotypes are "winning combinations," and, as demonstrated in the experiments of Mather and Harrison, selection to change them in order to meet a particular environmental stress is countered by selection favoring the retention of the successful integrated unit. The unusually sharp break between the two kinds of *Maniola jurtina* populations mentioned earlier in this chapter may represent the border between two such highly integrated units. Permanent directional progress is made only when the selective forces operating in favor of change are able to overbalance those operating in favor of retaining the successful combination. There is much to indicate that the phenomenon of overdominance with respect to adaptive value (selective advantage of the heterozygote over homozygotes) is one of the fundamental mechanisms contributing to genetic homeostasis in most animals and many plants. However, the term includes all methods of genetic autoregulation of populations.

GENETIC ASSIMILATION

When individuals of the plant *Achillea lanulosa* (Compositae) were transferred from localities at various altitudes in the Sierra Nevada to an experimental garden at sea level, the plants did not all grow to a uniform height (Fig. 7-21). In the now classic experiments of Clausen,

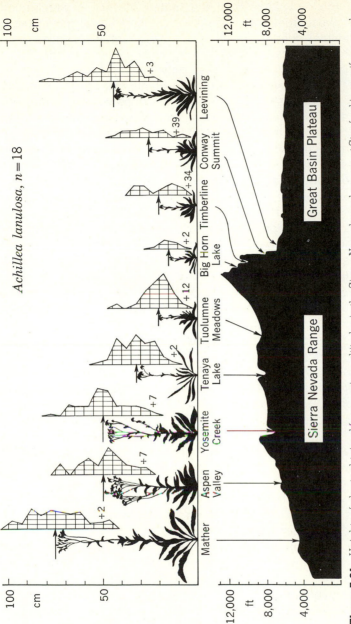

Figure 7-21 Heights of plants obtained from various altitudes in the Sierra Nevada and grown at Stanford in a uniform garden. Illustrated plants represent a population of about 60 individuals. The frequency diagrams show variation in height within each population. Horizontal lines separate class intervals of 5 cm according to the scale on the ordinate, and the distance between vertical lines represents two individuals. Numbers to the right of frequency diagrams indicate nonflowering plants. Arrows point to the mean of each distribution. (*From J. Clausen, D. Keck, and W. Hiesey, Carnegie Inst. Wash. Publ. 581, 1948.*)

Keck, and Hiesey, plants from the higher elevations were much shorter than those from lower elevations. Since all were grown under roughly identical conditions, those from the higher localities were shown to be genetically dwarfed; i.e., their genotypes tended to produce short individuals regardless of the environment in which they developed. However, when low-altitude plants were divided (giving genetically identical stocks) and these divisions were grown at sea level and at mid-altitude, the mid-altitude individuals were shorter than their identical twins at sea level. In other words, the low-altitude genotype interacting with mid-altitude environment developed into a plant similar to those with a mid-altitude genotype. Such forms, in which a phenotypic change simulates a genotypic change, have been termed **phenocopies.**

There seem to be many situations in nature where such phenocopying occurs, although rigorous demonstrations of the phenomenon generally are lacking. For instance, many butterflies have spring generations that are smaller and darker than their summer generations, the difference presumably being due to the seasonal variation in the environment. However, in more northern parts of their range, the butterflies have only a single summer generation, which is small and dark and resembles the spring generation of southern localities. In the northern populations, the individuals are presumed to have genotypes that produce the dwarfing and darkening. Although the critical transfer experiments have not been done, the greater constancy of the northern forms in the face of environmental changes supports these presumptions.

How can one account for the development of the high-altitude races of plants, the high-latitude races of butterflies, and similar phenomena? The environmentally produced changes cannot be directly transmitted to succeeding generations; the lamarckian idea of the inheritance of acquired characters has long been discarded. A method by which such acquired characters could, *through selection*, become assimilated in the genotype has been proposed by Waddington and is supported by a series of experiments by Waddington and others. This **genetic assimilation** is best explained by a brief example.

A strain of a wild-type laboratory population of *Drosophila melanogaster* was subjected to a high-temperature shock during the pupal stage. A few of the adults emerging from the treated pupae showed an abnormal break in the veins of the wings (the crossveinless phenotype). Only those individuals showing the acquired crossveinless phenotype were used as parents for the next generation. After more than a dozen generations, the frequency of crossveinless flies from treated pupae was over 90 percent, and a few crossveinless flies began to appear from untreated pupae of the selected strain. Crossing these latter flies produced strains that had a high frequency of crossveinless phenotypes in the absence of heat shock. It would appear that an acquired character had become heritable. Actually, in this experiment, selection seems to have favored those genotypes which had a low threshold for producing the favored phenotype. Eventually the threshold was lowered to a point at which no heat shock was necessary to move the developmental sequence to the crossveinless end point. Similar results have been obtained in studies of other characters in *Drosophila*.

It is most important to remember that the range of possible viable phenotypes is genetically determined and that selection may alter this range so that phenotypes that previously *could* be induced by altering the environment become genetically fixed as the *only* possible result of gene-environment interaction in the original environment. Phenotypic plasticity may be found where organisms normally face different environments in succeeding generations. Or, in contrast, a highly canalized development leading to a "winning" phenotype may evolve in situations where the environmental stresses are highly similar, generation after generation. Developmental aspects of genetic assimilation are discussed in Chap. 4.

POPULATION PHENETICS

In the early 1960s, members of the Population Biology group at Stanford University began to take a new approach to the study of variability in natural populations by performing careful multivariate statistical analyses of phenotypic variability. In a sense, these investigations of "population phenetics" represented a return to the morphological approach used by early evolutionists before the theory of population genetics was developed. On the one hand, however, the direct genetic explanation of patterns of phenetic (phenotypic) variation had remained an intractable problem because of the difficulties of studying many loci simultaneously. On the other hand, greatly enhanced understanding of both genetics and development and advances in statistical techniques and computing hardware gave population pheneticists enormous advantages over pioneers in the statistical study of phenotypic variation, such as Francis Galton.

Many fundamental questions remain to be answered about phenetic variation in natural populations, such as the following:

1. How are various kinds of phenetic variation related to the degree of genetic variation (note from the preceding discussion of

genetic assimilation that we can be sure that there is not a one-to-one relationship)?

2. Is variability in one character ordinarily related to variability in others?

3. What is the evolutionary significance of various kinds of asymmetry; is it related to developmental homeostasis?

4. Are there any general statements that can be made about phenetic variation in tropical vs. temperate populations, central vs. large populations; if so, what is their evolutionary significance?

Some progress has already been made in answering these questions. Soulé has attempted to investigate the **phenome,** which he defines as the phenotype as a whole. Phenotype is normally used to mean only the phenotypic expression of the alleles associated with a particular portion of the genotype, most commonly the alleles at a single locus. Soulé has studied intensively the variation in numerous characters of the side-blotched lizard (*Uta stansburiana*). He has found, for example, that on islands in the Gulf of California overall phenetic variation in any given population is representative of the variation of other local populations on the same island. He also finds strong evidence of what he calls a population variation parameter, in that populations highly variable in one character tend to be variable in others as well.

Soulé has been able to account for much of the variation statistically on the basis of island area, the number of related species of lizards on the islands, and measures of island isolation and the degree of phenetic divergence of the island populations from a mainland population. He has also determined, using electrophoretic techniques, the amount of heterozygosity at 20 loci in the populations inhabiting the islands. Variation in heterozygosity, like phenetic variability, is strongly affected by island size and the presence of related species. In contrast, however, heterozygosity is not significantly influenced by isolation or degree of phenetic divergence. The correlation between phenetic variability and level of heterozygosity at the 20 loci is significantly positive but far from perfect. Soulé concludes that underlying genetic variation is better estimated by phenetic variation than by the degree of heterozygosity at the 20 loci. The sample of 20 loci was small and the estimates of heterozygosity thus subject to large sampling error. In theory, the measure of overall phenetic variability "samples" a much larger portion of the variability in the genome, since many loci presumably affect each character. It clearly will be necessary to examine a larger number of loci electrophoretically to test this hypothesis, but this presents technical difficulties.

Soulé has proposed a verbal, i.e., nonmathematical, model consistent with his data in which the degree of phenetic (and thus genetic) variation is increased by long residence of populations in complex stable habitats, where variation has accumulated because of the developmental and physiological advantages of heterozygosity. The presence of competitors and the stable environment tend to subject the populations primarily to stabilizing selection, which, while setting a limit to variability, would not lead to the severe reduction in variance expected under a strong regime of directional selection.

Important work also is that of Roughgarden and others, who are attempting to build mathematical models to predict the degree of phenotypic variability displayed by populations with different genetic systems and under different evolutionary pressures. These general approaches, grounded in niche theory, show great promise for inducing broad generalities about the evolution of populations, and eventually of communities, which meld the ecological and genetic aspects of the problem, a satisfactory combination of which has long eluded classical ecologists and population geneticists.

ADJUSTMENT TO THE ENVIRONMENT

The diverse and ingenious ways in which organisms meet the problems of survival and reproduction are inferential evidence for the great efficacy of natural selection. This adjustment to the environment is usually called *adaptation*, but for reasons discussed in the final chapter we have avoided this ambiguous term whenever possible. One need not go into the details of the evolution of the bird's wing, the giraffe's neck, the vertebrate eye, the nest building of some fish, etc., as the selective origins of these and other structures and of behavioral patterns may be assumed to be basically the same in outline as those, e.g., industrial melanism, already been discussed. Even a slight advantage or disadvantage in a particular genetic change provides a sufficient differential for natural selection to occur. Thus the property of light sensitivity of unicellular organisms provides a starting point for the development, through selection, of the highly complex eyes found in vertebrates, insects, and certain mollusks.

The old anti-evolutionist argument that the vertebrate eye would be useless unless present in its modern complexity is nonsense. Many organisms with less complex and less specialized photo-receptors put them to good use, and it is easily seen that even a human being would be better off with a non-image-forming photo-receptor (one which gave information only on the amount of light present) than without any photoreceptor at all. Similarly, any nondetri-mental variation in a highly edible butterfly tending to make it look more like a sympatric distasteful species puts this deviant at a selective advantage.

The presence today of all degrees of refinement in the phenomena of mimicry and protective coloration argues strongly against the hypothesis that such resemblance must be virtually perfect before it is effective. Experimental evidence at hand indicates that less perfect copies of certain distasteful model butterflies also enjoy a degree of protection, though perhaps not as great as that of the more nearly perfect mimics. It is interesting to note that mimetic forms of various butterflies generally do not occur in areas where the models are absent, indicating that a selection pressure favoring mimicry is required to prevent regression to the wild type. Even such difficult to explain phenomena as the evolution of social behavior in bees are now yielding to investigation, e.g., Michener (1958). This problem is complicated because the unit of evolution is the colony, not the individual, since most members of a hive are nonreproductive. Selection in honeybees consists largely of differential reproduction of colonies rather than of individual genotypes.

Loss of features when they no longer confer selective advantage is one of the most widely observed evolutionary phenomena. Selectively neutral characters presumably are rare. The eye, very useful to most animals, may become an easily injured infection-prone liability to a cave fish. Body hair, which at one time protected human beings in certain environments against cold and injury, became a happy hunting ground for lice with the invention of clothing. Any "neutral" organ or appendage will represent an investment of energy to no purpose, and this actually will be a selective disadvantage.

SUMMARY

In this chapter examples of studies of evolutionary changes within populations have been chosen for the diversity of approach and material. In addition, inferential evidence bearing upon the efficacy of the selective process is discussed. It becomes apparent that although it is relatively simple to demonstrate changes, it is much more difficult to partition the responsibility for the changes among the various evolutionary forces. The problem is especially complicated because of complex interactions within genetic mechanisms and developmental systems. Most difficult of all to document is the role of genetic drift. It is nearly impossible to "prove" the efficacy of drift in natural populations, as one can always hypothesize the existence of some yet to be discovered selection pressure that could account for the observed phenomena. Indeed, there have been several cases, especially in snails, where differences at one time attributed to the action of drift have been

demonstrated to be caused by selection. However, the inevitability of drift, together with results of studies of gene frequencies in small populations of *Drosophila*, both in nature and in the laboratory, leads one to believe that drift, interacting with the other pressures, can be an important evolutionary force. Extensive studies still are needed on a wide variety of organisms before broad generalizations on the relative contribution of the various forces can be made with real confidence.

REFERENCES

Allison, A. C.: Metabolic Polymorphisms in Mammals and Their Bearing on Problems of Biochemical Genetics, *Am. Nat.*, **93:**5–16 (1959). Summary paper with extensive bibliography. (Allison, 1956, is the central reference on sickle-cell anemia.)

Camin, J. H., and P. R. Ehrlich: Natural Selection in Water Snakes (*Natrix sipedon* L.) on Islands in Lake Erie, *Evolution*, **12:**504–511 (1958). The principal paper on selection in *N. sipedon*.

Clarke, B.: Divergent Effects of Natural Selection on Two Closely Related Polymorphic Snails, *Heredity*, **14:**423–443 (1960). The bibliography lists most of the important papers on *Cepaea*.

Dobzhansky, T.: *Genetics of the Evolutionary Process*, Columbia University Press, New York, 1970. Essentially an updating and broadening of the author's classic work, *Genetics and the Origin of Species*, a must for every serious student.

————: Variation and Evolution, *Proc. Am. Phil. Soc.*, **103:**252–263 (1959). The bibliography will introduce the reader to evolutionary literature on *Drosophila*. The number of the *Proceedings* containing this paper has many articles of interest to the evolutionist.

————: Evolution and Environment, in Sol Tax (ed.), *Evolution after Darwin*, vol. 1, pp. 403–428, University of Chicago Press, Chicago, 1960. Contains a good discussion of drift in *Drosophila*.

Genetics and 20th Century Darwinism. *Cold Spring Harbor Symp. Quant. Biol.*, 1959, vol. 24. See especially Lamotte's paper on *Cepaea* for his side of the drift-selection controversy and the interesting papers by Carson, Dobzhansky, and Mayr.

Lerner, I. M.: *Genetic Homeostasis*, Wiley, New York, 1954. Highly theoretical and well worth reading (partly out of date).

————: *The Genetic Basis of Selection*, Wiley, New York, 1958. A comprehensive source on artificial selection.

Michener, C. D.: Distinctive Type of Primitive Social Behavior among Bees, *Science*, **127:**1046–1047 (1958). Evidence of the origin of worker-queen differentiation.

Sheppard, P. M.: *Natural Selection and Heredity*, Hutchinson, London, 1959. An excellent little book covering much of the material in this chapter.

————: Some Contributions to Population Genetics Resulting from the Study of the Lepidoptera, *Adv. Genet.*, **10:**165–216 (1961). Further information and references on the Lepidoptera examples and a guide to the literature on disruptive selection.

Stebbins, G. L.: Variation and Evolution in Plants: Progress during the Last Twenty Years, in Max K. Hecht and W. C. Steere (eds.), *Essays in Evolution and Genetics in Honor of Theodoseus Dobzhansky* (supplement to *Evolutionary Biology*), pp. 173–208, Appleton-Century-Crofts, New York, 1970. Surveys the work in the field of plant evolution since the author's publication of *Variation and Evolution in Plants*.

Waddington, C. H.: Genetic Assimilation, *Adv. Genet.*, **10:**257–293 (1961). Summary and discussion of the work and literature.

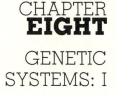

CHAPTER
EIGHT
GENETIC
SYSTEMS: I

The cytogenetic mechanisms discussed in the preceding chapters provide for the production of offspring sufficiently similar to the parental types to survive barring unduly rapid environmental change. When the environment changes gradually they provide also for sufficient variability for the organisms to change as well and thus survive. The persistence of an organism depends on the proper balance between these two phenomena: *fitness for the immediate environment and flexibility, in the long-range view, for whatever changes take place in the environment in the course of time*. It has already been pointed out that the genetic mechanism of most organisms provides for storing variability in unexpressed form, as well as for regulating the release of this variability. Diversity of means of storage and release of variability, which, like other traits, must be under genetic control, implies that these means change in the course of time; in other words, they evolve. The collective ways in which the amount and type of new gene combinations are controlled may be referred to as the **genetic system.** Thus one may speak of the evolution of genetic systems, which is the evolution of mechanisms effecting and affecting variability.

In this chapter the various sorts of genetic systems found in plants, animals, and microorganisms will be described. In the following chapter several specialized genetic systems which have become important in certain organisms will be discussed in greater detail. Elucidating the interrelationships and ultimate significance of these systems and integrating them into evolutionary theory are among the greatest challenges facing the modern evolutionist.

Although the actual course of evolution of genetic systems is not known, there has been considerable speculation about the main lines that it may have taken. Here genetic systems will be discussed, starting with the simplest and concluding with the most complex. This does not indicate an evolutionary sequence. The simplest biological phenomena often can be interpreted as having been reduced to this state from something more complex.

GENETIC SYSTEMS IN MICROORGANISMS

Transformation

Transformation is a phenomenon involving exchange of genetic material in some bacteria. For example, studies of strains of pneumococcus bacteria (*Diplococcus*) have shown that the DNA extracted from a strain with a particular trait can transform a strain lacking this trait into one that possesses it. Virulence in pneumococcus depends on the polysaccharide envelope of the bacterial cell. If a nonencapsulated, nonvirulent strain is grown with purified DNA extract from a virulent strain, virulent cells with a capsule will develop in the culture. Transformation may also occur in the opposite direction. Thus the DNA determines the polysaccharide coat. Transformation has been achieved for a large number of traits. Linkage of traits in the DNA material also has been found, since in experiments involving differing traits double transformations occur more frequently than would be expected for independent events. Genetic recombination involving transformation has been employed to some extent for genetic mapping of bacterial and viral chromosomes but has not been demonstrated in nature. *In vivo* transformation has been shown to occur in laboratory mice, however.

Recombination in viruses

A rather complex relationship exists between certain bacteria and bacterial viruses known as **bacteriophages.** A single phage consists of two major structural elements, a tail and an enlarged (often hexagonal) head. The protein coat surrounds a DNA core. When infecting a bacterial cell, the phage attaches itself by its tail to a specific receptor site on the bacterial cell wall. During the course of the next few minutes, the DNA leaves the phage and enters the bacterium. The cell eventually breaks open (undergoes **lysis**), releasing new phage. During the period of multiplication, new phage DNA is synthesized and new phage protein formed, the result being a hundred or so new phages of the same type as the infector. The original infector has, with its DNA, managed to preempt the synthetic processes of the bacterium and turn them to its own use, that of duplicating the phage.

Since they were first studied phages with many different traits have been found. It is possible to infect a culture of bacteria with a mixture of phages with different traits. When this is done, recombination of traits may occur in the phage progeny. With some characteristics, recombinants appear with equal frequency; with others there is a reduction in the number of recombinants, indicating that linkage exists between the genetic factors affecting the traits. What has happened then is recombination of the genetic material of the phage within the host cell of the bacterium. Radioactive-tracer studies have shown this genetic material to be the phage DNA. Such evidence, taken with that from experiments with bacterial transformation, in which the transforming agent is DNA, shows that in these forms the genetic material is DNA. It also shows that, in the absence of sexual reproduction, recombination of the genetic material can occur during the synthetic processes which produce more DNA.

Transduction

The bacteriophages discussed previously are virulent, and the host bacterium is killed as the cell undergoes lysis. However, there are also **temperate phages** capable of establishing a sort of symbiotic relationship which need not result in lysis of the host bacteria. These are called **lysogenic phages,** e.g., lambda phage. The noninfectious stage of the temperate phage is called prophage. Reproduction of the prophage and bacterium is so regulated and integrated that there is no known detrimental effect. Indeed, it seems clear that in the prophage stage, the viral DNA is physically incorporated (or integrated) within the bacterial chromosome. In this state it produces a repressor protein which prevents the expression of phage genes governing lytic processes.

Occasionally, however, at a rate of 10^{-2} to 10^{-5} per generation, prophage escapes repression and resumes the viral replication process. Lysis of the bacterial cell then occurs, and the released phages are able to adsorb onto other bacterial cells. In the course of excision from the bacterial chromosome, the prophage may carry along some of the genetic material of the host bacterium. When the phage is released and infects a new bacterium, it carries with it this genetic material of its former host. This, in turn, has a chance of being incorporated into the genotype of the recipient bacterium, a phenomenon known as **transduction.** The genetic material transferred usually is not a single genetic factor but groups of factors. Transduction thus is a special sort of genetic recombination, again involving not the sexual process but an infectious process. The phage acts as a vector for infectious transfer of bacterial genes. Transduction has been used for mapping very small segments of the bacterial chromosome.

Viruses, including phages which contain RNA instead of DNA are also known. Included in this group are bacteriophages of *Escherichia coli*, poliomyelitis and influenza of animal cells, and most plant viruses. The mode of replication of these viruses has been under intensive investigation.

Sexual recombination in bacteria

Sexual recombination also may occur in bacteria. Tatum, Lederberg, and others have shown that in *E. coli* **conjugation** of cells takes place. Genetic material is transferred from one cell to another, one of the conjugating cells being a recipient cell (−), the other a donor (+). This transference takes place through one of the structures called pili (singular, pilus), which serve as a connecting tube between the cells. The process of conjugation takes about 90 min; it can be interrupted at any time during this period by separating the cells by agitation. By interrupted-mating experiments, it can be shown that the amount of genetic material transferred is proportional to time. The process involves a linear, sequential transfer and thus is ideal for mapping the bacterial chromosome. Recombination of traits from different strains has been studied in some detail. The genetic material of the bacterium behaves as if it were on a single circular chromosome. Linkage maps of this chromosome and those of other bacteria, e.g., *Salmonella* and *Bacillus*, with similar chromosomes have been made. Spontaneous breakage of the transferring chromosome also occurs. Hybrids of *E. coli* and *Shigella dysenteriae*, as well as of *E. coli* and *Salmonella typhimurium*, have also been studied.

Microbial genetics and evolution

The frequency of occurrence and significance of such phenomena as transformation, transduction, and sexual recombination in many microorganisms are unknown. The extent to which these processes are distributed among microorganisms also is largely unknown. It seems obvious that the population genetics of these forms is apt to be rather different from that known in eukaryotic organisms.

The nature of the genetic relationships between very small replicating systems and larger organisms is now being extensively studied. Most of what has been observed in this category has been included under the term infection. At this level, of course, the distinction between genetics, infection, and development breaks down. Many viruses, particularly the RNA viruses of plants (which can be transmitted by sucking insects), produce morphological effects similar to those produced by chromosomal genetic material.

The role microorganisms have played in the evolution of the eukaryotic cell has become well documented. As discussed in Chap. 2, some evidence supports the notion that chloroplasts and mitochondria may once have been free-living prokaryotes, and there are similar reasons for arguing that other organelles have been derived in the same manner.

For example, some strains of *Paramecium* are known as killer strains, since they secrete into the medium in which they grow a substance poisonous to sensitive strains of *Paramecium*. It has been found that killer-strain individuals contain particles of DNA called **kappa particles.** For kappa particles to be maintained in the cell, the genotype of the strain must contain a dominant nuclear factor *K*. In the course of asexual fission, the kappa particles reproduce in such a way that all members of the resulting clone contain the particles. Should the rate of cell reproduction *exceed* that of particle reproduction, *KK* individuals *lacking* kappa may appear. Such individuals are not killers but sensitives. They are not able to initiate the formation of particles despite their chormosomal genotype. They can, however, regain kappa particles by contamination from a cell-free suspension of killer animals.

Sexual reproduction in *Paramecium* involves conjugation of cells and cytoplasmic exchange, during which kappa particles may or may not be transferred. In the cross between *KK* and *kk* individuals, the resulting *Kk* paramecia will be killers if conjugation has lasted long enough for kappa particles to be transferred. Paramecia that are homozygous for the recessive factor *k* may inherit kappa particles in the cytoplasm, but these are lost during subsequent generations of fission.

These particles are self-duplicating and contain DNA. Although their reproduction depends upon the genotype of the host, strains of *Paramecium* have been isolated with mutant kappa particles. The mutations in these particular strains are independent of nuclear control. Similar results have been obtained from work with chloroplasts in the alga *Euglena*.

An understanding of the nature and origin of various cellular organelles has shed new light on other problems of interest to the student of evolution. For example, diverse plant groups such as green algae (Chlorophycophyta), red algae (Rhodophycophyta), and golden algae (Chrysophycophyta) have chloroplasts which are characterized by different kinds of chlorophyll and related pigments.

Raven has suggested that these chloroplast differences represent independently derived endosymbiotic relationships between various photosynthetic protists (some of which may no longer be extant) and plant cells. Not all workers have accepted this conclusion, however.

This digression into what is usually termed **cytoplasmic inheritance** will serve to illustrate the difficulty of drawing distinctions between nuclear and cytoplasmic inheritance and between genetic recombination and infection at the level of cell and microorganism. It has become quite apparent that bacteria and other microorganisms, in addition to their role as reducers and decomposers, may have played an important part in the transmission of genetic information in the ecosystem. This puts the evolutionary significance of these organisms with regard to other members of the ecosystem or community in a new perspective.

GENETIC SYSTEMS OF OTHER ORGANISMS

The basic features of the genetic systems in the vast majority of living organisms are the same. There is an alternation of haploid and diploid phases, the result of alternating haplosis (meiosis) and diplosis (syngamy). As we have seen, in the few viruses and bacteria that have been studied, the genetic system is simpler and more variable. It has recently been demonstrated that sexual recombination similar to that found in bacteria also occurs in the blue-green algae (Cyanophycophyta), a group once thought to be unique in its lack of sexual recombination.

Sexuality and diploidy

The phenomenon of sexual reproduction is so widespread and its evolutionary significance is so immediately apparent that most biologists place its origin very early in the evolution of life. In many groups of plants and animals, every individual produces both female and male gametes. Even so, cross-fertilization is frequently the rule. Where the organisms are not hermaphroditic, the nature of sexual development may be affected by environmental factors such as light intensity, photoperiod, temperature, chemical composition of the medium, etc. It would seem that, in the course of evolution, this rather variable sex determination is replaced with more precise control systems or, at least, have such imposed upon them.

The first question to be considered is the origin of **diploidy.** It is usually assumed that diploidy has high selective value because of the opportunity it provides for the storage of recessive genes and thus of variability. It is generally felt that diploidy also provides necessary buffering in development and thus greater freedom from environmental effects. One would expect that this would be increasingly important as the complexity of organisms increases. Fusion of free cells occurs spontaneously in tissue cultures and cultures of unicellular organisms with such frequency that it is not difficult to imagine the origin of syngamy. Perhaps the resulting buffering effect in the diploid cell would have immediate selective value. Recent theoretical treatments have indicated that for genetic polymorphism to be maintained in haploid or asexual organisms, there must be either selection of the usual overdominant type on the diphophase or frequency-dependent selection on the haplophase, or both.

In many instances, fusion of cells leads to instability which is resolved by division. It was suggested in Chap. 1 that in early protoorganisms a sort of protorecombination might have taken place. Since a great many of the simpler algae and fungi are haploid for most of their lives (the only diploid cell being the zygote which immediately divides), one might conclude that recombination occurring with division immediately after syngamy has high selective value.

Sexual reproduction is a complicated process having many components that must be integrated in function. The stages in its evolution are not known. With the completely unstable genetic mechanism of protoorganisms, each "individual" would presumably have been different from all others. Fusion would occasionally combine complementary "genotypes," and this may have been the foundation of a selective advantage of fusion. On the basis of the speculations presented in Chap. 1, it may be suggested that the stabilization of the genetic mechanism of early protoorganisms involved loss of superfluous genetic material or its assumption of new functions, association of the DNA nucleotides with protein molecules to form chromosomes, and the restriction of gene function so that expressivity became less variable and control more precise. The variety of genetic systems in plants and animals includes many bizarre phenomena. Their common features suggest that, in nearly all instances, *the advantages of diploidy and of recombination have been joined*. Meiosis of the nuclear genetic material and fusion of cells are combined in a life cycle of varying degrees of complexity. This combination appears to have arisen independently in a number of different ways in plants and animals.

Stebbins has concluded that the wide occurrence of haploidy in the flagellates and filamentous algae is a result of their short and simple development and their rapid rate of reproduction, which makes the establishment of complex gene-developmental systems less important. The selective value of buffering and long-term storage of recessives would also be less important. With increasing complexity have come increased length of the developmental period and concomitant lengthening of the life cycle. The buildup of integrated gene complexes with ontogenetic buffering and genetic homeostasis is thus favored, and the diploid state has high selective value.

Diploid life cycles and alternation of generations

Some groups of Protozoa are predominantly diploid, with complex mating behavior effecting recombination. Certain groups of algae, notably some of the brown algae (Phaeophycophyta), diatoms (Bacillariophyceae), and some green algae (Siphonales), also are diploid during most of their life cycle. Meiosis results in the production of haploid gametes. The Metazoa, of course, have the same sort of genetic system.

Presumably this **diploid life cycle** arose independently several times from organisms with a predominantly haploid cycle. In addition, there also arose in plants and in fungi life cycles that involve **alternation of generations.** In these forms there is a regular cycle of haploid individuals that produce gametes and diploid individuals that produce asexual spores. Gametogenesis takes place by mitosis. Sporogenesis involves meiosis and resultant recombination. In some algae and fungi, the alternating generations are isomorphic (indistinguishable morphologically).

With increasing complexity, there seems to be a tendency for reduction of the haploid gametophyte generation in proportion to the diploid sporophyte. Thus in most ferns the gametophyte is a small thallus, usually a centimeter or less in diameter, whereas the sporophyte may be quite massive. In gymnosperms and flowering plants the gametophytic generation is reduced to relatively few cells (the pollen grain is a male gametophyte) and the sporophyte is the conspicuous stage. In mosses, on the other hand, the gametophyte is the conspicuous stage. It is perhaps better to regard the bryophytes (mosses and liverworts) as a specialized offshoot of ancient terrestrial plants and not on the main phyletic line of the vascular plants.

No process comparable to alternation of generations is known to occur in animals, with the possible exception of some Sporozoa. In other animals the haploid phase is represented by the gametes only. The Coelenterata have the so-called alternating generations of medusae and polyps, but these are morphologically, not cytologically, different.

The fungi, as a group, have a number of distinctive genetic systems of interest here insofar as they shed light on the selective forces involved in the evolution of genetic systems. These highly specialized organisms are poorly understood cytogenetically. The occurrence of somatic crossing-over and systems of multiple-mating types attests to the selective advantage of recombination. The water molds (Phycomycetes), which are filamentous and without cross cell walls, build up numerous haploid nuclei in the common cytoplasm of the filaments, or hyphae. Within the mycelium of these fungi genetically different populations of nuclei may arise, producing the so-called heterokaryotic state.

Mushrooms and toadstools (Basidiomycetes), the most complex of the fungi, and some of the sac fungi (Ascomycetes) show an interesting parallel with the hypothesized evolution of diploidy in plants and animals. Haploid mycelia, often with specific mating types, develop in the soil. The cellular hyphae of different mycelia, coming into contact, may fuse. Eventually a mycelium results in which each cell has two haploid nuclei—one from each haploid mycelium—which do not fuse until reproductive structures are formed. This special sort of diploidy is known as **dikaryosis.** Just as diploidy is associated with developmental and structural complexity in plants and animals, dikaryosis appears to be requisite for great morphological complexity in the fungi.

Recombination and genetic systems

To recapitulate, when the many different genetic systems that have evolved are compared, certain common features stand out. Except where simple unicellular structure and rapid reproduction are found, it seems that diploidy has had selective value. Simple, rapidly reproducing organisms have great flexibility for these reasons alone, and mutation is the chief source of variability. Such organisms as yeast and *Paramecium* are exceptions. With increased complexity, integrated combinations of genes controlling the developmental pathways are built up. The diploid state (or dikaryosis) provides the necessary buffering. It also permits the storage of variability in the form of recessive genes and of polygenes in balanced systems. Meiosis and syngamy, resulting in recombination, produce the release of variability as new gene combinations.

It is usually assumed that, without environmental change, most new gene arrangements will have lower selective value than existing ones. Thus a certain wastage of zygotes occurs in addition to the fortuitous wastage of gametes in sexual reproduction. Nevertheless, the wastage in diploid outcrossing organisms is surely much less than that in haploid organisms. The wastage of recombinants in the bacteria-bacteriophage systems is probably very much higher still. This suggests that there has been selection, also, for genetic systems that not only provide for buffering and the buildup of gene complexes and for storage and gradual release of variation but also reduce the wastage of biological materials and thus energy in the course of evolution. It must not always be assumed, however, that what appears to be wastage is disadvantageous.

Reduction of recombination

In nearly all genetic systems, modifications that reduce the amount of recombination have occurred. The result is reduction of wastage, through the preservation of superior gene combinations, and fitness is thereby increased. It has been generally assumed that the most primitive organisms were asexual, and these modifications are usually referred to as *reversions to asexuality*. Since recombination mechanisms have been found in nearly all organisms, and since the ability to effect recombination appears to be a fundamental property of DNA, this is perhaps an inappropriate designation. Sexuality and diploidy probably evolved relatively early in time. Reduction of recombination takes many forms. In unicells there may be absence of sexual reproduction. In the more complex multicells, reduction of recombination may occur as a result of reduction or elimination of crossing-over, assumption of specialized mating systems, inbreeding, self-fertilization, or loss of sexuality. As Stebbins has pointed out, such modifications are found most often in pioneer forms whose populations experience pronounced fluctuations in size. Under conditions of environmental change, such forms can exploit newly opened habitats through rapid duplication of closely similar genotypes. Often there is reduction in body size and in developmental complexity, as well as increased reproduction rate. Cytogenetic mechanisms affecting recombination are discussed in Chap. 9; systems of mating will be considered here.

Since the amount of recombination has selective value, the genetic system of nearly any organism is always in a state of flux. The simultaneous existence of variable mechanisms producing increase and decrease in recombination provides a system buffered against short-term environmental change but able to respond to long-term change. Most of the mating systems are bivalent in this sense or are combined with other mechanisms to produce this bivalence.

Mating systems and recombination

The basic type of mating against which the others may be compared is **random mating.** If individuals that are alike in phenotype and/or genotype are more apt to mate than would be expected by chance, the system is said to be **positively assortative.** If unlike individuals mate more frequently than expected by chance, the system is **negatively assortative.** The nature of reproductive mechanisms in plants and animals makes it unlikely that truly random mating ever occurs. It is difficult to specify precisely the extent of deviations from randomness. Mating systems do not, of course, fall into discrete classes, and any population in nature may show several of the arbitrarily delimited types. Furthermore, mating-system type is affected by such things as selective advantage of particular characters and population size.

As pointed out in Chap. 6, random mating without selection results in constant gene frequency with no change in variability. As the population size reaches lower limits, the effects of sampling error lead to genetic drift. Small population size also leads to inbreeding and a deviation from randomness of mating when the species is considered as a whole.

Inbreeding systems

Inbreeding may be referred to as positive genetic assortative mating, for it increases the chance of mating by organisms with like genotypes. Mating systems leading to inbreeding result in the breakup of a population into smaller groups that only rarely exchange genetic information. The heterozygosity of the population is reduced as fixation occurs in the subgroups. Genetic variance is increased in the population as a whole unless there is strong selection for particular homozygotes or heterozygotes. Morphological and physiological mechanisms leading to inbreeding are common in plants. Their degree of restriction ranges from facultative self-pollination to obligate self-fertilizing types with cleistogamous flowers (see Chap. 9).

Detailed studies by Stebbins and others have shown that the degree of restriction of recombination in plants is closely correlated with their growth form and habitat. As discussed in Chap. 9, populations often combine genetic systems that have opposite effects on recombination. Herbaceous plants, which have short generations and thus more recombination, tend to have low chromosome numbers. Perennial plants, with longer generation time, have higher chromosome numbers. Plants that occur in ecological communities usually thought of as "closed," i.e., those in which most offspring do not survive to maturity, tend to have genetic systems that promote genetic recombination. Oaks are an example. On the other hand, pioneer organisms, wherever they may occur, are members of "open" communities. In order for zygotes to survive, they must have the proper genotypes; there is no time for the organism to experiment with recombinants. It is not surprising, therefore, to find that plants of desert regions, grasslands, and cleared areas in the tropics generally have genetic systems that result in reduced recombination.

When inbreeding is imposed on populations that are usually outbreeding, a loss of fitness referred to as **inbreeding depression** occurs. The relationship between fitness, heterozygosity, and outbreeding is not yet well understood. Many groups of plants have successfully employed inbreeding as a mating system for long periods of time with only occasional outbreeding. It cannot be assumed that all organisms necessarily maintain developmental and genetic homeostasis through outbreeding and heterozygosity.

Outbreeding systems

Negative assortative mating is the mating of unlike individuals with a frequency greater than that expected under random mating. The differentiation of sexes in animals generally assures that self-fertilization cannot occur. In addition, most animals have developed systems of varying degrees of complexity which influence the degree of outbreeding. These are briefly discussed in Chap. 5. There is evidence that genetically controlled components of dispersal affect outbreeding and gene flow in insects. Behavioral mechanisms in both invertebrates and vertebrates often operate to reduce the frequency of nearest-neighbor matings. In *Homo sapiens* such ethological mechanisms reach an extreme.

A diversity of mating types affects recombination in microorganisms. Dispersal mechanisms in plants, as in animals, make nearest-neighbor matings less frequent than would be expected if chance alone determined the pairings. Flowering plants have floral pollination mechanisms which also function to determine the amount and type of recombination. It is commonly assumed that there has been a general evolutionary trend from open flowers, composed of numerous parts, which are pollinated more or less indiscriminately, to flowers with the few stamens and stigmas positioned in such a way that pollen is precisely applied to the body of the pollinator and withdrawn from it.

Other plants have physiological **incompatibility systems** that ensure outbreeding. Since reproduction in the higher plants requires pollination by male gametophytes, as well as fertilization by male gametes, the process can be interrupted at many steps. A common system has a multiple allelic basis, the gene for incompatibility, S, existing in many states. The various possible genotypes affect fertilization; the pollen-tube growth is very slow in a style which has the same allele of S as the male gametophyte. A polymorphism involving a form of frequency-dependent selection results. Other more complex incompatibility systems have been studied in the flowering plants.

SUMMARY

Viruses and bacteria show simple, variable genetic systems producing recombination with great wastage of recombinants. In flagellates and filamentous algae, the life cycle may be predominantly haploid with a diploid zygote that divides by meiosis immediately after formation. From this situation, diploid life cycles and cycles with alternation of generations seem to have evolved independently a number of times in plants and animals. In the fungi, specialized mechanisms producing recombination and diploidy have developed. The selective advantage of diploidy and sexuality seems to lie in their provision for developmental buffering and the storage and release of variability, as well as in the reduction of the amount of gamete and zygote wastage. Recombination may be modified in multicellular organisms by mating systems leading to predominantly inbreeding or predominantly outbreeding individuals, or by cytological changes.

REFERENCES

The references are given at the end of Chap. 9.

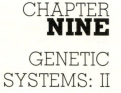

CHAPTER NINE

GENETIC SYSTEMS: II

As we have seen in the consideration of evolutionary change in the quantitatively measured characteristics of organisms, recombination often is a more significant factor than mutation as a source of variability. Since the genotype of an organism evolves under conditions determined, among other things, by the constitution and organization of its chromosomes, the investigation of the chromosomal mechanism is particularly important in evolutionary studies. The basis of variation lies in the genetic code, but changes in the structure and number of the chromosomes bearing the code may directly or indirectly affect the amount of recombination of the code.

Genes are defined here as those regions of the chromosome between the closest points of crossing-over. They are therefore the smallest units of recombination in higher organisms. (Although recombination within genes occurs in prokaryotes, it may be disregarded in population studies of eukaryotic organisms.) Larger units of recombination may involve particular portions of the chromosome in which crossing-over or its effects are restricted; these may be called **supergenes.** There are whole chromosomes, e.g., the sex chromosomes, which behave as recombinational units. The *entire nucleus* is the unit of transmission and recombination in organisms that reproduce asexually (some kinds of apomictic organisms).

The cytological mechanisms that determine, in part, the amount of recombination in a population are an important aspect of the genetic system of the organism. Often they are visible in the phenotypic appearance of the chromosomes, the **karyotype.** However, structural and genic changes may occur without any obvious changes in the karyotype. The cytological mechanisms about to be discussed often are referred to as aberrations or mutations. Since they appear to occur with measurable frequency in organisms, and since they may become characteristic of entire populations or taxonomic groupings of animals and plants, they are best thought of as aspects of the cyto-genetic repertoire of organisms. Like any other characteristic, they arise spontaneously and are maintained as long as they confer a reproductive advantage.

MEIOTIC DRIVE

In a variety of organisms, not all alleles of a locus have an equal chance of inclusion in the gametes. This phenomenon has been called **meiotic drive,** since one allele is favored or driven. As part of the genetic system, it may reduce the amount of recombination, but Thomson and Feldman have shown, using models due to Prout and others, that the opposite may occur. A case in point is the T-allele system in mice, where in males the gametic output from a hetero-zygotic male is almost entirely of a deleterious allele. This allele in homozygous form leads to sterility. The balance of these opposing forces may be the underlying cause of the polymorphism.

Mechanisms that affect the distribution of alleles have been studied in some detail in the fruit fly *Drosophila melanogaster*. A case in point is the segregation distorter factor (SD), which is located on the second chromosome. Meiotic drive occurs only in males who are heterozygous for the SD factor and only then if synapsis of the second chromosome pair has taken place. Although meiosis is normal, functional sperm generally are those containing the chromosome which bears the SD factor. Thus this example also shows that an allele could increase in a population as a result of meiotic drive even if it had a deleterious effect. Eventually, selection presumably would result in the accumulation of factors that reduce the effect of the gene responsible for meiotic drive. It is possible that meiotic drive might result in the spread of a beneficial gene. The extent to which this has occurred in nature is unknown. Various examples of meiotic drive have been reported in other plant and animal species.

CHANGES IN CHROMOSOME STRUCTURE

The usual cytological mechanisms of haploid and diploid organisms already have been described. The student should review the material on chromosome behavior in Chap. 3 before reading the following discussion of derivative genetic systems, those in which cytological changes operate to increase or (more often) to decrease what is usually thought of as the standard amount of recombination.

Inversions

Inversions are of evolutionary importance when a combination of genes which have a selective advantage are included within the inverted chromosome segment. Such a group of genes may be referred to as a supergene. The net effect of an inversion is to prevent a breakup of a supergene as a result of recombination. It should be recalled from Chap. 3 that when an individual is heterozygous for an inversion (either paracentric or pericentric), recombination can be effectively restricted since crossover products are nonviable (Fig. 9-1). However, it must be pointed out that multiple crossovers within the inversion loop can result in viable crossover gametes (Fig. 9-2).

Often inversions with no apparent selective advantage have been found in plant and animal populations. They are referred to as floating inversions. If they are of an appropriate length, and if mutations occur which offer a selective advantage, these inversions will increase in frequency within the population and perhaps even reach complete fixation. The genus *Drosophila* provides numerous examples (see Chap. 7).

It has been found that concomitant with inversions regularly associated with certain chromosomes, there may be marked increase of crossover frequency in other chromosomes. The selective advantage of the inversion must lie in the formation of supergenes rather than simply in the reduction of the amount of recombination generally. Otherwise, the increase in crossover frequency in other chromosomes probably would not be observed.

When individuals from geographically isolated populations or from different species are crossed in the laboratory, they are often found to differ by very long inversions (perhaps of an entire arm). Several such inversions prevent the formation of fertile offspring; thus they may be thought of as isolating mechanisms. Their importance in nature is not as clear-cut as that of shorter inversions. It seems clear that spontaneously occurring inversions might rather rapidly differentiate karyotypes of populations that become separated.

Translocations

Reciprocal translocations likewise may function in setting up supergenes in populations heterozygous for the chromosomal change. As discussed in Chap. 3, viable meiotic products can be produced in translocation heterozygotes only if there is an alternate disjunction of the chromosomes during meiosis. When this occurs, only two classes of viable gametes ultimately result (see Fig. 3-5). This pattern of chromosome segregation is possible because of the pairing of

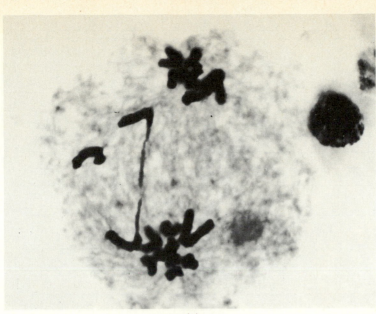

(a)

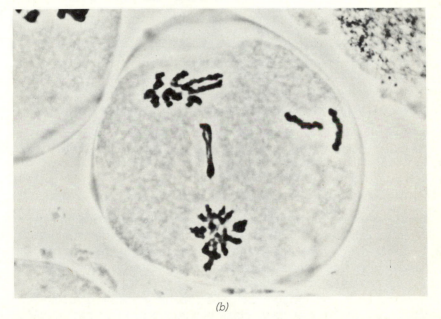

(b)

Figure 9-1 (a) An inversion heterozygote in the plant *Urginea* (Liliaceae). The bridge and fragment at the first meiotic anaphase indicate that a crossover occurred within the inversion loop (see Chap. 3), (b) a double bridge (center) and two fragments (right) in *Urginea* indicate a double crossover within an inversion loop. In either case inviable meiotic products will result. (*Courtesy of K. Jones.*)

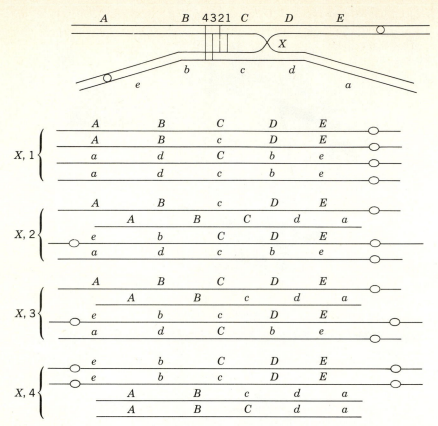

Figure 9-2 The chromatid and genetic products which are the consequences of double crossing-over within a paracentric inversion. The crossover at X is assumed to be constant, with the second crossover being at the 1, 2, 3, or 4 position. X,1 would be a two-strand double, X,2 and X,3 would be three-strand doubles, and X,4 would be a four-strand double. (*From C. P. Swanson, T. Merz, and W. J. Young, Cytogenetics,* © *1967 p. 108; reprinted by permission of Prentice-Hall, Inc.*)

homologous regions of the translocated chromosomes and the subsequent crossing-over in the distal portions of the chromosome arms. Crossing-over, on the other hand, rarely occurs in the regions near the centromere. This means that recombination of the genes located near the centromeres is effectively inhibited due to the lack of both crossing-over and independent assortment of the chromosomes. It also then means that each class of meiotic product has included within it what may be regarded as a supergene, i.e., a particular combination of genes that is transmitted as a unit.

The size of a supergene depends upon the number of chromosomes included in the rings at meiosis (Fig. 9-3). Rings of more than four chromosomes are found at meiosis if three or more pairs are involved in interchanges (Fig. 9-4). In some plants, e.g., *Rhoeo* and several species of *Oenothera* (evening primrose), all the chromosomes exchange segments and all are united, forming one great ring at meiosis (Fig. 9-5). In these organisms there is then in fact only a single linkage group.

Translocations occur in a large proportion of diploid plants, e.g., *Campanula*, *Paeonia*, as well as in scorpions and cockroaches. In many invertebrate animals, translocations are involved in the sex-chromosome mechanism. When an XX female and XO male chromosome mechanism exists, one of the autosomal chromosomes may become translocated to an X. This leaves one of the autosomes with its homolog as part of a modified sex chromosome (which may be called a neo-X). This autosome subsequently behaves like Y chromosomes or heterochromatic regions of chromosomes in general. It is said to have become heterochromatized and is called a neo-Y. Apparently this process has occurred in several grasshoppers and mantids, as well as other invertebrates. Presumably some readjustment of gene function takes place. The genes in the autosome translocated to the X are now present in half their previous number in the males. This is because heterochromatization probably results in

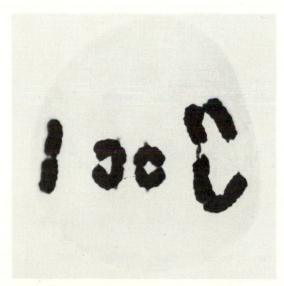

Figure 9-3 A translocation heterozygote in the plant *Gibasis pulchella* (Commelinaceae). At the first meiotic metaphase there are three bivalents and a ring of four chromosomes (beginning to separate due to terminalization of chiasmata). (*Courtesy of K. Jones.*)

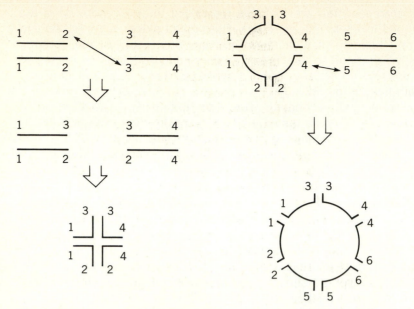

Figure 9-4 How a translocation heterozygote is formed and how the size of a chromosome ring may increase as a result of an additional reciprocal translocation. (*From I. H. Herskowtiz, Genetics, Little, Brown and Company, 1962.*)

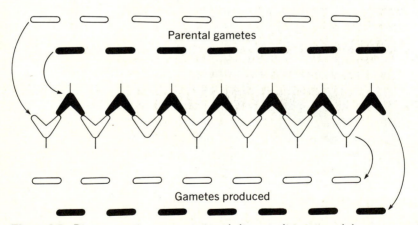

Parental gametes

Gametes produced

Figure 9-5 Diagrammatic representation of alternate disjunction of chromosomes in a complex structural heterozygote, indicating that the parental chromosomes form a linked group that is passed intact from one generation to the next and that there is no independent assortment of chromosomes. (*From I. H. Herskowitz, Genetics, Little, Brown and Company, 1962.*)

their loss in the homologous autosome (the neo-Y). Multiple sex-chromosome mechanisms (with several X's and Y's) may evolve by this process, as has been the case apparently in *Drosophila miranda*. Karyotype evolution in the *Drosophila virilis* group also involves translocations between autosomes and sex chromosomes. An extreme case of multiple sex-chromosome mechanisms is found in the Palestinian beetle *Blaps polychresta*. In the males there are 12 X chromosomes and 6 Y chromosomes. The females have 24 X chromosomes.

Genetic systems based upon regularly occurring structural heterozygosity for reciprocal translocations are characteristic of a few groups of plants. The behavior of translocation systems in nature is not well understood for most plants, although rings are reported for a number of genera, e.g., *Rhoeo*, *Paeonia*, *Datura*, *Hypericum*, and many Onagraceae, such as *Clarkia* and *Oenothera*. Although the selective advantage of translocation systems in nature is not well understood, it is commonly held that heterozygote superiority must be a factor. Species of *Oenothera* and *Clarkia* have been studied in some detail cytologically. It has been estimated that in *Clarkia* more than 20 percent of the plants in nature have rings of chromosomes, indicating the presence of translocations.

Some plants, such as the evening primroses, *Oenothera* subgenus *Euoenothera*, have evolved translocation systems of amazing complexity. The development of these systems seems to have involved selection for heterozygosity. This group is well known through the work of Cleland and others. The mutation hypothesis of de Vries was based upon studies of *Oenothera erythrosepala* (*O. lamarckiana*). Some species of the subgenus, for example, *O. hookeri*, have seven pairs of chromosomes (Fig. 9-6) or small rings of chromosomes plus pairs. The rings are the result of **floating translocations;** i.e., they occur in varying number and are not stabilized as a fixed part of the genetic system of the species. In others, such as *O. biennis*, all the chromosomes are in a ring of 14 at meiosis (Fig. 9-6). In addition to the reciprocal translocations that involve all the chromosomes, there are balanced sets of lethal genes, the Renner complexes.

In such ring-forming types, crossing-over rarely occurs in the regions of the centromeres. In these regions of the chromosome, recessive mutations may accumulate. Eventually each haploid set of seven chromosomes develops mutations which result in lethality when homozygous (Fig. 9-7). The lethal mutations are part of the Renner complexes of genes in the centromere region. Since there is always alternate disjunction in these plants, heterozygosity is always maintained. Union of gametes with like lethal alleles results in death

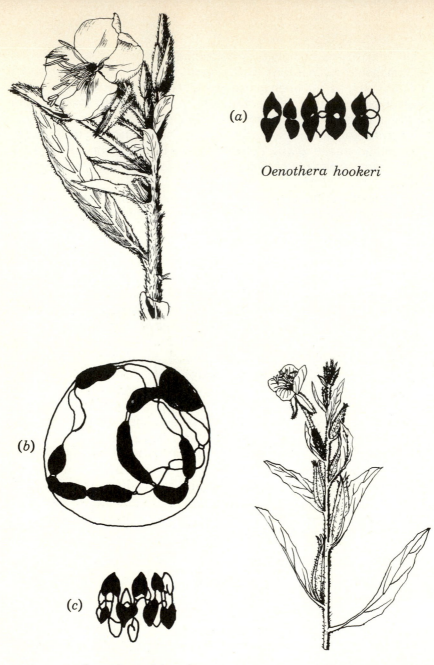

(a)

Oenothera hookeri

(b)

(c)

Oenothera biennis

Figure 9-6 Chromosome behavior in two species of *Oenothera*. (*a*) Seven bivalents in *O. hookeri*; (*b*) ring of 14 in meiotic prophase of *O. biennis*; (*c*) ring of 14 in alternate disjunction at anaphase I in *O. biennis*. [*After R. Cleland, Bot. Rev., vol. 2 (1936) and after L. Abrams, Illustrated Flora of the Pacific States, Stanford University Press (1951).*]

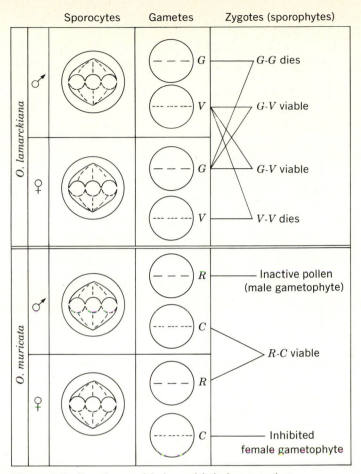

Figure 9-7 Two forms of balanced lethal system that ensure permanent translocation heterozygosity in the genus *Oenothera*. In *O. lamarckiana* zygotic lethality removes the homozygote from the population; in *O. muricata* homozygosis is prevented through gametic lethality. (*From C. P. Swanson, T. Merz, and W. J. Young, Cytogenetics,* © *1967, p. 119, reprinted by permission of Prentice-Hall, Inc.*)

of the zygote, and no homozygous individuals are produced. More complex systems involving gametophytic lethals have also evolved. The ring-forming species have small flowers and are largely self-pollinated (in contrast to *O. hookeri*, for example). They consequently form a very large number of highly heterozygous, mostly true-breeding, and partially isolated races. Occasional outcrossing between the races leads to the origin of new ring systems and new racial types. The great number of such races, which are the primary evolutionary units in the ring-forming oenotheras, makes the taxonomy of the group very difficult.

As Cleland has pointed out, these ring-forming oenotheras illustrate that various types of "mutations" which individually might be considered harmful may, *in combination*, produce a workable genetic system. The races form a widespread and even weedy group. Reciprocal translocations (which usually lead to sterility), lethal genes, and self-pollination combine to form a system in which heterozygosity of all the chromosomes is preserved and the plants are highly fertile. There are what appear to be disadvantages of such a genetic system. Recombination is reduced, and there may be wastage of gametes and zygotes. The failure of any part of the system destroys the whole. Polyploidy apparently is excluded since it is incompatible with a complete ring system and lethal gene complexes.

Changes in chromosome size and shape

By their nature, inversions and translocations also change the size and appearance of the chromosomes. Pericentric inversions may change the position of the centromere and thus alter the relative length of the arms. Unequal reciprocal translocations also may change the chromosome length. Since the number of crossovers is proportional to the length of a chromosome arm, there may be a change in the amount of recombination simply as a result of size changes. As is discussed below, if a chromosome is very small, for example, because of translocation and does not pair properly, it may be lost and the basic chromosome number changed.

CHANGES IN CHROMOSOME NUMBER

One way of expressing the amount of recombination afforded by the genetic system is to use the recombination index of Darlington. This simple measure is equal to the haploid number of chromosomes plus the average number of chiasmata. The larger the recombination index (RI), the greater the number of new gene combinations formed by recombination and segregation. The index does not take into account the existence of supergenes formed as a result of structural changes or apomixis, but its use leads to interesting comparisons.

An increase in the basic number will increase the RI; for this reason, chromosome-number change may be regarded as an aspect of the genetic system related to the balance between immediate fitness and long-range flexibility. Stebbins has made an analysis of the distribution of RI as indicated by chromosome number in flowering plants. In woody plants (trees and shrubs) the basic gametic number is significantly higher than that of herbaceous genera. This suggests that among long-lived plants a high RI has been favored by selection. In short-lived rapidly reproducing organisms, the genetic system usually seems to favor immediate fitness at the expense of flexibility. There is, however, a certain amount of flexibility inherent in a short life cycle with its rapid turnover of genes. Within strictly cross-pollinated herbaceous flowering plants, Stebbins finds that annual species tend to have a lower RI than perennials. Most annual species are either cross-pollinated with a low RI or predominantly self-pollinated, with a high RI. They appear to be specialized for rapid occupation of uniform habitats.

POLYPLOIDY

The basic chromosome number clearly appears to have selective value. How are changes in the basic number accomplished? The minimum number of chromosomes, though all different, that function as a harmonious and integrated unit is a **genome.** For purposes of discussion, the genomic number may be symbolized as x. The total number of chromosomes in any nucleus is its chromosome complement. It is easy to see that the complement may include one or more genomes or parts of genomes. In discussing the regular alternation of gametic and zygotic chromosome numbers, it is convenient to use a different symbol n. The haploid and diploid numbers of all organisms are n and $2n$. Organisms that have experienced an increase in chromosome number are called **polyploid.** Here n includes more than one x or portions of x. In polyploid organisms n and $2n$ usually are called haploid and diploid, even though in a tetraploid, for example, $n = 2x$ and $2n = 4x$. The commonest type of change is **euploidy,** which is increase by whole genomes. The oldest members of a euploid series generally are those with the lowest numbers. Thus if a plant is found to have $2n = 22$ ($x = 11$) chromosomes, its closely related tetraploid derivative would have $2n = 44$ ($x = 11$). The number of chromosomes in one genome may also change, leading to **aneuploidy.** Strictly speaking, an aneuploid series is a series of numbers, such as 11, 12, 13, 15, not a series of organisms; aneuploidy is reversible and may arise in several ways.

Aneuploidy

Genetically active chromosomes cannot be added or subtracted from the genome. Nor is there cytogenetic evidence that centromeres arise

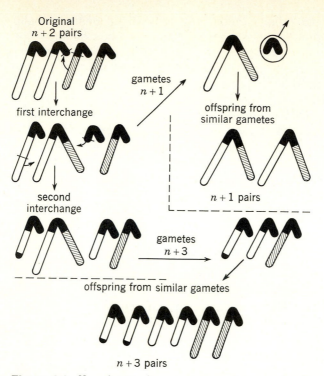

Figure 9-8 How basic chromosome number can be increased or decreased by reciprocal translocation of unequal chromosome segments. Nonhomologous chromosomes are white and hatched; black indicates supposedly inert segments. (*From G. L. Stebbins, Variation and Evolution in Plants, Columbia University Press, 1950, and after C. D. Darlington, Recent Advances in Cytology, used by permission of McGraw-Hill Book Company.*)

anew in populations in nature. Nevertheless, conditions favoring the loss or gain of chromosomes can be brought about by unequal reciprocal translocations. In many organisms the region of a chromosome adjacent to the centromere is genetically inert (heterochromatic). Darlington has suggested that if the active euchromatic arms are translocated to other chromosomes, the remaining heterochromatic centric fragment usually will not pair properly and may be lost. A chromosome is removed from the genome, but the *same amount of genetic material remains* (Fig. 9-8). Thus a class of spores or gametes will result at meiosis which will have one less chromosome in its genome. If this genome is spread through a population in such a manner that it eventually occurs in both the male and female germ lines, offspring homozygous for the chromosome reduction will occur.

Such an individual will be isolated cytogenetically from its parent; if it forms a new population, further differentiation may be expected to occur.

On the other hand, a second translocation might apportion some active material back to the centric fragment, *dividing* the genetic material of another chromosome. Again pairing will be upset and a variety of gametes will be produced. This will lead to a situation similar to that involving an aneuploid reduction in chromosome number, except that offspring may result which are homozygous for a chromosome increase. The original amount of genetic material must be present or the combination will not be viable.

Aneuploid change in chromosome number has been studied extensively in plants. The correctness of the above model has been shown by Kyhos, who demonstrated that *Chaenactis fremontii* ($n = 5$) and *C. stevioides* ($n = 5$) were independently derived from *C. glabriuscula* ($n = 6$). The derivative species, *C. fremontii* and *C. stevioides* are restricted to the rather recently formed deserts of California and adjacent states. *Chaenactis glabriuscula* is ecologically more diverse and is found in mesic sites throughout California. Both derivative species are morphologically similar to *C. glabriuscula* and could be considered to be segregates of this species if it were not for the difference in chromosome number. In addition to the unequal reciprocal translocations and associated loss of centromeres involved in the formation of these two species, other chromosomal interchanges have occurred (Fig. 9-9). The exact sequence of these events is currently under study.

There are numerous examples suggestive of progressive increase in chromosome number but unfortunately little experimental evidence. The genus *Clarkia* apparently is one in which chromosomes have been added to the genome. This increase may be associated with the formation of supernumerary chromosomes (see below). Aneuploidy that simulates progressive increase may result when loss or gain of one chromosome is followed by amphidiploidy (doubling of the chromosome number following hybridization of two diploids). Part of such a series has been produced in *Brassica* (mustard), where $x = 8$, 9, and 10 may represent a phylogenetically ascending series. The numbers known in nature or experimentally produced are $n = 17$, 18, 19, 27, and 29. The genus *Carex* (sedges) has the most extensive aneuploid series known. Haploid numbers ranging from $n = 6$ to $n = 56$ have been reported, and every number from 12 to 43 is represented by one or more species. Presumably structural changes and polyploidy have produced some of the numbers in this series.

Structural rearrangements in which two acrocentric chromo-

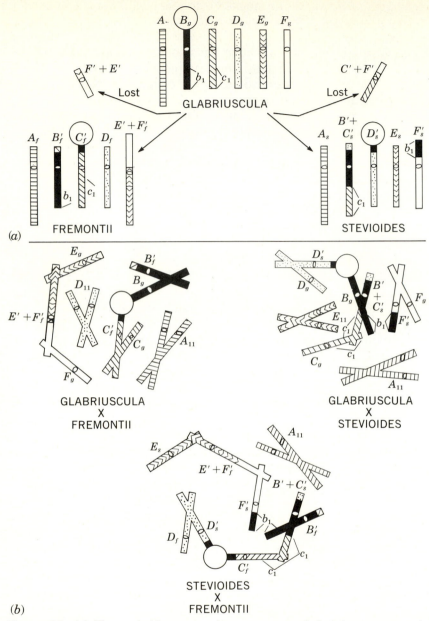

Figure 9-9 (*a*) The probable structural arrangement of the chromosomes of *Chaenactis glabriuscula, C. stevioides,* and *C. fremontii.* The approximate positions of the centromeres are indicated by ovals in the chromosomes. The nucleoli are indicated by larger circles attached to the ends of the chromosomes. (*b*) Modal pairing configurations of the chromosomes in the hybrids between *Chaenactis glabriuscula, C. stevioides,* and C. fremontii. The maximum configurations occur in

somes give rise to a large metacentric chromosome and a minute fragment, which subsequently disappears, are common in *Drosophila*, grasshoppers, and reptiles. The process is known as **centric fusion** and represents a special sort of reciprocal translocation. In many families or genera, the number of long arms remains constant while the relative number of acrocentrics and metacentrics fluctuates. Many examples could be given. An interesting one is the cricket genus *Nemobius*. *Nemobius fasciatus* has a metacentric X chromosome and seven acrocentric autosomes. Other species have additional metacentrics and fewer acrocentrics, presumably the result of structural rearrangements.

Supernumerary chromosomes

In addition to the basic number of chromosomes in the genome, both plants and animals may have extra chromosomes called **supernumerary chromosomes.** It has been verified experimentally that extra chromosomes ordinarily are not tolerated, for they cause genetic imbalance and upsets in meiosis. This is why, in general, only reciprocal translocations can change the basic number. *The genetic material remains the same; only its distribution among the centromeres is changed.* When supernumerary chromosomes are found, it is evident that they must be neutral in some sense or have a special function. Often they are variable in number from cell to cell, or individual to individual. Nevertheless, it seems unlikely that they are completely inert, since they may in some cases remain in the population.

These supernumerary chromosomes in plants commonly are called B chromosomes, and they are surprisingly frequent. In general, the B chromosomes are smaller than the others and pair only among themselves. In most instances they are heterochromatic. They vary in number among individuals, they may be in odd or even numbers or absent, and their presence usually cannot be detected in the phenotype of the plant. B chromosomes may have physiological effects, however, the evolutionary importance of which is unknown.

In *Clarkia*, there is strong evidence that supernumeraries have arisen as the result of structural changes in the chromosomes. These plants often have ring or chain arrangements of chromosomes in

the hybrids *C. glabriuscula* × *C. stevioides*, and *C. stevioides* × *C. fremontii* when chiasmata are also formed between the areas designated as B_1. The approximate positions of the centromeres are indicated by ovals in the chromosomes. The nucleoli are indicated by larger circles attached to the ends of the chromosomes. [*From D. Kyhos, Evolution, vol. 19 (1965).*]

translocation heterozygotes, and unequal separation may lead to the presence of supernumerary chromosomes (trisomics). For some reason, in this genus extra chromosomes do not readily disturb the genetic balance or reproduction. Lewis, Juhren, and Mathew have performed an extensive study of supernumeraries in *Clarkia williamsonii*. They have found plants with up to seven supernumeraries in nature and have experimentally produced plants with up to 17 extra chromosomes. However, plants with more than three supernumeraries showed a significant reduction in fertility. These chromosomes pair in varying degrees with the normal chromosome complement. It is most likely that pairing is frequently partial or completely eliminated because of mutations or structural rearrangements which prevent synapsis.

It is interesting to note that possession of B chromosomes in *Clarkia williamsonii* appears to be deleterious under all conditions tested. Lewis has concluded that the reproductive characteristics of the chromosomes is the reason for the increase of individuals with B chromosomes rather than any advantage conferred upon the individuals in which these chromosomes occur.

Supernumerary chromosomes are found among many invertebrates. They appear to be largely heterochromatic. Some are mitotically stable; others show nondisjunction.

Euploidy

Euploidy as a genetic system occurs frequently in plants but seems to be rare in animals. It has been suggested that increase in chromosome number in animals would upset the sex-determination mechanism and for this reason it has been selected against. Only in parthenogenetic animals are there polyploid series comparable to those in plants. Chromosome numbers have been sampled, however, for relatively few organisms, and conclusions about the occurrence of polyploidy are dangerous. On the basis of available numbers in plants, polyploidy appears to be most common among the vascular plants. For this reason, the following discussion will be limited to that group.

The study of chromosome numbers and behavior in wild forms and in hybrids made in the laboratory usually enables one to identify polyploid organisms. It is even sometimes possible to synthesize in the laboratory a species recognized in nature. If the diploid hybrid can be made, chemicals, such as colchicine, which upset the spindle mechanism can be employed to double the chromosome number. If a plant has a high chromosome number and is impossible to cross with other forms, one can only infer its polyploid nature. If pairing is perfect in such a plant, and if its fertility is high, it must be assumed that it has become functionally diploid. Only the techniques of comparative morphology (including, in some instances, comparisons of proteins) offer hope of deducing the origin of the polyploid.

It was, for example, with the aid of comparative morphology that Ownbey established the origin of two polyploid species in the genus *Tragopogon* (Compositae). *Tragopogon mirus* ($n = 12$) is a morphological intermediate of *T. dubius* ($n = 6$) and *T. porrifolius* ($n = 6$); likewise *T. micellus* ($n = 12$) is morphologically intermediate between *T. dubius* and *T. pratensis* ($n = 6$). Both polyploids are fertile, and although they appear morphologically similar to the sterile diploid hybrids, they are larger and more robust in habit.

In discussing this sort of genetic system it is customary to make a distinction between polypoids in which all the genomes are alike, **autoploids,** and polyploids in which the genomes are different, **alloploids.** In practice, this distinction may be difficult to draw. Clearly, organisms are autoploid when they are the result of somatic doubling of chromosome number (unless the parental organism is a diploid hybrid). Usually plants resulting from the fusion of haploid or polyploid gametes of the same species are regarded as autoploids. The genomes, if not the genes, are presumably much the same. The difficulty arises when wider crosses are involved or when hybrids experience doubling, leading to complex intermediate states between autoploidy and alloploidy.

Among the higher plants sporophytic haploids with the gametic number of chromosomes (*monoploids*, where $n = x$) in their somatic tissues are unknown in natural populations. They may be produced in culture and are of great interest cytogenetically, but they are of no direct evolutionary significance. On the other hand, sporophytic haploids derived from eupolyploids (*polyhaploids*) have been found in nature. Chromosome pairing is quite regular in these plants since there is a high degree of genome homology. Although the evolutionary significance of polyhaploidy is still rather speculative, it does seem reasonable to assume that a reversible shift between ploidy levels might serve as a mechanism for adjusting the balance of fitness and flexibility for the organisms of the group through time.

Strict autoploids are found in nature. The presence of more than two similar genomes in an autotriploid ($3x$) or autotetraploid ($4x$) organism leads to difficulties at meiosis. Only two homologous chromosomes can synapse at any one point. Instead of the usual bivalents, multivalents and/or univalents may be found, depending upon chance and the length of the chromosomes as related to chiasma number. Although numerous autoploid organisms may show enhanced "vigor" or other physiological properties considered advantageous by the plant breeder, they sometimes are unable to reproduce sexually and are selected against in nature unless they acquire an efficient mode of sexual or asexual reproduction. The reduction in fertility may not be complete (it may be very slight); furthermore, most of the higher plants have one or more modes of vegetative propagation. Many of the most important horticultural and pomicultural plants are triploids (such as most bananas, many apples, some cherries, Japanese iris, tiger lilies, tulips). In addition, both autotriploids and autotetraploids occasionally occur spontaneously in animals, e.g., in certain salamanders. Autoploids rarely become established as populations in nature, however.

Alloploidy, on the other hand, is known in virtually every phylum of the plant kingdom. Well-known examples are the allotetraploid of radish and cabbage, *Raphanobrassica*, and the allotetraploid of two horticultural species of primrose, *Primula kewensis*. The first involves a cross between different genera, the second different species; therefore the two genomes in the hybrid would most certainly be different.

In diploid hybrids between species and genera, chromosome behavior is variable. For example, in *Raphanobrassica*, there seem to be so many small differences between the chromosomes of the parental species that pairing does not take place and only univalents are found at meiosis. The same is true of the cross between *Allium fistulosum* (scallion) and *A. cepa* (onion). The univalents are distributed at random, and only rarely will cells with balanced chromosome complements be produced. At the opposite extreme, in the *Primula* example and in the cross between the grasses *Festuca pratensis* (meadow fescue) and *Lolium perenne* (perennial rye), bivalents are formed in the diploids. This suggests that there are fewer differences between the genomes involved. Nevertheless, following disjunction, the diploid hybrids are sterile. Doubling of the chromosome number in each hybrid provides the chromosomes of each genome with the appropriate homologs at meiosis. Bivalents are always formed, and fertility is restored.

The resulting allotetraploids are sometimes referred to as amphidiploid since, in effect, they are diploid for each parent. That is, $2n = 2x_1 + 2x_2$. Allotetraploids of this sort occur in nature. Observation of the chromosomes of artificial hybrids between the suspected tetraploid and its possible parents will suggest the relationship involved. In a backcross to a genuine parent, a mixture of bivalents and univalents occurs. It is apparent, however, that there is an evolutionary disadvantage inherent in the makeup of a strict allotetraploid. The genomes may be so differentiated that when they are combined in the fertile tetraploid hybrid, genome recombination does not occur. The parental genomes presumably continue to exhibit the same amount of recombination as they did in the diploid. Because pairing is perfectly normal, genes of different *genomes* cannot be recombined and segregated.

Probably more common in nature than a strict allotetraploid is a tetraploid in which the genomes are partly differentiated but still sufficiently similar for multivalents to be found in which crossing-over between chromosomes of the two parents may occur to produce genetic recombination. This is referred to by Stebbins as **segmental alloploidy,** because only some segments of the chromosomes are different. While segmental alloploids are of the greatest significance for evolution, they are difficult to recognize in nature. They may result from hybridization of morphologically similar parents, and because of multivalent formation they simulate autoploids. Furthermore, as a result of their partial sterility, they are unstable. Selection, acting to increase fertility, may tend to favor further chromosome differentiation. Bivalents may be formed and a strict alloploid result.

There are many groups of plants in which the major evolutionary differentiation has involved alloploidy. One of the best examples is the fern genus *Asplenium* (spleenwort), studied by Wagner and his associates. As shown in Fig. 9-10, there are three primary species with $n = 36$. Meiosis is regular, and 36 bivalents are formed in each. By a study of morphology and by analysis of chromosome behavior in other species, it could be shown that *A. bradleyi*, *A. ebenoides*, and *A. pinnatifidum* are allotetraploids involving the primary species in pairs. The sterile diploid hybrid of the same parentage as *A. ebenoides* is known; in it there are 72 univalents instead of 72 bivalents. *Asplenium trudellii* is a backcross of *A. pinnatifidum* to one of its parents, for it shows 36 bivalents (*montanum* genome) and 36 univalents (*rhizophyllum* genome). *Asplenium gravesii*, on the other hand, is a hybrid between two different *allotetraploids* that share *A. montanum* as a parent. This is revealed by the behavior of the chromosomes also, for the two *montanum* genomes form bivalents, while the *platyneuron* and *rhizophyllum* genomes remain unpaired (36 bivalents plus 72 univalents).

In other groups, much more complicated polyploid complexes are known. The diagram of Stebbins (Fig. 9-11) shows the combinations of genomes and of autoploidy and alloploidy that may occur (although higher levels have been found). In some groups, allotetraploids are able to form partially fertile hybrids with autoploid forms of either of their parents. In plants, particularly those with any degree of vegetative propagation, the result may be to blur or obliterate the morphological limits of the taxa originally involved. A group of populations with different degrees of ploidy and varying morphology,

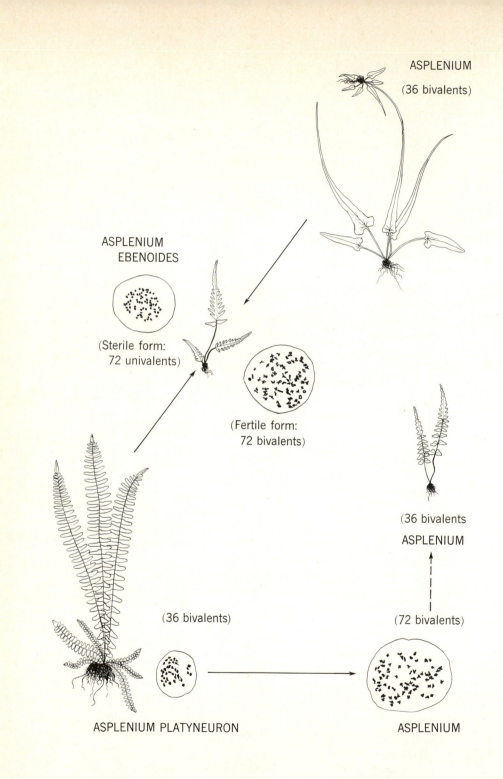

ASPLENIUM
(36 bivalents)

ASPLENIUM
EBENOIDES

(Sterile form:
72 univalents)

(Fertile form:
72 bivalents)

(36 bivalents
ASPLENIUM

(36 bivalents)

(72 bivalents)

ASPLENIUM PLATYNEURON

ASPLENIUM

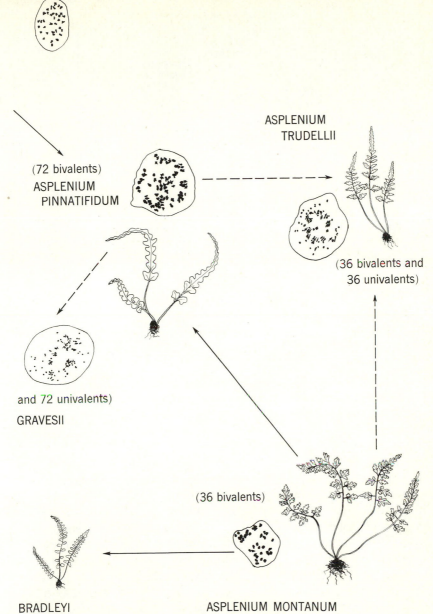

RHIZOPHYLLUM

ASPLENIUM
TRUDELLII

(72 bivalents)
ASPLENIUM
PINNATIFIDUM

(36 bivalents and
36 univalents)

and 72 univalents)

GRAVESII

(36 bivalents)

BRADLEYI   ASPLENIUM MONTANUM

Figure 9-10 Polyploid complex in the fern genus *Asplenium*. Discussion in text. [*In part after W. H. Wagner, Evolution, vol. 8 (1954).*]

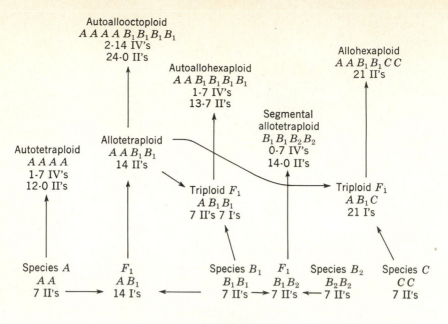

Figure 9-11 Various levels and types of polyploidy, their genome constitution, and their mode of origin. Letters represent genomes; roman numerals indicate pairing behavior; I = univalent, II = bivalent, etc. Thus 24–0 IIs = 24 to 0 bivalents. (*From G. L. Stebbins, Variation and Evolution in Plants, Columbia University Press, 1950.*)

as well as ecological preferences, may arise. Only by cytological analysis can such complexes be resolved and then sometimes only partially. The situation may be further complicated if the plants reproduce apomictically, perpetuating individuals that are sexually quite sterile (see below).

It is difficult to specify the general importance of polyploidy as a genetic system. It has been estimated that about one-half of the flowering plants are of polyploid origin. Similarly, the ferns and their allies show much polyploidy, as do the algae and mosses. On the other hand, polyploidy is rare in some families of gymnosperms. Also, the fungi seem to form an exception, but they have been very poorly sampled cytologically. Unfortunately very little is known of tropical species, so that the sampling is far from thorough or representative. Whole families or subfamilies, e.g., the subfamily Pomoideae of the Rosaceae, the apples and their relatives, as well as genera, have numbers that suggest a polyploid origin. Despite the existence of regular progressions of chromosome number in animals, polyploidy generally is not invoked in explanation. As has been seen, centric fusion is thought to be the cause. It is generally felt that polyploidy would upset the sex-determination mechanism of animals in which sex depends upon the ratio of autosomes to sex chromosomes. There are various groups of nonvascular and vascular plants with XX:XY sex chromosomes. In some of these, e.g., the dock *Rumex*, polyploids are known. The taxonomy of this genus is not generally agreed upon, but in *Rumex acetosella* there are four polyploid levels, from diploid to octoploid. In octoploid forms, "male" plants have one Y, and all but one of the X chromosomes become "autosomal." Thus it is clear that it is possible to evolve ways of combining polyploidy with sex-chromosome systems.

Other work suggesting polyploidy in animals has involved not only the counting of chromosomes but also determinations of the amount of DNA in cells. Ohno and coworkers have presented evidence that although the chromosome numbers of all placental mammals vary from 17 to 80, they have nearly identical amounts of DNA per diploid nucleus: 7.0 pg (picograms; 1 pg $= 10^{-12}$ g). Similar amounts are found in marsupials, monotremes, and one group of reptiles. According to Ohno, this suggests that the reptilian ancestors of birds and mammals had stabilized the amount of DNA per nucleus. Further evolution of the genetic material in these groups did not involve polyploidy.

Polyploidy among vertebrates, Ohno maintains, should be sought in fishes and amphibia. The first bisexual tetraploid species discovered in nature is a South American frog *Odontophrynus*. Two other frogs, *Ceratophrys* and *Hyla*, have naturally occurring polyploids, and polyploidy has been reported in three genera of tailed amphibia, *Ambystoma*, *Eurycea*, and *Notophthalmus*. In *Ambystoma*, polyploidy is combined with parthenogenesis, as discussed below.

Ohno has proposed that carp (*Cyprinus*) and goldfish (*Carassius*) with $2n = 104$ are $4x$ to other members of the family Cyprinidae. A diploid-tetraploid relationship also exists between clupeoid fishes (anchovy, herring) and salmonoid fishes (trout, salmon). Here, although centric fusions confuse an analysis made by counting chromosomes, DNA determinations show that the salmonoids have about twice as much DNA per nucleus as the clupeoids. And, in fact, supposed diploids have two loci for the lactate dehydrogenase gene, while the supposed tetraploids have these duplicated.

The effects of polyploidy as a genetic system are varied and at times opposed. For example, autoploidy may severely restrict or eliminate genetic recombination in one organism, only to act as a bridge between an alloploid and one of its parents. A strict amphidiploid is limited in recombination potential to that of its parents, but in segmental alloploidy recombination of the parental characteristics, including ecological preferences, may occur. In general, one might say that polyploidy may act to increase the scope of evolutionary units, enhancing the occasional interbreeding through time, and thus occasionally the production of more fit individuals. The situation resembles that postulated by Wright for most rapid evolution: subunits partially isolated in space which occasionally exchange genetic material. It makes possible further exploitation of the advantages of hybridization at the diploid level.

APOMIXIS

The ultimate in the restriction of recombination is the elimination of sexual reproduction altogether. In extreme cases, recombination is completely eliminated. This appears to be relatively rare, and there are many types of asexual reproduction with varying amounts of recombination. The term **apomixis** describes all types of asexual reproduction, i.e., those types of reproduction which tend to replace or act as substitutes for sexual reproduction. The classification of the types of apomixis is exceedingly complex, since a variety of situations intermediate between sexuality and obligate apomixis may occur. In most plants, either the gametophytic or the sporophytic generation or both may be involved. Terminology relating to apomixis also has become unfortunately complex; it will be simplified as much as possible in the following account.

Apomixis may be **facultative** or **obligate.** Some organisms reproduce sexually at times and asexually at other times, usually under different environmental conditions. However, in animals, a regular cyclic change from sexual to asexual reproduction is not strictly facultative apomixis, since the asexual portion of the cycle is obligate.

The simplest sort of apomixis is **vegetative reproduction;** it may function as the only mode of reproduction. For example, some aquatic plants are known to reproduce only asexually in northern Europe, e.g., *Elodea*, where only pistillate plants are found, and sexually in other parts of their range. The self-sterile triploid day lily (*Hemerocallis fulva*) reproduces asexually in America. The effect of such vegetative propagation, whether by rooting and further growth of broken-off branches or by the production of specialized propagules such as gemmae or bulbils (which are modified branches or buds), is to create a population of genetically identical organisms. Such a population is called a **clone.** Clones may be formed by animals that reproduce by fission (Protozoa) or budding (Coelenterata, Annelida, etc.). From the standpoint of genetics, the clonal population is one genetic individual. Since they usually separate from one another eventually, members of a clone are individuals physically and play the role of individuals ecologically.

Among both plants and animals, vegetative reproduction may occur in individuals that also reproduce sexually and thus have a dual breeding system. This usually is not referred to as apomixis, except where it assumes the entire reproductive function during a portion of the life cycle. Ordinarily, as in some species of *Viola* (violets), sexual reproduction and asexual reproduction occur at the same time (Fig. 9-12). For example, individual plants of *Viola* grow runners, or stolons,

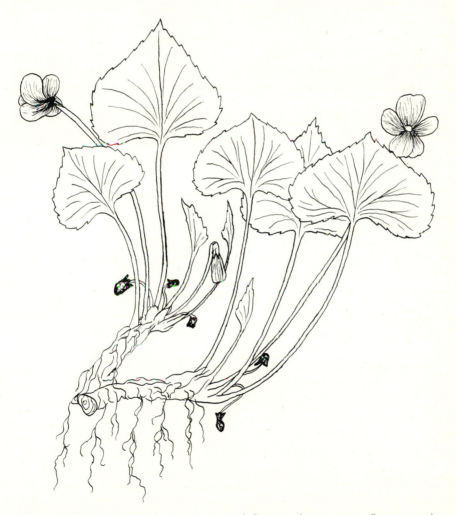

Figure 9-12 Violet (*Viola*) showing typical flowers, cleistogamous flowers, and vegetative propagation (creeping underground stems).

at the base; these eventually produce a series of new rosettes which may become separated from the parent plant. Typical violet flowers also occur; they are pollinated by insects, producing heterozygous seeds. At the same time, however, cleistogamous flowers may be formed. These flowers, close to the ground among the leaves, never open and are self-pollinated within the bud. Seeds from these flowers, virtually sown at dehiscence, produce plants genetically similar to the parents. Thus three modes of reproduction, providing different amounts of genetic recombination, occur in the same plant at the same time.

Somewhat more cryptic sorts of vegetative reproduction, without the formation of seeds, also are known. These include some instances of so-called "vivipary" in plants. That is, the propagules are tiny bulbils which occur in the inflorescences and may replace the flowers. Some species of onion (*Allium*), for example, as well as some *Poa* (bluegrass) species reproduce in this fashion. From a developmental point of view, one may say that individual flower primordia, which are modified spore-producing branches, develop into specialized vegetative branches that fall from the plant, subsequently to root and grow. Clone formation is the result.

Besides these relatively simple modes of apomictic reproduction, there are sorts involving seed production whose complexity is amazing and whose effects are greatly puzzling. They are known collectively as **agamospermy** (Table 9-1). Considering only the higher plants as examples, one must first remember that their life cycle has a gametophytic period, which separates meiosis and fertilization and which alternates with the sporophytic generation. The

TABLE 9-1. MODES OF APOMIXIS IN PLANTS INVOLVING SEED PRODUCTION: AGAMOSPERMY

Gametophytic apomixis: gametophytes produced
 Apospory: production of a gametophyte from somatic cells of the reproductive structures of the *sporophyte* through mitotic divisions
 Apogamety: production of a sporophyte by mitotic division of cells of the *gametophyte* other than the egg
 Parthenogenesis: development of the egg without fertilization, often occurring in apospory or diplospory
 Diplospory: production of diploid spores, and thus diploid gametophytes, through mitotic or only partially meiotic division of those cells of the sporophyte which ordinarily undergo meiosis
Sporophytic apomixis: no gametophyte produced
 Adventitious embryony: cells of nucellus or integument produce new sporophyte directly through mitotic divisions

---

TABLE 9-2. MODES OF APOMIXIS IN ANIMALS INVOLVING GAMETE PRODUCTION

Arrhenotoky: haploid males from unfertilized eggs, diploid females from fertilized eggs (also known as haplodiploidy or haploid parthenogenesis)

Thelytoky: females only (or males very rare) from unfertilized diploid eggs resulting from ameiotic or partially meiotic divisions or cell fusion (also known as diploid parthenogenesis)

Cyclical thelytoky: alternation of sexual and asexual forms

---

gametophyte may consist of only a few cells, e.g., eight in the female gametophyte of many flowering plants, or it may be a relatively massive multicellular plant, physiologically independent of the sporophyte (as in most ferns). In either event, apomixis may occur in this generation if the egg develops without fertilization. The four basic types of apomixis in which the gametophytic generation is involved are known collectively as **gametophytic apomixis.**

In the various modes of gametophytic apomixis, development may depend upon the influence of a pollen tube or on nutritive tissue (itself a product of fertilization), called endosperm, that surrounds the apomictically produced embryo sporophyte. Thus pollination is necessary, but fertilization of the egg does not take place. This is called **pseudogamy** and is generally one of the most difficult aspects of apomixis to detect.

Apomixis in the higher plants may progress to the point where the gametophytic generation is suppressed completely. A variety of agamospermy known as **adventitious embryony** results. Here cells of the integument or nucellus of the ovule produce the new sporophyte by mitotic divisions. Thus sporophytic tissue gives rise directly to a new sporophyte that is enclosed in the usual seed coats and is superficially indistinguishable from the usual situation. Adventitious embryony also may be pseudogamous. In ovules where adventitious embryony occurs, gametophytes and sexually produced embryos may also occur.

It will be seen from the above that apospory and adventitious embryony produce plants genetically identical with their parent. Since diplospory may involve divisions that are partly meiotic, the possibility of genetic recombination exists. The nature and possible results of partial recombination of this sort are discussed below under thelytoky in animals.

Just as apomixis in the higher plants is complicated by the existence of alternation of generations, so is it complicated in higher animals by sex (Table 9-2). In a number of groups of animals, males arise from unfertilized eggs and are haploid. Females are produced from fertilized eggs and are diploid. This type of apomixis, known as haplodiploidy, haploid parthenogenesis, or **arrhenotoky,** obviously may be a means of sex determination. Apparently arrhenotoky has arisen about seven times in the Metazoa (among the insects, arachnids, and rotifers). In the insects, the Hymenoptera as an order are characterized by this form of apomixis and sex determination, which also is known in the Homoptera, Coleoptera, and Thysanoptera.

In arrhenotoky, no true meiotic division occurs in the male, the sperm being produced by a mitosis or a simulated meiosis and thus genetically identical (except for mutation) with the parental genotype. The evolutionary genetics of haplodiploid organisms is by no means well understood. For example, it can be seen that recessive genes must be immediately expressed in males of species with this genetic system, unless some special mechanism prevents this. It has been suggested that by virtue of some sort of "repeat" mechanism of gene duplication the organisms may be functionally diploid. It is well known that most highly differentiated tissues of an organism are endoploid (often to a high degree); this might protect the soma from harmful mutations. On the other hand, haploid males could act beneficially as a sort of screen for recessive lethals.

A rather different mode of apomixis in animals, called **thelytoky,** is a kind of parthenogenesis in which females are produced from unfertilized eggs. It appears to have arisen in most major animal groups, including the vertebrates, under natural conditions. In some organisms thelytoky is complete: Males are very rare or unknown, and every individual arises from unfertilized eggs. More common is **cyclical thelytoky,** where there is an alternation, usually in an annual cycle, of sexual and asexual forms. This mode is found in aphids, gall wasps, cladocerans, many parasitic worms, and rotifers.

As in plants, reduction division may be entirely suppressed, and the divisions resulting in the egg are mitotic. The egg is then diploid. On the other hand, meiosis may occur essentially as normally, but doubling takes place at a later stage, restoring the diploid condition (Fig. 9-13). This is comparable to the results of diplospory in plants. The usual products of meiosis are four haploid cells. In the higher plants these would be megaspores, and in animals an egg and two (or three) polar bodies. If the first division of meiosis occurs, crossing-over may take place and the basis for recombination established. The daughter cells, following the second division, may fuse in pairs and their nuclei combine to form **restitution nuclei.** Only two cells are formed, and meiosis and recombination are partially suppressed. In animals this may be the result of the fusion of the egg with the second polar body or, following the cleavage division of the egg, the fusion of the cleavage nuclei.

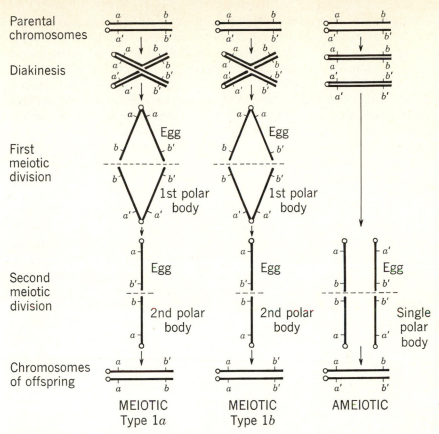

Parental chromosomes

Diakinesis

First meiotic division

Egg

1st polar body

Second meiotic division

Egg

2nd polar body

Chromosomes of offspring

MEIOTIC
Type 1a

Egg

1st polar body

Egg

2nd polar body

MEIOTIC
Type 1b

Egg

Single polar body

AMEIOTIC

Figure 9-13 Diagram of three types of parthenogenesis and their genetic consequences. Type 1a, somatic number restored by fusion of second polar body nucleus with egg nucleus; type 1b, somatic number restored after cleavage division of egg by fusion of cleavage nuclei; ameiotic type, divisions resulting in egg are mitotic. (*From White, Animal Cytology and Evolution, 2d ed., Cambridge University Press, 1954.*)

White, Darlington, and others have pointed out that since no segregation can occur in ameiotic thelytoky, recessive mutations and structural rearrangements of the chromosomes tend to *accumulate.* Daughters will resemble their mothers, but the line will tend to become more and more heterozygous. This may account for the vigor with which some apomictic organisms are endowed. On the other hand, in partially meiotic thelytoky, where segregation can occur, existing heterozygosity will be *reduced* in time, without the fusion of gametes to restore it.

The effects of apomixis on the organism are varied and complex. In both plants and animals, apomixis commonly is associated with both hybridization and polyploidy, although there is no reason to infer a direct causal relationship between the three. If a major factor in the relative absence of polyploidy in animals is the sex-determination mechanism, it is clear that thelytokous animals are free to become polyploid. Also, the various cytological components of apomixis are controlled by a large number of genes that are largely of negative selective value when separated. For example, mutations leading to upsets in spindle formation would be deleterious unless combined with mutations leading to the formation of restitution (fusion) nuclei and to the development of eggs without fertilization. Hybridization may serve to bring these genes together in a functional system. It has been suggested also that some sort of buffering may take place if the organism is polyploid.

In those plants in which apomixis has been adopted as a genetic system, a common effect is the origin of what are called **agamic complexes,** which are the despair of the taxonomist and the cytogeneticist alike. The genera *Rubus* (blackberries and raspberries) and *Poa* (bluegrass) are familiar examples in which relationships are so blurred that they prevent any clear understanding of the group as a whole. The American species of *Crepis* (false dandelions) form the only agamic complex of size that has been studied in its entirety from both the systematic and cytological points of view. It is interesting that of the 196 species of *Crepis* (which have been grouped into 27 sections), apomixis and polyploidy are known only in five species of the section Pyrimaches in southeastern Asia and in nine species of the section Psilochaena in America.

The ten North American species of section Psilochaena make up a group more heterogeneous than any other. Their basic chromosome number, $x = 11$, is not found in any other section, and it may have originated as a result of interspecific hybridization of Asiatic species with $n = 7$ and $n = 4$, or with $n = 5$ and $n = 6$. *Crepis runcinata* is the only member of the section that shows neither apomixis nor polyploidy, having a chromosome number of $2n = 22$. Seven of the remaining species each have a 22-chromosome diploid form, with which are associated numerous polyploid apomictic forms. The other two species include only polyploid apomictic types derived by hybridization between two or more of the above-mentioned seven.

Thus there are seven primary diploid types involved in polyploidy and apomixis. Five have greatly restricted geographic distribution and very different ecological preferences. No localities are known in which two diploids occur together, and no diploid hybrids have been found. The polyploids, with chromosome numbers of 33, 44, 55, 77, and 88, show combinations of characteristics of two or more of the diploids and all are apomictic (some facultative). In general, the apomicts exceed their diploid ancestors in geographic distribution.

The analysis of this agamic complex may serve as a case study for such problems in general. The unveiling of relationships and the construction of a useful classification depended upon recognition of the primary sexual diploids. Around these could be grouped the derivative autoploid and alloploid asexual forms. Although the variation at first appears baffling and overwhelming, it can be made sensible; in so doing, interesting facts are revealed. For example, diversity is greatest in the regions where the sexual species are found and is less in the peripheral areas of the range occupied only by apomicts. The latter appear to have radiated, subsequent to their formation by hybridization and polyploidy, from the more central area of their sexual ancestors.

Agamic complexes in plants other than in *Crepis* are less well known. They may be quite small or exceedingly large and complex. The chromosome numbers may become very high and unbalanced. Study of other genera has proved to be quite difficult where the ancestral sexual diploids have become extinct (*Rubus*) or there may be several agamic complexes in one genus (*Poa*).

Large agamic complexes are less common in animals. The case of *Artemia salina* (brine shrimp) with sexual diploids and thelytokous triploids, tetraploids, pentaploids, octoploids, and decaploids, is well known. In the Curculionidae (weevils) and Lumbricidae (earthworms) larger complexes have been found; these may include diploids, triploids, tetraploids, and pentaploids, or even hexaploids and decaploids. In the weevils there is reason to believe that occasional fertilization of usually parthenogenetically developing eggs by sperm from bisexual races or species takes place.

Thirteen species of thelytokous earthworms have been studied cytologically; all are polyploids ranging from triploids to a possible decaploid. Some sexual species are polyploid as well. Since the Lumbricidae generally are hermaphroditic, thelytoky involves modification or loss of the male organs. Oogenesis is complex, there being chiasma formation and bivalents even in odd-numbered polyploids. The apparent success of these forms, if their wide geographic distribution is to be a criterion, may be accounted for by postulating heterosis as a result of alloploidy.

Apomixis has been reported in the vertebrates, but its extent and evolutionary importance are not well known. An interesting situation has been studied in some detail in the fish genus *Poecilia* (formerly *Mollienesia*). Several populations of *P. formosa* have been sampled in southern Texas, where they occur in streams and drainage ditches. The fishes have also been raised in the laboratory and their genetic similarity studied by means of tissue transplants. Grafts of donor tissue are rejected by the host fish, because of tissue antigens produced by the host genes, in a length of time roughly proportional to the degree of genotypic similarity between the host and donor.

Poecilia formosa is parthenogenetic, but eggs do not develop without the stimulation of sperm. Since males of *P. formosa* are exceedingly rare in nature, sperm from related species (in this instance *P. latipinna*) is necessary to initiate development. This mode of reproduction, in which the genetic information of the sperm is not incorporated into the zygote, is known as **gynogenesis.** As in other instances of apomixis, the genetically identical progeny of a female form a clone. Tissue-transplant studies have shown that two clones of *P. formosa* make up about 80 percent of the population in one drainage ditch near Olmito in the valley of the Rio Grande. The remainder belong to a third clone or cannot be identified. Clones sampled in 1961 were the same as those found in 1960.

In 1954, several dozen *Poecilia formosa* were taken from the Olmito ditch and released in the San Marcos River, some 250 mi to the north. The species has become established, as has *P. latipinna*, which was introduced into the area many years earlier. The San Marcos population thus was available for comparison with the Olmito fishes; it was also sampled in 1960 and 1961. Only two clones were found; these were the common clones at Olmito. Thus the clones of these poecilias probably have remained relatively unchanged (within the limits of tissue-transplant discrimination) for at least 7 years and possibly for much longer.

Polyploidy and parthenogenesis are combined in naturally occurring populations of the salamander *Ambystoma*. A complex of four species has developed with two diploid forms, *A. laterale* and *A. jeffersonianum*, and two parthenogenetic triploids, *A. trimblayi* and *A. platineum*. Uzzell and others have analyzed this situation in great detail with populations extending from the Midwest to the northeastern United States.

The parthenogenetic triploids are thought to have arisen following the Wisconsin glaciation some 10,000 years ago. Each of the triploid species is the result of hybridization between the two diploid species. Genomic analysis of the triploids was made using morphological and serum protein phenotypes. *Ambystoma platineum* is thought to have two sets of *A. jeffersonianum* chromosomes and one set of *A. laterale* chromosomes. On the other hand, *A. trimblayi* probably has two sets of *A. laterale* chromosomes and one set from *A. jeffersonianum*. In both triploids, activation of meiosis by sperm from a diploid male is required: *A. jeffersonianum* sperm for *A. platineum* and *A. laterale* sperm for *A. trimblayi*. The chromosome complement of the activating sperm is not incorporated into the zygote; thus these salamanders are gynogenetic. Oocytes in the triploid salamanders become 6n before activation and subsequently undergo a meiosis in which sister chromosomes pair.

Asher and Nace have analyzed the genetic and evolutionary implications of this sort of parthenogenesis. Only mutation or segregation resulting from rare quadrivalent formation introduce genetic variability. They have produced mathematical models to quantify the influence of mutation, segregation, and selection upon genetic variability in parthenogenetic populations. According to their one-locus, two-allele models, apomictic species (ameiotic) should not become completely heterozygous (as predicted by Darlington, White, and others) because of mutation alone. Rather, an equilibrium state consisting of both homozygotes and heterozygotes should occur; furthermore, selection and segregation should drastically alter this equilibrium. Also, species which are partially or completely meiotic need not become completely homozygous, as previously predicted, if selection favors heterozygosity. Thus, parthenogenetic species may not be evolutionary "dead ends."

The formation of the two triploid species of *Ambystoma* is hypothesized to have occurred 10,000 years ago. This means the 3n forms would have had some 5,000 generations to diverge. In fact, there is evidence that the postulated ancestral heterozygous serum protein phenotypes are present in the contemporary triploids. Asher and Nace have suggested that either mutation and segregation are not affecting the genome or selection is maintaining the constancy of the genome. Since both segregation and mutation are rare events, heavy selective pressures are thought to be maintaining the ancestral genotype.

In several species of lizards, males have never been observed. This suggests that parthenogenesis occurs in these species. There are 19 all-female species and subspecies distributed among five genera and three families. According to Maslin, who has worked extensively with the genus *Cnemidophorus*, actual proof of parthenogenesis has been found in only seven forms.

It would appear, from various lines of evidence, that parthenogenesis generally arises following hybridization, and in some genera this leads to triploidy. The advantage of parthenogenesis is that it makes possible the colonization of new areas by individuals with advantageous combinations of genes but sexually sterile. Zweifel, Taylor, and others have studied variation in parthenogenetic forms, comparing them with sympatric sexual species. In general, the bisexual species are more variable than the parthenogenetic forms. Zweifel found, however, that the parthenogenetic *Cnemidophorus tessellatus* was just as variable over its entire range as two subspecies of the bisexual *C. tigris*. Within local populations of *C. tessellatus* there was reduced variation, however.

There remains to discuss only cyclical thelytoky, a peculiar genetic system found in aphids, gall wasps, Cladocera, and rotifers. The cytological mechanisms differ from group to group, but we may single out a species of aphid as an example. In *Tetraneura ulmi*, which produces galls on elm leaves, there is a sequence of generations which have been given names descriptive of their behavior, fundatrices, emigrantes, exules, sexuparae, and sexuales. In the spring, females of the fundatrix generation become adults within the elm leaf galls. There each produces, parthenogenetically, female offspring which later develop wings and fly away to feed on the roots of grasses. They are the emigrantes, which produce, also parthenogenetically, several generations (females) of exules (also living on grasses). Eventually, the exules give rise to the sexuparae, winged females which fly back to the elm and there parthenogenetically produce both males and females called sexuales. The latter pair, and from fertilized eggs appear once more the female fundatrices, the gall-making generation.

Cytological investigation shows that female sexuales have $2n = 14$, while males have $2n = 13$; there is evidently an XX:XO sex-chromosome system. The fundatrices, emigrantes, and exules types of thelytokous females have a diploid set of 14 chromosomes. There is a single maturation division in oogenesis. The eggs produced are diploid because the division is not reductional and they develop into females that are identical genetically (except for mutation). The sexuparae produce eggs of two kinds. In those eggs which will be female-determining, all the chromosomes split, as in mitosis. In those which will give rise to males, the chromosomes behave similarly, except for the X chromosomes, which pair and are reduced as in meiosis. One X remains in the egg; the other goes into the polar body.

In spermatogenesis of the male sexuales, the X chromosome is apportioned to one of the secondary spermatocytes in normal fashion. However, those cells without the X chromosome eventually degenerate, and only X-containing cells produce sperm. Thus the sexuales males can have only daughters, which complete the cycle as fundatrices.

Other aphids have similar cycles in which the number of generations may differ and in which there are two kinds of sexuparae: male-producing and female-producing, etc.

In such complexes as those described for plants and animals, the usual concept of species is very difficult to apply. The sexually reproducing diploids may be comparable to species in other organisms. But the autoploids and alloploids combine the characteristics of two or more diploids in asexually reproductive and therefore very fertile organisms. This breaks down the utility of criteria based upon morphological intergradation, gene exchange, and geographic distribution. Combining the classic techniques of taxonomy with the methods of cytogenetics, however, the biologist may be able to identify the major evolutionary units within the complex. To these he customarily gives the rank of species, while the multitude of apomictic forms may be described with or without formal taxonomic recognition, whichever appears most useful.

Aside from greatly complicating the work of the biologist, what are the effects of apomixis as a genetic system? It is obvious that *apomixis makes possible the survival of many genotypes that are vigorous and well-adapted but sexually sterile* for one reason or another, e.g., in unbalanced polyploids. Apomixis also permits building up of large numbers of genetically similar individuals for the rapid colonization of newly available habitats. One finds apomixis often to be the genetic system of weedy or pioneer organisms and of those in habitats subject to frequent or regular catastrophe, such as sand bars, lawns, etc.

It is also true that apomixis limits the genetic variability of the organisms that have adopted it as their sole mode of reproduction. For this reason, it generally is found to be an alternative or secondary genetic system. Apomixis usually is not combined with other systems that reduce the long-range flexibility of the organism for the sake of immediate fitness, e.g., self-fertilization. It is interesting that, even in those groups, such as *Poa*, where apomixis and high polyploidy are carried to what appear to be extremes, the situation is not, as usually described, dead end. The pollen of obligate apomicts may be functional, and pollination of an apomict may occasionally result in the segregation that leads to an escape from asexuality.

SUMMARY

Populations of plants and animals often exhibit cytogenetic mechanisms controlling the amount and nature of genetic recombination. These mechanisms, along with others previously mentioned, make up the genetic system of the population that determines how many new gene combinations are produced in a unit of time. They range from inversions and translocations, which produce relatively small groups of linked genes, through polyploidy, with its diverse and variable effects, to apomixis, in which recombination is largely or completely eliminated. Such mechanisms are often considered disadvantageous in the very long-range view. However, they are extremely common in both plants

and animals and must result in a selective advantage. The bizarre and complicated genetic systems of some organisms discussed are poorly understood and have not been satisfactorily integrated into evolutionary theory.

REFERENCES

Darlington, C. D.: *The Evolution of Genetic Systems*, 2d ed., Basic Books, New York, 1958. A remarkable attempt to unify cytology and genetics in evolutionary terms.

Garber, E. D.: *Cytogenetics: An Introduction*, McGraw-Hill, New York, 1972. An excellent paperback discussion of cytogenetics.

Oliver, James H., Jr. (ed.): Symposium on Parthenogenesis, *Am. Zool.*, **11:**239–398 (1971). A great variety of papers on many aspects of parthenogenesis in invertebrates and vertebrates.

Stebbins, G. L.: The Comparative Evolution of Genetic Systems, in Sol Tax (ed.), *Evolution after Darwin*, vol. 1, *The Evolution of Life*, pp. 197–226, University of Chicago Press, Chicago, 1960. A thorough account of genetic systems, together with speculations concerning their origin.

———: *Chromosomal Evolution in Higher Plants*, Addison-Wesley, Reading, Mass., 1971. A concise, paperback review of chromosomal variation in higher plants.

Stent, G. S.: *Molecular Genetics*, Freeman, San Francisco, 1971. A recent source of material on the genetic systems of microorganisms.

White, M. J. D.: *Animal Cytology and Evolution*, 2d ed., Cambridge University Press, New York, 1954. The standard reference for genetic systems in animals, though now somewhat out of date. Other references to animals will be found in *Evolution after Darwin* Vol. 1 cited above.

Wilson, Edward O.: *The Insect Societies*, Belknap Press, Harvard, Cambridge, Mass., 1971. This very comprehensive and beautifully illustrated book discusses the genetic systems of social insects.

PART
THREE

POPULATIONS: DIFFERENTIATION

The process of evolution is sometimes divided into microevolution (changes within populations) and macroevolution (the origin of major variation patterns). Where to draw the distinction is an arbitrary decision, which we prefer not to make. In the preceding section we have considered primarily changes within populations. In this section the ways in which evolving populations change and interact to produce the diverse life forms on the earth are presented.

Chapter 10 deals with the basic splitting process of evolution, the ways in which a single evolving entity becomes two or more entities. This subject is discussed first by comparing different patterns of diversification which have been observed and then attempting to explain how they may have come about.

Chapter 11 is concerned with the patterns produced over long periods of time by populations evolving and dividing and also becoming extinct. No special factors are postulated to account for the evidence derived from a study of the fossil record, which is accepted as fragmentary and biased in various ways. The same processes that produce elaboration of different populations across a diversified habitat are viewed as being responsible for the elaboration of populations through time. The apparent problem of how "higher" taxonomic categories arise is considered an artifact created by the taxonomic method applied to situations where much extinction and loss of data have occurred.

CHAPTER
TEN

THE
DIFFERENTIA-
TION OF
POPULATIONS

It is obvious to anyone observing the variation of living things in nature that organisms do not vary continuously. Variants of one type of organism may be arranged in a continuum, but there are gaps in the variation from continuum to continuum. Plants and animals, viewed by our usual techniques of studying organisms, seem to be aggregated into discrete or nearly discrete clusters usually called **species.** Certainly the living world may be structured by the scientist in many ways different from this customary taxonomic one; some of these may be of considerable interest to the evolutionist. In the last chapter of this book some of the problems involved in perceiving and describing structure and pattern in nature are discussed. Nevertheless, it is possible to recognize taxonomic units and to classify them; this has led biologists to attempt to understand the origin of such units in nature. This generally has been studied from the point of view of how a single supposedly interbreeding population can differentiate into discrete clusters. The processes presumed to be involved make up what is frequently referred to as **speciation.**

Elucidating the mechanisms of speciation often has been regarded as the central problem of evolution. Darwin's classic work was entitled *The Origin of Species* . . . , and many monographs in both botany and zoology in recent years have emphasized the so-called species level of recognizable biological difference. This emphasis may have had the effect of obscuring some exceedingly important and interesting problems usually thought of as falling within the province of ecology, e.g., the nature and evolution of communities of plants and animals. However, in discussing here the question of how recognizable aggregates of similar organisms arise in nature, we for the moment shall accept the commonly used analyses and designations.

One usually gains the impression from even a casual study of living things that there is a spectrum of degree of similarity among organisms. Some forms appear to be very distinct from all others; some appear to intergrade almost imperceptibly with others that are closely similar. In approaching the problem of how populations become differentiated, it will be useful to consider the nature and size of the gaps in variation between clusters of similar organisms. In this chapter, examples from the spectrum of variation will be discussed, examples in which the degree of differentiation is relatively small. To put it another way, we shall examine situations that seem to be close to branch points in the evolutionary tree—organisms that seem to be on the verge of fragmenting into multiple entities, and multiple entities that appear to be of rather recent origin. A series of examples is presented first, to give the reader some feel for the types of patterns that occur. The probable causes of these patterns are then discussed, illuminated with further brief examples. The very distinct forms will be dealt with in Chap. 11.

In what follows, the term **character** will denote any trait that varies in the overall group under discussion. Thus the presence or absence of plastids is a character when one considers all organisms. For any given organism one can determine whether or not it possesses plastids. Their presence or absence is not a character in roses for they are uniformly present. Femur length is a character in man because it varies within a group and can be measured for any individual. Femur length is also a character when adult mice and adult men are compared, but the ranges of observed variation in this character are not overlapping. Such discontinuities in variation (in single characters or in constellations of characters) are here referred to as gaps.

RELATIVE LACK OF GEOGRAPHIC VARIATION

In some cases populations of the same species which are isolated from each other by great distances show relatively little geographic variation. For instance, individuals of the small lycaenid butterfly *Lycaena phlaeas* in the Sierra Nevada of California are very similar to *Lycaena phlaeas* in eastern North America and Europe. The sand crab *Emerita analoga* has a strongly disjunct Northern-Southern Hemisphere distribution, but there is no obvious differentiation of the populations. Reef fishes of tropical seas often show little geographic differentiation, and similarly the plants which grow on the numerous low atolls scattered through the Pacific seem virtually identical everywhere.

EXAMPLES OF DIFFERENTIATION

Continuous geographic variation

In many instances, variation is so continuous that no dividing lines between segregates are obvious. Variation in some characters may occur in gradients. These gradients in single characters are called **clines,** and the variation is then called **clinal.**

Color, pattern, and size variation in animals.
Geographic variation in color, pattern, and size is one of the most widely studied of all biological phenomena. This variation is often of the sort already described (*Biston, Cepaea, Natrix*), in which populations differ primarily in the frequency of different types of individuals present. Another example is the North American tiger swallowtail butterfly (*Papilio glaucus*), some populations of which are composed of yellow and black striped (tiger) males and females and other populations of tiger males, tiger females, and uniformly dark-brown females. In southern Canada and the extreme northern United States, the populations of *P. glaucus* are composed only of tiger individuals. In south-central Florida, the proportion of dark-brown females is very low (6 to 8 percent), and in southern Florida, dark females may be completely absent. In most of the southern United States, however, populations show high frequencies (up to 50 percent) of dark-brown females.

In many cases, variation is not in *frequency* of types (partially intrapopulational). Individuals within a given population may all be closely similar (little intrapopulational variation), but color or pattern may change from population to population in broad geographic trends. Among mammals and birds the tendency for populations in colder, drier parts of the range to be lighter than those in the warmer, more moist parts is so common as to have been dignified as Gloger's rule. Other so-called ecological rules deal with variation (not necessarily continuous) in size and shape. One (Bergmann's) states that homoiothermal vertebrates in warm areas tend to be smaller than those from cool areas. Another (Allen's) states that projecting parts and appendages (wings, legs, noses, etc.) tend to be shorter in cooler areas than in warm ones.

Ecotypic variation in plants.
Botanists have attached more importance than zoologists have to the local population as a basic unit, perhaps because of the greater ease with which the less motile plants can be studied. The Swedish botanist Turesson began the study of local populations in the early years of the twentieth century. By collecting plants from a variety of habitats and growing them in

experimental gardens under constant conditions, he was able to discover the effects of environmental factors on phenotypes. He found, for example, that plants from woodlands, sandy fields, and dunes retain their characteristic phenotypes under garden cultivation. Each local race with habitat-correlated morphology is called an **ecotype.** Turesson was able to distinguish races where soil factors were of greatest importance, which he called **edaphic ecotypes. Climatic ecotypes** are those in which climatic factors are important selective agents.

Turesson's work set the stage for extensive studies of plant ecotypes. The work of Clausen, Keck, and Hiesey over many years was directed to an analysis of the variation within and between populations of plants widely distributed in California. Making use of field growing stations at Stanford (sea level), Mather (4,600 ft), and Timberline (10,000 ft), they were able to separate environmental and genetic components of variation to a large extent. Perennial plants that can be propagated vegetatively can be grown at all these locations and their physiological responses to environmental factors thus investigated. In effect, the same genetic individual can be studied simultaneously in three different ecological situations. Such studies have led to recognition of many ecological races, or ecotypes of plants.

The genus *Achillea* (yarrow) in the sunflower family has already been mentioned. By means of transplant studies, Clausen, Keck, and Hiesey analyzed the *A. millefolium* complex in some detail. The plants are found throughout the Northern Hemisphere, where they grow from sea level to timberline. There is a continuous morphological variation, from plants some 6 ft high in the San Joaquin Valley to alpine plants only a few inches in height. Other morphological traits also intergrade from population to population, so that taxonomic distinctions are difficult to determine.

Adjustment of *Achillea* plants to their environment depends on the proper integration of many physiological processes, such as rates of photosynthesis and respiration, resistance to cold, and time of dormancy and other periodic phenomena. Each local population is composed of many different genotypes. Depending upon the level of study, these can be viewed as being aggregated into groups of varying size. Clausen, Keck, and Hiesey concluded from transplant studies along a 200-mi transect of California that the genotypes and local populations are arranged into at least 11 physiological races. Two taxonomic species are represented along this transect, where they occur in different habitats. *Achillea lanulosa*, of the higher elevations, is primarily a species of continental habitats, whereas *A. borealis* occurs at lower altitudes and is a coastal species, in the main. It is interesting to note that in the northern portion of its range, where *A. lanulosa* comes to the coast, it has developed coastal ecotypes that are similar to those of *A. borealis*.

It may well be that plants, being rooted, become adjusted to the local conditions with a precision that would not be of selective value in animals.

Ecotypes of relatively recent origin in Britain have been studied by Bradshaw and his associates. Lead-tolerant populations of the perennial grass *Agrostis tenuis* have been found growing around abandoned mines on workings which have a high lead concentration. It has been established that the lead tolerance has a genetic basis. In one study, populations not tolerant to lead have been shown to exist in close proximity to the lead-tolerant populations. The transition zone between these two races is less than 20 m wide.

Although *A. tenuis* is self-incompatible, no hybrids between the tolerant and nontolerant races have been observed. Bradshaw has suggested that severe selection against the hybrids accounts for their absence. It is clear that the lead-tolerant populations arose within the pollination range of the nontolerant populations, but since hybrids were not viable, the populations could easily become established and maintained.

Clinal variation in plants.   The butterfly weed (*Asclepias tuberosa*) also shows geographic variation, but it has been studied in a very different manner. The subspecies occurring in the eastern two-thirds of the United States have been studied in great detail by Woodson. The distribution of these subspecies is shown in Fig. 10-1. Only *A. tuberosa tuberosa* and *A. t. interior* will be discussed here. In most parts of their range these subspecies can be distinguished by the shape of the leaf. Fortunately, two important components of leaf shape can be quantified and the change in shape studied geographically. These components are angle *A*, a measure of the taper of the apex of the leaf, and angle *B*, which measures the shape of the base of the leaf (Fig. 10-2). The two subspecies meet along a broad front in the eastern United States, and there is a zone of intergradation, as can be seen in Fig. 10-1. Woodson has studied geographic variation by dividing a map of the country into equal-area quadrats and measuring the herbarium specimens collected in these areas. He has also studied local-population samples and has measures of variation within and between individuals and colonies.

By comparing the measurements of specimens collected in 1946 along a 1,200-mi transect from Kansas to Virginia with the available herbarium specimens from the quadrats in which the

Figure 10-1 Distribution of subspecies of *Asclepias tuberosa*. Each symbol represents a county record. Large dots, *A. t. interior*; small dots, *A. t. tuberosa*; hollow circles, putative hybrids between subsp. *tuberosa* and subsp. *interior*; half circles, *A. t. rolfsii*. [*From R. E. Woodson, Ann. Mo. Bot. Gard., vol. 34 (1947).*]

transect falls, Woodson was able to study the effect of time. The herbarium specimens, collected over a period of many years, represent a sample which is, on an average, older than the 1946 transect. It was clear that characteristics of *A. t. interior* were moving eastward while those of *A. t. tuberosa* were moving westward but at a much slower rate. When, in 1960, samples were once again collected along the transect, the changes that had occurred in the 14-year interval could be accurately measured. Apparently reciprocal diffusion of the eastern and western genotypes of both angle *A* and angle *B* has occurred. Woodson interprets the eastern subspecies as being in the process of genetic submergence by the western one, since its western movement is proportionately less. Nevertheless, its effects on the western leaf shape can be clearly seen.

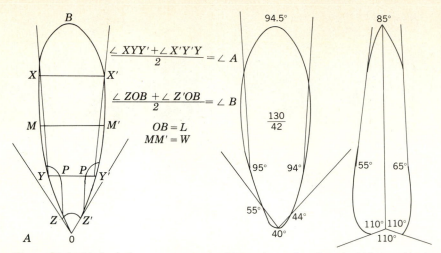

Figure 10-2 Method of measuring angle A (apical taper) and angle B (shape of leaf base) in the butterfly weed (*Asclepias tuberosa*). [*After R. E. Woodson, Ann. Mo. Bot. Gard., vol. 34 (1947).*]

Clinal variation in animals. In some cases, although the geographic variation is continuous, experimental evidence indicates that a considerable amount of differentiation has occurred. Variation in the leopard frog (*Rana pipiens*) is extensive and discordant. The variation in 12 characters is summarized in Table 10-1. No overall pattern of variation is evident; indeed, many of the characters seem to vary completely independently. Moore's detailed studies of variation in developmental processes have yielded abundant provocative data. For instance, northern and southern populations of *R. pipiens* show different temperature-tolerance ranges for normal embryological development (Fig. 10-3). These differences parallel those found between northern and southern frogs belonging to clearly distinct species. For example, the northern *R. sylvatica*, which ranges from the subarctic to the central United States, can develop normally between 2.5 and 24°C, whereas *R. catesbiana*, living from southern Canada to Mexico, has a range of 15 to 32°C; where these two overlap, *R. sylvatica* breeds in the early spring, *R. catesbiana* in midsummer.

Recent experimental data have shown the pattern of population differentiation in *Rana pipiens* to be quite complex. For example, normally developing embryos of diploid hybrids involving Vermont females have been obtained with males from population representing a broad geographic range in the United States and Mexico (Fig. 10-4). However, when the development of haploid hybrids involving enucle-

TABLE 10-1. POPULATION FORMULAS FOR MEADOW FROGS OF EASTERN NORTH AMERICA* †

	A	B	C	D	E	F	G	H	I	J	K	L
Quebec	A	B	C	D	E	F	G	H	I	J	K	L
Maine	A	B	C	D	E	F	g	H	I	J	k	L
Vermont	A	B	C	D	E	F	G	H	I	J	K	L
N. New York	A	B	C	D	e	F	G	H	I	J	K	L
Massachusetts	A	b	C	D	E	F	G	H	I	J	K	L
Rhode Island	A	B	C	D	E	F	G	H	I	J	K	L
S. New York	A	b	c	d	E	f	g	h	i	J	k	l
New Jersey	A	b	c	d	e	f	g	h	i	J	k	l
Maryland	a	B	c	D	e	F	g	h	i	J	k	l
North Carolina	A	b	c	D	e	F	G	h	i	J	k	l
South Carolina	—	b	—	D	e	f	g	h	i	J	—	—
Georgia	a	b	c	d	e	f	g	h	i	J	k	l
Florida	a	b	c	d	e	f	g	h	i	j	K	l
Ontario	—	B	C	D	E	F	G	H	I	J	K	—
Michigan	—	B	C	D	e	F	G	H	I	J	—	—
Wisconsin	A	B	C	D	E	F	g	H	i	J	K	L
Minnesota	—	B	C	D	e	F	g	H	i	J	K	L
South Dakota	A	B	C	D	e	F	g	H	i	J	K	L
Nebraska	—	B	C	D	e	F	g	h	i	J	k	L
Indiana	A	B	C	D	e	F	g	H	i	J	k	l
Kentucky	a	B	c	D	e	F	g	H	i	J	—	—
Illinois	a	B	C	D	e	F	g	H	i	J	k	l
Missouri	A	B	C	D	e	F	g	H	i	J	k	L
Kansas	A	B	C	D	e	F	g	H	i	J	k	L
Arkansas	A	B	c	D	e	F	g	H	i	J	k	l
Oklahoma	a	B	C	D	e	F	g	H	i	J	k	l
Mississippi	a	b	c	D	e	F	G	h	i	J	k	l
Louisiana	a	B	C	d	e	F	G	h	i	J	k	l
Texas	a	B	C	D	e	F	g	H	i	J	K	L

*

A	head width/head length 0.92 or greater	a	head width/head length less than 0.92
B	50% or more with tibia bars	b	less than 50% with tibia bars
C	average number of tibia bars (when present) 1.4 or greater	c	average number of tibia bars (when present) less than 1.4
D	50% or more without femur bar	d	less than 50% without femur bar
E	50% or more without tympanic spot	e	less than 50% without tympanic spot
F	50% or more with light reticulum	f	less than 50% with light reticulum
G	number of dorsal spots less than 13	g	number of dorsal spots 13 or more
H	number of lateral spots 12 or more	h	number of lateral spots less than 12
I	more lateral than dorsal spots	i	lateral spot number equal to, or less than, dorsal spot number
J	50% or more without lateral reticulum	j	less than 50% without lateral reticulum
K	50% or more of males with oviducts	k	less than 50% of males with oviducts
L	50% or more of males with no, or poorly developed, external vocal sacs	l	less than 50% of males with no, or poorly developed, external vocal sacs

† After J. Moore, *Bull. Am. Mus. Nat. Hist.,* vol. 82 (1944).

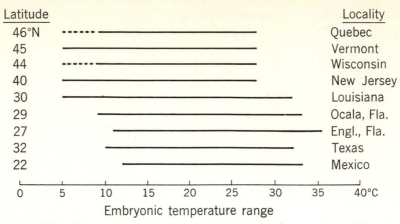

Figure 10-3 Temperature tolerance ranges for normal embryological development of *Rana pipiens* from different localities. The lower limit for Quebec and Wisconsin has not been determined but is believed to be identical with that for Vermont. [*From J. Moore, Evolution, vol. 3 (1949).*]

ated eggs of *R. pipiens* from Vermont and sperm from other localities (haploid hybridization) are scored, the range of genetically compatible populations is much narrower (Fig. 10-5). In addition, similar kinds of data show that local populations have undergone considerable genetic differentiation. This conclusion is drawn from investigations of groups of populations in Oklahoma, Texas, the Isthmus of Tehuantepec, and the Florida peninsula, which are geographically close to one another but are genetically distinct when tested by haploid hybridization. Moore has stated that haploid hybridization is the most sensitive measure of genetic relatedness used to date on population studies of the leopard frog and accurately reflects the pattern of population differentiation in this species.

It has also been found that populations of *Rana pipiens* in different regions have markedly distinctive mating-call types. Since these call types lead to reproductive isolation in areas of overlap, some biologists feel that *R. pipiens* is actually a complex of species. Four distinctive call-types, western, eastern, northern, and southern, have been recognized across the range of the so-called species. It is unfortunate that the common leopard frog, used so extensively in biological research, may in fact be not one but four species and that, in some research, two or more species may have been combined as subjects.

The British satyrine butterfly *Coenonympha tullia* shows a pattern of differentiation reminiscent of *Rana*. Crosses between individuals from widely separated populations resulted in some broods in which a number of "females" were intersexual, indicating some genetic incompatibility. This result follows Haldane's rule that inviability or sterility in hybrids will most likely appear in the heterogametic sex, in this case the females. In crosses between less

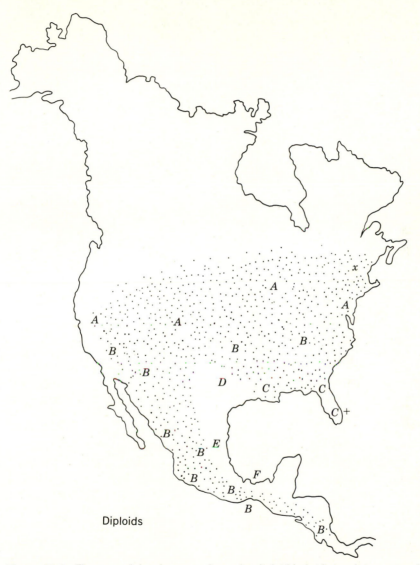

Diploids

Figure 10-4 The type of development shown by diploid hybrids involving eggs of *Rana pipiens* from Vermont and sperm from other localities. The degree of abnormality is shown by the letters *A* to *F*. The dotted area shows the known extent of populations that can be crossed and the hybrids show no or only slight defects. (*From J. Moore, Recent Results in Cancer Research, Springer-Verlag, 1969.*)

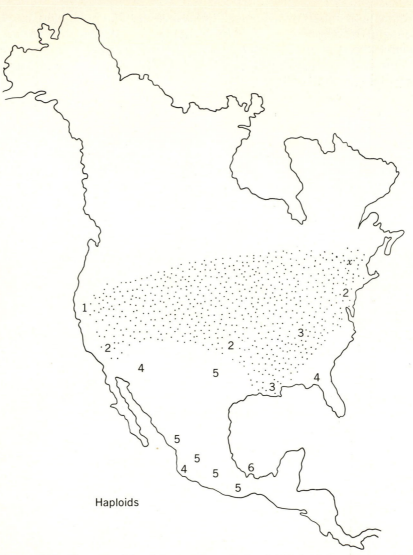

Figure 10-5 The type of development shown by haploid hybrids involving enucle-ated eggs of *Rana pipiens* from Vermont and sperm from other localities. The degree of abnormality is shown by the numbers 1 to 6. The dotted area shows the known extent of populations capable of forming haploid hybrids that show typical haploid development or only mild defects. (*From J. Moore, Recent Results in Cancer Research, Springer-Verlag, 1969.*)

distant populations no abnormalities were found. However, this butterfly has not been as intensively studied as the gypsy moth (*Lymantria dispar*), for which Goldschmidt has described many degrees of intersexuality in crosses between populations of various levels of differentiation. For those interested in details, this work is well summarized by Dobzhansky.

Variation and genetic isolation. Populations of some animals that are connected by a series of intermediate forms may occur together and remain distinct. For instance, Stebbins has described an interesting pattern of variation in the plethodontid salamanders of the genus *Ensatina*. These animals live along the western coast of North America from southwestern British Columbia to southern California. In California they are confined to coastal areas, the Sierra Nevada, and southern interior mountain ranges. There is considerable geographic variation in color pattern and, to a lesser extent, in size (Fig. 10-6). The coastal populations are brown or reddish brown above, while the Sierra and interior populations become progressively more spotted with orange, yellow, cream, and finally orange again, as one travels southward. In the Sierra Nevada, throughout the western central foothills, there are populations similar to those of the central coast, and hybrid individuals intermediate between the Sierra Nevada and coastal types are found there.

In the characters studied (and with the exception just mentioned) there seems to be rather continuous north-south variation, although taxonomists have broken the continuum into a series of subspecies and zones of intergradation. However, where the southern coastal and inland types meet south of the Central Valley, there is a rather sharp discontinuity in the variation. Strikingly different uniformly colored and blotched forms occur widely and are known to coexist at two localities, on the south side of the San Bernardino Mountains and on Mount Palomar. Occasional hybrids are found in these areas, but the frequency of successful interbreeding seems to be less than in the Sierra Nevada.[1]

A similar situation has been reported for neotropical fruit flies called *Drosophila paulistorum*. Here, the pattern is more complex than that described for *Ensatina*, there being three areas where two groups occur together without interbreeding. In these areas not only is hybridization not detected, but in laboratory tests where the forms were denied the opportunity of mating with their own kind, not even cross-insemination (let alone the production of viable hybrids) was found. However, in laboratory tests it was possible to exchange genetic information between these forms by using a series of intermediate "bridging" cage populations sampled from other ge-

---

[1] We are deeply indebted to Dr. R. C. Stebbins for keeping us informed of the progress of this most interesting work.

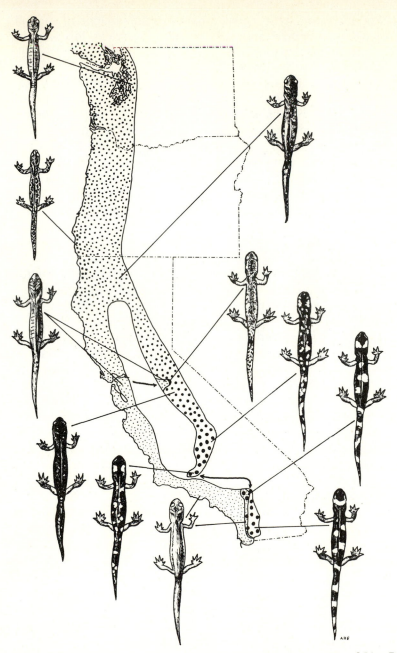

Figure 10-6 *Ensatina* in western North America. Discussion in text. [*After R. C. Stebbins, Univ. Calif. Publ. Zool., vol. 48 (1949)*.]

ographic areas. How much actual exchange takes place in nature through such bridging populations is an open question. Such complex situations are found in more and more organisms as detailed studies are made.

Closely related isolates

Species swarms in fishes.   The East African lakes, Victoria, Tanganyika, and Nyassa, support a large number of closely related fish species of the family Cichlidae. For example, in Lake Victoria are found some 70 endemic and 6 nonendemic species of the genus *Haplochromis* living in three different ecological zones. One group consists of deep-bodied forms with short snouts, horizontal mouths, equal jaws, and bicuspid outer teeth (Fig. 10-7). These fishes are found inshore and are bottom feeders. The cichlids of a second group have more slender bodies and longer snouts, their mouths are slightly oblique, their lower jaws prognathous, and their outer teeth conical and caniniform (Fig. 10-7). The members of this group are fish-eating predators, hunting the middle depths of open and inshore waters. A third group of *Haplochromis* are slender and long-snouted. They have very oblique mouths (in two forms almost vertical), extreme prognathism of the lower jaw, and caniniform outer teeth (Fig. 10-7). These are predaceous surface feeders, eating principally other fishes and insects.

There is only a moderate amount of diversity in this large complex of closely related distinct clusters. Although some of the forms are virtually indistinguishable morphologically, they have been found to be ecologically differentiated and to have distinctive breeding coloration. The greatest morphological variation is in the teeth and structures of the head, which is hardly surprising in view of the diverse feeding habits within the group.

Sibling species of alpine butterflies.   In some cases, superficial similarity may disguise a rather large amount of diversity. Lorkovič has shown that the holarctic butterflies of the *Erebia tyndarus* group, although very much alike in outward appearance, have wide divergence in chromosome number ($n = 8$, 10, 11, 15, 24, 25, 51, and perhaps 52) and (to a lesser extent) in the morphology of the male and female genital structures. Such outwardly similar forms are often called sibling species. In the western Alps (Fig. 10-8) two forms, *Erebia cassioides* ($n = 10$) and *E. nivalis* ($n = 11$), occupy two barely overlapping ecological zones, the former in the subalpine (1,400 to 2,400 m) and the latter in the alpine (2,300 to 2,700 + m). Although *E. cassioides* and *E. nivalis* share a narrow border strip, there is little

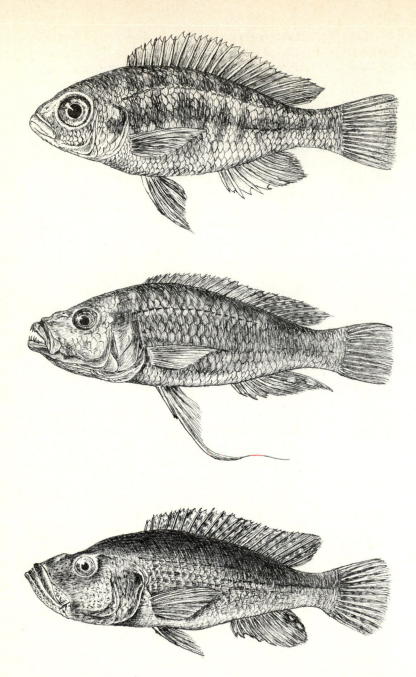

Figure 10-7 Cichlid fishes (*Haplochromis*) from Lake Victoria, Africa. Top, *H. macrops*; center, *H. bayoni*; bottom, *H. cavifrons*. These fishes are approximately 100, 180, and 145 mm long, respectively. [*After G. A. Boulenger, Catalogue of the Fresh-water Fishes of Africa in the British Museum (Natural History), vol. III, 1915.*]

evidence of hybridization. Only 2 of 400 specimens examined were not unequivocally assignable to one species or the other. The two forms have quite distinct life cycles, *E. cassioides* completing its development in 1 year, *E. nivalis* requiring 2. Laboratory crosses indicate strong behavioral isolation (females not responsive to males of the wrong form; copulation, when induced, abnormally brief), and hybrids, when produced artificially, are wholly sterile and unlike any individuals found in nature. In short, the two forms seem to be completely isolated from each other genetically.

Erebia cassioides, however, has a wider geographic distribution than *E. nivalis*, and in regions where *E. nivalis* is absent it extends to altitudes as high as those occupied by *E. nivalis* where both are present. Conversely, in some areas where the mountains do not reach great heights, *E. nivalis* lives at lower elevations, with *E. cassioides* correspondingly lower or absent.

Erebia tyndarus ($n = 10$) occurs in essentially the same life zone as *E. cassioides*, but, as one can see from Fig. 10-8, the two are not sympatric. *Erebia tyndarus* occurs in the central Alps, with *E. cassioides* on the east and west. Experimental crosses (*E. tyndarus* females × *E. cassioides* males) showed little behavioral isolation, and about 15 percent of the eggs from these crosses hatched with 25 percent survivorship among the young larvae. Although the ranges of *E. tyndarus* and *E. cassioides* adjoin very closely, both in the east and

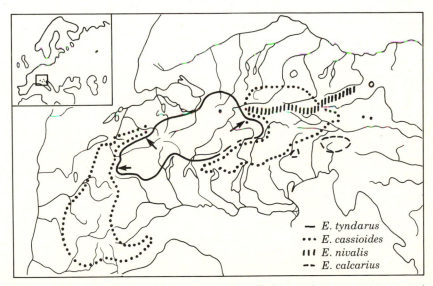

Figure 10-8 Distribution of butterflies of the *Erebia tyndarus* group in central Europe. Arrows indicate points of contact. [*From Z. Lorković, Biol. Glas., vol. 10 (1957).*]

TABLE 10-2. SPECIES OF DARWIN'S FINCHES (GEOSPIZINAE)*

Name	Description and habits	Number of islands on which species is permanent resident
Geospiza magnirostris	Large; forages on ground and in bushes and trees; feeds on small variety of very hard, generally large seeds	14
fortis	Medium; habits as above; feeds on large variety of moderately hard, small to large seeds	12
fuliginosa	Small; forages on ground more than *G. fortis* or *magnirostris*; feeds on large variety of soft, generally small seeds	14
difficilis	Medium; forages on ground; presumably takes soft seeds; known to eat ticks on iguanas and on Wenman Island to take blood from boobies	7
scandens	Medium; rests mostly on *Opuntia* cactus; feeds on a small variety of moderately hard seeds, also soft plant tissues and nectar	11
conirostris	Large; habits poorly known but similar to *G. scandens*	3
Platyspiza crassirostris	Large; mostly in dense brush and high trees; feeds mainly on fleshy fruits, soft to moderately hard seeds, young leaves, and flowers	10
Camarhynchus psittacula	Medium; forages in trees, brush, and occasionally on ground; primarily insectivorous, excavating fairly deeply into woody tissues, usually on larger branches	11
pauper	Same as *C. psittacula*, which it replaces on Charles Island	1
parvulus	Small; forages in trees, brush, on cactus, and on ground; primarily insectivorous, excavating less deeply than *C. psittacula*, usually on smaller twigs and in lichens	12
Cactospiza pallida	Medium; tanager-like habits; probes with stick or cactus spine; primarily insectivorous	7
heliobates	Medium; habits poorly known, primarily insectivorous; restricted to mangroves	2

TABLE 10-2. *(cont.)*

Name	Description and habits	Number of islands on which species is permanent resident
Certhidea olivacea	Small; warbler-like; forages at all levels in trees, occasionally at ground level; takes only animal food, especially small larvae	16
Pinaroloxias inornata	Medium; reported feeding on ground and in trees; presumably takes insects, nectar, and some fruits	1

*   Modified from Bowman (1961).

west, there seems to be no significant overlap. Indeed the three species, *E. tyndarus*, *E. cassioides*, and *E. nivalis*, show a striking aversion to coexistence. A fourth alpine species, *E. calcarius* ($n = 8$), is found in the Julian Alps, but its exact spatial relationship with *E. cassioides* is not known at this time.

The Galápagos finches.   A large cluster of distinct closely related groups is found in the subfamily Geospizinae of the finch family (Fringillidae). These birds, known collectively as Darwin's finches, are restricted, with a single exception, to the Galápagos Islands. One member of this group is found on Cocos Island. First studied by Darwin, they have been the subject of brilliant monographs by Lack and Bowman. There are some 14 distinct kinds of finches, considered by ornithologists to represent six genera (Table 10-2).

The 14 species are distributed in various patterns over the islands, individual islands within the Galápagos having between 3 and 10 species each (Fig. 10-9). The birds differ primarily in size and in the form of the beak and in other structures related to their feeding habits (Fig. 10-10). There is almost a complete continuum in the amount of differentiation. The cluster known as *Platyspiza crassirostris* is distributed over eight of the islands but shows almost no interisland differences in the characters studied. The warbler finch (*Certhidea olivacea*) is found on all the Galápagos and shows considerable variation in color from island to island. For instance, the upper parts of both sexes vary from gray-brown (James Island) to very pale gray (Barrington Island). Superimposed on this is variation in the amount of olive tinge. The under parts range from pale olive-buff

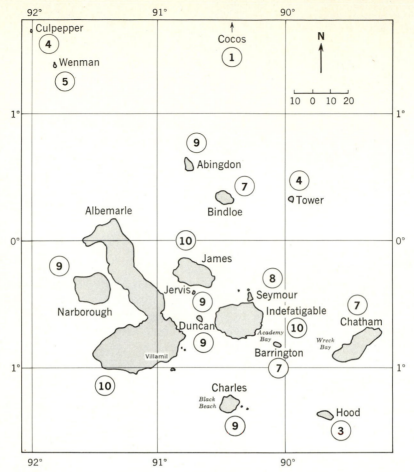

Figure 10-9 Map of Galápagos Archipelago showing main islands. Numbers give the total of different kinds of geospizine finches recorded from each island. [*After R. I. Bowman, Univ. Calif. Publ. Zool., vol. 58 (1961).*]

(James Island) and wash-brown (Culpepper, Wenmen, and Charles Islands) to white (Barrington Island).

The peripherally distributed *G. conirostris* is, except for size, quite similar to the central *G. scandens*, the two forms being completely allopatric. Three ground finches, *G. magnirostris*, *G. fortis*, and *G. fuliginosa*, are widely sympatric and differ superficially in overall size and relative beak size. Where the three forms occur together, there is almost no overlap in measurements and observations indicate that individuals recognize and mate only with members of the proper form. The three forms sometimes take the same food; their feeding habits are overlapping but not congruent, as shown in Table 10-2.

Bowman and Billeb have reported that a fourth species of ground finch, *G. difficilis*, includes a population on Wenman Island which have become blood eaters. After landing on the tail of the red-footed booby (*Sula sula websteri*) or masked booby (*S. dactylactra granti*), the finches bite the host bird and feed on the blood as it moves along the Booby's feathers. Blood eating is not the sole means of obtaining food for these finches; they, like other populations of *G. difficilis*, ground-forage. Blood eating, however, is an efficient means of obtaining a high-protein food source, and since the blood of both birds are nearly isotonic, feeding on blood in the absence of a fresh drinking water is no problem. It has been suggested that initially these finches probably fed on ectoparasites removed from the host birds and that this mutualistic relationship led subsequently to the derivation of the parasitic blood-eating habit. They are known to pick ticks

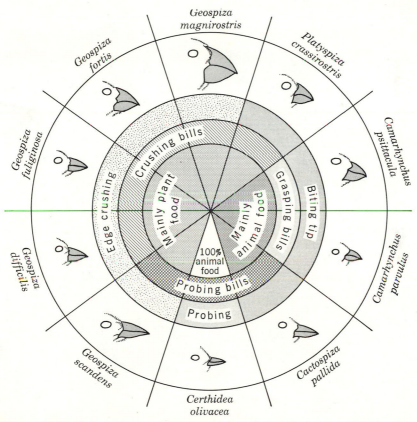

Figure 10-10 Schematic representation of the relationship between bill structure and feeding habits in 10 species of Geospizinae from Indefatigable Island. [*From R. I. Bowman, Univ. Calif. Publ. Zool., vol. 58 (1961).*]

from iguanas. Feeding on ectoparasites of other animals is not uncommon in birds.

The two members of the genus *Cactospiza*, *C. pallida* (the tanager-like finch) and *C. heliobates* (the mangrove finch), are very similar. Both forms are primarily insectivorous, *C. pallida* having the unique habit of excavating for beetles and other insects with its beak and then probing the excavations with a cactus spine or twig. This remarkable behavior compensates for the lack of a long tongue to use as a probe and is one of the few instances of birds using a "tool." On the only island occupied by both *C. pallida* and *C. heliobates* (Albemarle), their ecological differences isolate them (*C. heliobates* in the coastal mangrove belt and *C. pallida* inland). Therefore they are allopatric, although in one case they both live on the same island. The extremes of differentiation within the genus can be seen (Fig. 10-11)

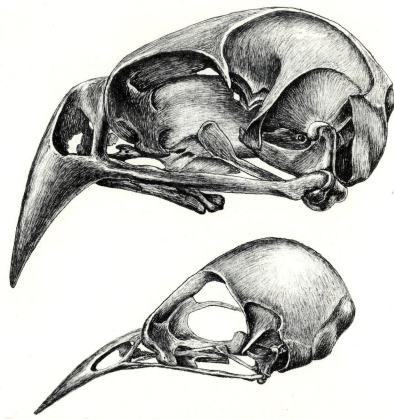

Figure 10-11 Extremes of differentiation in skulls of Galápagos finches. Upper, *Geospiza magnirostris*; lower, *Certhidea olivacea*. [*After R. I. Bowman, Univ. Calif. Publ. Zool., vol. 58 (1961).*]

by comparing the skulls of the broadly sympatric *Geospiza magnirostris* (the large ground finch) with *Certhidea olivacea* (the warbler finch).

Darwin's finches show a pattern of differentiation opposite that found in most groups of birds. The most closely related forms of these finches differ primarily in the size and shape of the beak, whereas closely related forms of other birds are usually differentiated most strongly by plumage color. The very closely related Asiatic nuthatches, *Sitta tephronota* and *S. neumayer*, are clearly differentiated by plumage pattern as well as bill length but only where they are sympatric (Fig. 10-12). Where their ranges are separate, they are almost indistinguishable. This exaggeration of differences in an area of sympatry is an example of **character displacement,** a phenomenon common in birds.

Breedlove has demonstrated a striking example of character displacement in the plant genus *Fuchsia* (Onagraceae). Two Mexican species, *Fuchsia encliandra* subsp. *encliandra* and *F. parviflora*, show a remarkable similarity in floral morphology where their ranges do not overlap (Fig. 10-13). Floral morphology determines the kind of pollinator. In these nonoverlapping areas both species are pollinated by bumblebees and hummingbirds. On the other hand, where their ranges overlap, the flowers of these two species are quite different and each species has a different pollinator. *Fuchsia encliandra* subsp. *encliandra* is pollinated by the hummingbird *Amazilia beryllina*, whereas *F. parviflora* is pollinated by *Bombus formosus*, a bumblebee.

Other instances of character displacement have been described in ants, beetles, crabs, fishes, and frogs. Two kinds of termites have been reported to swarm at the same time of day where they are allopatric and at a different time of day where their ranges overlap. Possible causes of character displacement will be discussed later in this chapter.

Host preference in parasitic organisms

Differentiation in host preference is widespread among parasitic organisms. This phenomenon has been studied in such diverse organisms as cuckoos, human lice, and nematodes. Cuckoos have developed an unusual form of differentiation. They show what has been called brood parasitism; i.e., they lay their eggs in the nests of other kinds of birds. In most cases the cuckoo egg is incubated by the foster parents, and the voracious cuckoo hatchling crowds its pseudosiblings out of the nest, eventually monopolizing the food brought by its adopted parents. Usually the egg laid by the cuckoo bears a

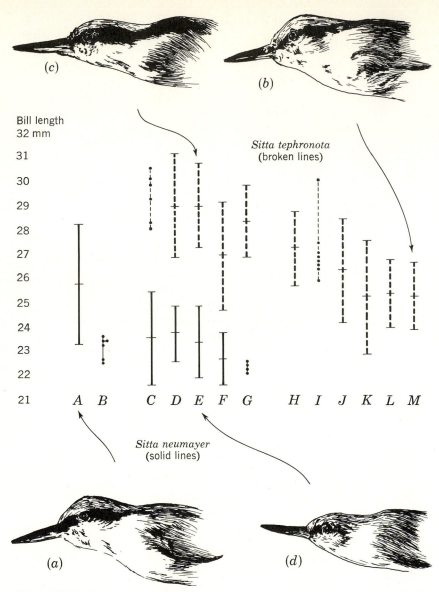

Figure 10-12 Character displacement in Asiatic nuthatches. Bill length and facial stripe in the two species are very different in areas where they occur together but are quite similar where they occur alone. Populations west of the zone of overlap (*Sitta neumayer*): *A*, Dalmatia and Greece; *B*, Asia Minor. In the zone of overlap: *C*, Azerbaijan and Northern Iran; *D*, Kermanshah; *E*, Luristan and Bakhtiari; *F*, Fars; *G*, Kirman. East of the zone of overlap (*S. tephronota*): *H*, Persian Baluchistan; *I*, southern Afghanistan; *J*, Khorasan; *K*, north-central Afghanistan north of the Hindu Kush; *L*, northeastern Afghanistan (Pamirs); *M*, Ferghana and western Tian Shan. [*After C. Vaurie, Am. Mus. Novit. 1472 (1950).*]

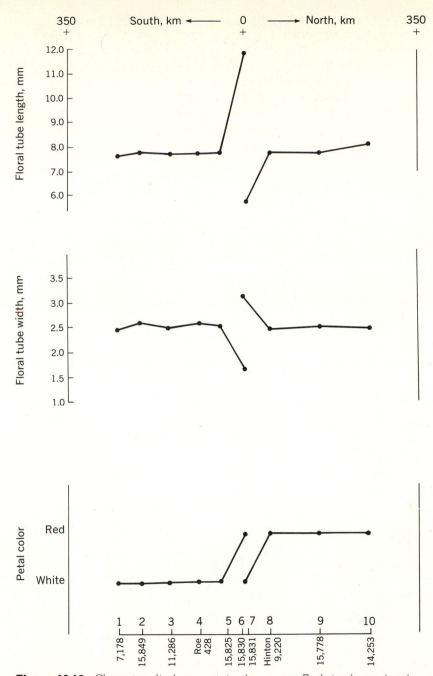

Figure 10-13 Character displacement in the genus *Fuchsia* shown for three characters in a linear north-south series of 10 populations from Mexico. [*From D. Breedlove, Univ. Calif. Publ. Bot., vol. 53 (1969).*]

remarkable resemblance to that of the host bird (Fig. 10-14). Cuckoo species seem to be subdivided into groups, each of which tends to lay its eggs in the nests of only one kind of host bird. A subdivision is called a **gens** (plural gentes); they are not geographically isolated from each other, individuals of one being found in close proximity to individuals of one or more of the others.

There are two distinct forms of human lice (*Pediculus humanis*), head lice and body lice, which differ strongly in their "ecology" but are nearly identical morphologically. Head lice, as the name implies, are found primarily in the relatively fine hair of the head. Their eggs are glued to hair. Body lice, on the other hand, live in the clothes, sucking blood where the clothes are in contact with the body. Body lice attach their eggs to the clothing.

The common human nematode parasite (*Ascaris lumbricoides*) is morphologically and serologically indistinguishable from the pig parasite (*A. suum*), but in most cases eggs from one will not develop properly in the host of the other.

Figure 10-14 Egg mimicry. Upper row, eggs of Asiatic crows, *Corvus coronoides*, *C. splendens insolens*, and *C. s. protegatus*; lower row, eggs of the cuckoo *Eudynamys scolopacea* laid in the same nests as eggs of the three crows. [*After E. C. S. Baker, Proc. Zool. Soc. Lond.* (*1923*).]

DISCUSSION OF OBSERVED PATTERNS

The basic reason for the diverse patterns of differentiation described in the preceding section is that the physical environment is, and always has been, heterogeneous. This heterogeneity has meant that evolutionary forces, especially selective forces, have operated differentially. In turn, this has produced a heterogeneous biotic environment and further differentiation in the forces operating in any portion of that environment.

It is all too easy to fall into the mistake of assuming that differentiation of populations into genetically isolated forms is somehow the "goal" of evolution. Presumably, differentiation often takes place, but usually those instances where incipient species become submerged again are only very rarely recorded. When morphological divergence of two populations has progressed to a certain point, often we no longer say that they interbreed, we say that they hybridize. To some this implies that they have somehow or other made a "mistake." The fact that occasional genetic interchange enriches the variability of both populations is too often forgotten. Differentiation is not necessarily a step toward the formation of isolated populations. It is merely one of the many things that happen to populations in nature.

Geographic variation in selection pressures

Geographic variation, in most cases, may be attributed primarily to the different selection pressures prevailing in different areas. Thus *Achillea* growing at high altitude is under selection pressure that favors dwarfed forms physiologically adjusted to rigorous mountaintop conditions. Lowland *Achillea* are not subjected to the same pressures but to others. *Natrix sipedon* populations have adjusted both to "normal" swamp habitats and to the special conditions present on the Lake Erie islands. *Cepaea* populations face selection pressures partially determined by local vegetation types, and *Biston* populations to pressures varying with the relative presence or absence of industrial pollution. The frequency of the sickle-cell gene in human populations varies geographically; it is high in regions where malignant tertiary malaria is present and low elsewhere.

Biologically sophisticated readers will be familiar with myriad examples in which differences due to different selective pressures because of different environments have been inferred. For the student a very few more are added. Arctic foxes (following Allen's rule) have short ears and snouts, whereas tropical foxes have long ears and snouts, presumably because the low surface-volume ratio in the former helps conserve body heat, whereas the high ratio in the latter permits the heat to dissipate more rapidly. Indeed, both Allen's and Bergmann's rules seem to be simply functions of the surface-volume ratio problems concerned with heat retention and dissipation. A great many homoiotherms show clinal variation in conformance with these rules, but physiological work to support their validity is mostly lacking.

Populations of the swallowtail *Papilio glaucus* have high frequencies of dark females in certain areas, supposedly because these dark-brown females resemble the *Aristolochia* swallowtail (*Battus philenor*), also found in these areas. *Battus philenor* is distasteful to birds, and birds have been observed feeding on adult *P. glaucus* in the field. Selection apparently favored the development and maintenance of the mimetic form of *P. glaucus* in these areas.

In the high Sierra Nevada of California, two forms of the butterfly *Oeneis chryxus* are found, a light form in areas of granite rock and a dark form in areas of basaltic rock. The selection pressure involved has not been discovered, but the correlation suggests the work of an unknown visual predator. Similar examples of geographic variation in color and in the habitat of many groups of animals have been reported and could be multiplied indefinitely; geographic variation is ubiquitous, and selection has been shown to play a major role in differentiating most populations that have been studied thoroughly. It is a truism to state that populations of organisms in different places will, under most circumstances, be genetically different.

The variation described in *Asclepias* seems attributable to a combination of differing selection pressures and genetic drift. *Asclepias tuberosa* is a long-lived perennial growing in colonies of several to many plants. The effective population size is estimated to be less than the median census size of 11 plants. One would expect that in populations of this small size genetic drift would become an important factor in evolution, and the data suggest that it is. Nevertheless, there appears to be a strong selective component in the changes that have occurred along the transect during the 14 years. It is not possible to discuss these in detail, but it is clear that the western genotypes for angle A and angle B are selectively better off than the comparable eastern genotypes. In *A. tuberosa* subsp. *interior* the genotype for angle B has a selective value about three times that for angle A. During the 14 years there has been a decrease of about 30 percent in variability for both angles.

In both years, about midway along the transect there appeared to be a zone of heterozygosity expressed as greater variability as well as greater size and apparent vigor. It is probable that hybridization between these subspecies, which were separated from one another during the Pleistocene glaciation and which have subsequently come together, has resulted in an increase in vigor and an extension of range. This study is one of the few in which populations have been investigated over a period of years. Unfortunately, however, *Asclepias tuberosa* is not a good subject for genetic studies. The cytogenetic bases for the phenomena that have been detected biometrically are not known. Many other interesting aspects of variation in this species of *Asclepias* also are poorly understood. These include the apparently centrifugal variation of *A. tuberosa interior*, which has resulted in a concentrically distributed peripheral subspecies in the western and northern parts of the range, and an interesting variation pattern with respect to color of the flower.

Exchange of genetic information

The degree to which gene exchange (gene flow) limits the amount of differentiation between populations is unknown. However, data from plant and animal populations suggest that gene flow is not as important as it was once thought to be. Studies of migratory behavior in various animal populations, as well as pollination and seed-dispersal studies in plant populations, indicate that on the whole the exchange of genetic information among populations is *quite limited*. Many cases are known in which populations have been totally isolated for hundreds of thousands of generations and yet show little differentiation. In other cases gene flow between distant populations may be so slow as to be practically indistinguishable from mutation as a source of variability. Ehrlich and Raven have argued that it is selection itself that is both the primary cohesive force and the primary disruptive force in evolution and that the selective regime determines what influence gene flow has on observed patterns of differentiation. Populations will differentiate if they are subjected to different selective forces and will tend to remain similar if they are not.

A classic example of an extremely limited exchange of genetic information is found in populations of a single species of a California desert plant. Around the southern and western edges of the Mojave Desert a small annual *Linanthus parryae* (Polemoniaceae) shows an interesting pattern of variation. With good rainfall, the plants form an essentially continuous carpet over large areas. In 1941 *L. parryae* was studied intensively between Palmdale and Lucerne Valley, where it had developed a practically continuous population over some 700 mi. In most areas investigated, the plant samples consisted of white-flowered forms, but in three isolated sections samples with varying numbers of blue-flowered individuals were collected. The composition of the samples taken from the westernmost of these areas between Palmdale and Llano is shown in Fig. 10-15. The central

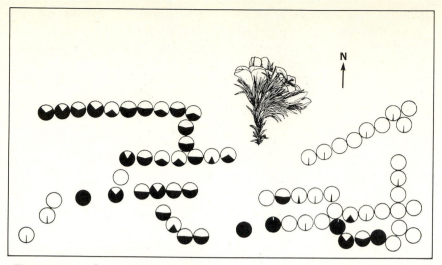

Figure 10-15 Composition, with respect to flower color, of samples of *Linanthus parryae* from an area in southern California. Black sectors, blue flowers; white sectors, white flowers. [*From C. Epling and T. Dobzhansky, Genetics, vol. 27 (1942) and after Abrams, Illustrated Flora of the Pacific States, Stanford University Press, 1951.*]

variable area is separated from the western one by a 25-mi gap and from the eastern one by an 8-mi gap.

The frequency of blue-flowered individuals in samples from these variable areas is shown in Fig. 10-16. This frequency distribution of phenotypes is reminiscent of the theoretical gene-frequency distributions in small populations subjected to some selection pressure (see Fig. 6-9). The distribution in Fig. 10-16 cannot be fully interpreted because the genetic bases of the observed variation are unknown. However, the U shape is strongly suggestive of populations sufficiently small for considerable random fixation and loss of genes producing blue flowers to have occurred. It seems likely that selection, drift, and highly restricted gene exchange all interact to produce the observed pattern of variation in *Linanthus*. Some unknown selective factor probably gives a small advantage to blue-flowered plants in the variable areas, while the small effective population size (calculated by Wright to be 14 to 27 plants, although recent evidence indicates that these numbers may be too small) permits considerable drift. In addition, the usual desert year (much drier than 1941) would produce only scattered *Linanthus* populations. This, combined with a negligible gene flow in the species (caused by pollinators transferring pollen mostly between nearby flowers), would help to maintain the established pattern.

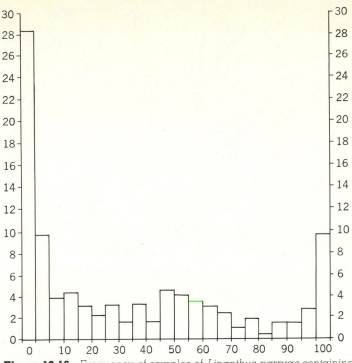

Figure 10-16 Frequency of samples of *Linanthus parryae* containing different proportions of plants with blue flowers. Ordinate, percentage of samples, abscissa, percentage frequency of blues. [*From C. Epling and T. Dobzhansky, Genetics, vol. 27 (1942).*]

Isolation

When the distribution of an organism becomes broadly discontinuous, populations normally will be found in different environments and subjected to different selection pressures. Such situations may develop in many ways. Emerging land may divide a marine habitat. The Isthmus of Panama has been repeatedly submerged through geologic time. With each new emergence some populations of marine organisms, previously more or less continuously distributed, become isolated in different environments. There is considerable evidence that glacial advances during the Pleistocene repeatedly fragmented the ranges of many organisms. Repeated changes of continental seaways (Fig. 10-17) have isolated and reconnected portions of the continents, causing manifold changes in the continuity of the distributions of organisms.

Climatic changes have profound effects on the distributions of plants and animals. Trends toward aridity produce desert or steppe

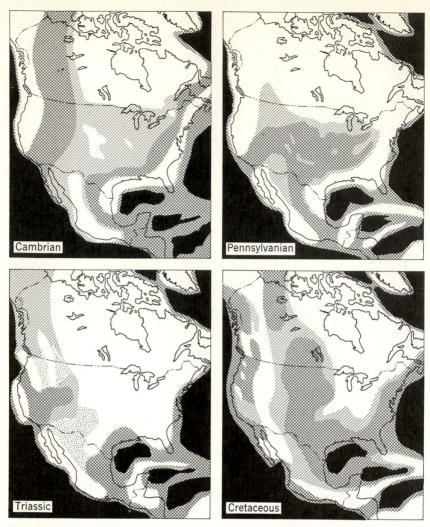

Figure 10-17 Distribution of seas in the northern Western Hemisphere during four geologic periods. Black, oceanic areas; dark gray, most persistent seaways on the continental platform; light gray, areas of temporary flooding; white, persistent land areas. (*From R. C. Moore, Introduction to Historical Geology, copyright 1949, Mc-Graw-Hill Book Company; used by permission.*)

barriers to the passage of organisms requiring high humidity or dense plant cover. Drought may cause large lakes to divide into numerous smaller lakes and rivers to be reduced to isolated series of pools. Increasing rainfall, on the other hand, tends to reunite isolated bodies of water, making gene exchange at least theoretically possible in aquatic organisms while forming barriers for terrestrial organisms. Belts of high humidity form barriers for desert and steppe organisms.

It is interesting to note that discontinuities in the distribution of an animal are not necessarily the result of barriers which the individuals are *unable* to cross. In many cases behavior patterns prevent dispersal across areas that could easily be traversed if the attempt were made. Thus rivers may serve to isolate bird populations on opposite banks, or a narrow strip of woods may effectively separate two meadow populations of butterflies.

Whatever its cause, physical isolation probably accelerates the differentiation of populations of sexually reproducing organisms when selection favors differentiation. Isolation is often referred to as if it were the *cause* of differentiation. It is not, of course; recombination between isolates is prevented, and each isolate responds to the selection pressures of its own environment. By definition, the environments in which two new isolates find themselves cannot be identical. Thus the two isolates are often subjected to different selection pressures. Because of sampling error, the two new isolates will be of different sizes and will have gene pools that are not identical, so that evolutionary forces will be operating in unlike genetic environments.

Since mutation and recombination are random processes, it is not to be expected that identical mutations or recombinations will show up in the two isolates. Thus aided by physical isolation, a single evolving unit may become two or more evolving units. As long as such evolutionary units remain isolated, they are free to respond independently to evolutionary forces. But isolated units need not differentiate. Indeed selection may favor a "winning combination" genotype which is capable of producing different highly adaptive phenotypes in different environments. In this case selective agents will operate to prevent differentiation. In some cases a balance may be struck between selection tending to differentiation and selection against that tendency.

Through the course of time, there have occurred innumerable environmental changes that might have brought previously isolated populations back together. The effects of ancient cataclysms have been modified or completely erased from the record. Two relatively recent overlapping events have undoubtedly produced effects which are conspicuous today. The first of these events, the Pleistocene glaciations, changed climate on a vast scale, causing plants and animals to migrate or perish. The glaciers also scoured the earth, mingling soil types, creating lakes, and rearranging drainage systems.

The advent of man has had similar effects. It is difficult to overestimate the importance of man as a factor in evolution. First with fire and then with domesticated plants and animals, he has modified the environment of many organisms. Many of the instances of inter-breeding of previously separated forms are the result of man's conscious or unconscious intervention. By breaking into the vast stored-energy reserves of climax ecological communities, man has diverted energy for his own purposes and grossly modified what we think of even yet as the "natural" (prehuman) environment.

Fusion of populations.   At one extreme, populations may have been isolated for a short time, subjected to such similar conditions or subjected to selection pressure to maintain a winning combination genotype, so that divergence has been minimal. When the populations reunite, individuals mate at random and the offspring from matings between parents of different populations are as successful as those from members of the same isolate. The two previously isolated populations then fuse into a single population. This is seen whenever *Drosophila* lines, isolated for a few generations in the laboratory, are combined or when guppies kept for years in one aquarium are dumped in with those that have been kept in another. The fusion of populations is certainly very common in nature also; it has not been given very much attention by zoologists.

Meeting with no gene exchange.   At the other end of the spectrum, populations that have diverged a great deal may come together and all sexual attraction between individuals of the different populations may have been lost, resulting in no interpopulation matings. This means that gene exchange between the evolutionary units is no longer possible, although, as in *Drosophila paulistorum*, genetic information may possibly still be exchanged by such isolates via the connecting series of populations. The pattern in *Ensatina* is of great interest, since it is now apparent that the terminal populations are differentiated to such an extent that fusion of these populations is most unlikely and thus a typical ring-of-races pattern has been produced.

Situations where two or more similar kinds of organisms live in close proximity without apparent hybridizing are numerous, but there seem to be no observations of segregates lacking sexual interest in each other at the time of rejoining. This does not indicate that such situations never develop, merely that they have not been observed. It is also important to note that gamete wastage is reduced between populations that cannot hybridize. Levin and Kerster have demonstrated that color dimorphism between populations of species in the plant genus *Phlox* which do not hybridize has resulted from selection pressures preventing interspecific pollen flow. *Phlox pilosa* includes populations essentially all of whose members possess pigmented corollas, as well as populations in which individuals with white corollas predominate or are the only form. Pollen grains of the two species can be distinguished on the basis of large, discontinuous differences in pollen diameter ($P.\ glaberrima$, $\overline{X} = 55\ \mu$m; $P.\ pilosa$, $\overline{X} = 30\ \mu$m). It was shown that in areas of sympatry which include both population types of $P.\ pilosa$ and populations of $P.\ glaberrima$, 30 percent of the pigmented form of $P.\ pilosa$ bore the pollen of $P.\ glaberrima$ while only 12 percent of the white form bore pollen of this species. In addition, 4.8 times as many grains were found on the pigmented plants as were found on equal numbers of flowers of the white form. It was concluded that the color dimorphism aids pollinator discrimination and conserves reproductive potential.

Perhaps the most likely candidate for such a situation in which two closely related kinds of animals have become sympatric and have not required selection against hybrids to reinforce isolating mechanisms (as described below) is in the flycatcher genus $Empidonax$. Johnson has studied an area in eastern California, where two forms, $Empidonax\ wrightii$ (gray flycatcher) and $E.\ oberholseri$ (dusky flycatcher), are sympatric. It is possible that this sympatry is of very recent origin, the result of habitat changes brought on by logging operations in the middle of the last century. $Empidonax\ wrightii$ breeds in sagebrush or small trees and tends to forage in open areas, whereas $E.\ oberholseri$ is a forest-chaparral bird. Where they overlap in an area of mixed clearings and broken forest, the birds retain their habitat separation but defend their territories $interspecifically$. The two kinds are so similar that they can be distinguished with assurance only by careful wing measurements and wing-tail ratios. Johnson has been able to detect distinct differences in vocalizations, presumably concerned with pairing and pair-bond reinforcement, but the challenge calls and appearance are so similar that $E.\ oberholseri$ territories are defended against $E.\ wrightii$ and vice versa. In spite of this similarity, several dozen mated pairs collected by Johnson all showed positive assortment (no $oberholseri$-$wrightii$ pairings), indicating that the birds have no trouble in properly choosing mates. Perhaps the recognition signs (whatever they may be) were originally weaker and were reinforced by selection, but if this was the case, the process must have been completed very rapidly.

Limited gene exchange.   In a similar case of sympatry, presumably permitted by human disturbance of the environment, two kinds of Mexican towhees, $Pipilo\ erythrophthalmus$ and $P.\ ocai$, have come into contact and are now hybridizing. Previous ecological isolation ($P.\ erythrophthalmus$ mostly in oaks and brushy undergrowth, $P.\ ocai$

mostly in coniferous forest and associated undergrowth) seems to have broken down when lumbering and agriculture produced second-growth situations suitable to both forms. Indeed hybridization at one level or another is widespread among birds, e.g., gulls, ducks, grackles, grosbeaks, honeyeaters, birds of paradise, and is usually interpreted as occurring in zones of secondary contact. Whatever its interpretation, it is evident that differentiation of bird populations may be a much more complicated process than one would assume from the neat arrays of "species" and "subspecies" found in bird checklists.

Similarly, it has been shown rather clearly that in many groups of plants considerable genetic interchange is possible after the so-called specific level of differentiation has been reached. This may be entirely at the diploid level, or it may involve polyploidy and apomixis. Examples of the latter have been discussed in Chap. 9. Most instances of such hybridization fall into the category of what has been called **introgressive hybridization** by Edgar Anderson. The changes of an F_1 hybrid offspring crossing with another such hybrid in the early stages of hybridization are much less than the probability of crossing with one or the other of its parents. The hybrid derivatives are almost always intermediate with respect to their ecological requirements, just as they are intermediate with respect to morphological traits. Backcross individuals from one parent or another will thus be more likely to find an appropriate ecological niche than F_1 or F_2 individuals. The result is that genetic submergence of the two hybridizing entities is unlikely to occur. Rather, portions of the germ plasm of one species will infiltrate the genotype of the other. Variability of the parental types will be increased in the direction of the hybridizing entity, and the species or subspecies may be able to increase its range and to move into habitats previously unoccupied.

An interesting example of introgressive hybridization in the sunflower genus (*Helianthus*) in California has been studied by Heiser. *Helianthus annuus* is a common weedy species found in most of the United States; there is considerable evidence that it was introduced into California by Indians in relatively recent times. *Helianthus bolanderi* occurs in California in two races. One is almost completely restricted to serpentine soils, an ecological situation in which relatively few and specialized plants are found. The other subspecies is a weedy form. Heiser has shown that the weedy race of *H. bolanderi* probably originated from the introgression of genetic material of the widespread weedy *H. annuus* into the serpentine race of *H. bolanderi*. The details of how this came about need not concern us here, but by making the appropriate crosses the derived subspecies can be synthesized in the laboratory. It is interesting that the introgression has been reciprocal; in addition to creating a larger, weedier form of *H. bolanderi*, it has resulted in the formation of a smaller form of *H. annuus*, apparently with extended ecological amplitude.

Most species of plants seem to be strongly restricted ecologically. Where the ecological barriers are strong, genetic interchange does not often take place and the hybrids with their intermediate ecological preferences do not survive. But when the ecological barriers break down, as in habitats subject to erosion or disturbance by man or glaciers, for example, the intermediate types may suddenly find suitable habitats and become common. When crossing between two species does occur, the result usually is not the swamping of the original species but the enrichment of the variation of the parental forms. Indeed, this appears to be very common in many genera of perennial plants studied in the United States.

Selection against hybrids. It seems likely that when highly differentiated populations rejoin, selection operating against hybrid individuals usually reinforces factors tending to prevent hybridization. The exchange of genetic information between the isolates often becomes negligible. It is entirely possible that when more is known about the processes of differentiation, it will be discovered that hybridization between individuals of rejoining segregates is almost universal, in other words, that mechanisms preventing exchange of genetic material between differentiated forms usually arise only through relatively unsuccessful hybridization after sympatry has been reestablished. Selection against hybrids and the subsequent evolution of mechanisms which prevent hybridization is the basis of the phenomenon of character displacement, discussed earlier in this chapter. Mechanisms which tend to prevent hybridization are referred to as **prezygotic barriers** to crossing since they prevent the formation of zygotes.

An interesting experimental demonstration of this mechanism was obtained by Koopman, who synthesized artificial mixed populations of *Drosophila pseudoobscura* and *D. persimilis* and held them at low temperatures (16°C) at which sexual isolation between the two is at a low level: hybrids are formed more readily at low than at high temperatures. Under the experimental conditions, the hybrids were extremely unsuccessful, but Koopman intervened to produce complete failure of hybridization by removing all hybrid individuals before they could reproduce (hybrids were identified by genetic markers).

Over a period of several generations the proportion of hybrids formed showed a marked decrease, indicating a reinforcement of whatever factors were operating to prevent hybridization. Koopman was able to show that the isolating mechanism was at least in part sexual; i.e., males "preferred" to mate with females of their own kind.

In nature, *D. pseudoobscura* and *D. persimilis*, although occurring in the same geographic areas, presumably do not hybridize for two reasons. First, there is considerable ecological isolation, *D. persimilis* usually occurring higher in the mountains and preferring cooler, shadier spots than *D. pseudoobscura*. Sexual isolation must also play a part, for except at low temperatures newly captured flies show little tendency to hybridize. It is suspected that other undetected factors also help to keep the two entities apart in nature. In the experiments just described, the two known factors were removed by crowding the flies together at low temperatures. In a very short time the action of natural selection established a barrier that was at least partly sexual where one had not existed previously.

Patterns of differentiation

The Galápagos finches and African cichlids. The complexity of differentiation patterns must not be underestimated. A great many forces seem to have interacted to produce the complicated pattern observed today in the Galápagos finches. It is likely that the Geospizinae are all descended from a small flock (perhaps a single pair) of fringillid ancestors which accidentally reached the islands from the South American mainland. It is highly unlikely that the immigrants would have represented a large and random sample of the parental population; indeed, they were most likely a small and biased sample, containing only a restricted segregate of the parental gene pool. This sampling error means that the selective forces on the islands, even if they were similar to those of the mainland, would operate differentially on the island birds, because they are operating in a genetic environment quite different from that on the mainland. Remember that in discussing the operation of evolutionary forces a single locus cannot be considered in isolation; the effects of pressure on one locus will depend in part on the composition of the entire genotype. Sampling error and the resultant change in genetic environment that commonly occurs when new colonies are established have been described as the **founder principle.** It is, of course, a special case of genetic drift.

Once the finches were established on one island, it was only a matter of time before migrants reached others in the group. Populations on different islands, being subjected to different selection pressures and the effects of the founder principle, probably differentiated rapidly. When migrations and remigrations brought differentiated forms into contact, selection in many cases must have operated against the tendency to hybridize, as described above. Size and shape of the beak of Galápagos finches are related to the kinds of food eaten and also are used by the birds for identifying mates. Abundance of different food sources varies between the islands, and it is likely that selection caused some differentiation in the beaks and associated structures in isolation. When contact was reestablished, selection probably caused greater differentiation of these structures because they were important in recognition. Such differentiation also seems to have an additional advantage in reducing the types of individuals eating the same kind of food. Such selection for "reduced competition" obviously occurs, but its exact mode of operation is unclear.

The history of the *Haplochromis* swarm must have been similar to that of the Galápagos finches. Migrants from river systems colonized ancient Lake Victoria. Multiple colonizations (separated by appropriate time intervals) alone could account for the observed diversity. New arrivals found old immigrants already partially differentiated; selection against hybridization finished the job. In addition, repeated cycles of drying and inundation may have alternately fragmented and reunited segregates, permitting the mechanisms of differentiation to act. It is also possible that areas of different types of bottom, shoreline, water depth, etc., acted (and act today) as intrinsic barriers to the dispersal of the various forms within the lake.

Sibling butterfly species. The *Erebia tyndarus* group is another excellent example of the subtle interactions possible between differentiated forms. For instance, in spite of their different life-cycle adjustment to altitude, *E. cassioides* and *E. nivalis* seem to be strongly influenced by each other's presence or absence, the altitudinal restriction appearing where the two forms occur together. The present distributional picture seems to represent the results of differentiated populations interacting over a varied field. As with Darwin's finches, the exact nature of these interactions is difficult to specify. "Selection to avoid competition" is not an explanation in itself. It is necessary to know exactly how differential reproduction of genotypes came about.

Presumably there was some differentiation before meeting. For instance, when the populations ancestral to today's *E. nivalis* and *E. cassioides* populations came together, the *E. nivalis* may have, on an average, lived more toward the upper limit of the joint range and *E. cassioides*, on an average, nearer the lower limit. Then there must have been some advantage accruing to the *E. nivalis* genotypes that "preferred" the high altitude location and those *E. cassioides* which tended to remain low. Perhaps the relative scarcity of the other form reduced the possibility of wasting gametes through unsuccessful hybridization. Then again, maybe unlike larvae tended to cannibalize each other. Perhaps waste products of the *E. nivalis* larvae tended to inhibit the growth of *E. cassioides*, and/or vice versa.

The food requirements of the various isolates may have been too similar or the number of available niches too restricted to permit the sort of habitat specialization seen in Darwin's finches.

The result is that each form now occupies the areas to which it is best suited. The historical details will probably never be known. Some forms, perhaps, lacking the genetic variability to differentiate further, became extinct. This could happen when severe conditions in an area greatly reduced the food supply and larvae of another *Erebia* species proved to be much more efficient at utilizing the restricted food supply. There is a growing body of information indicating that population extinction is a common occurrence in butterflies. Some forms, e.g., *E. calcarius* and western *E. cassioides*, may never have met. It is interesting to note that in the *E. tyndarus* group hybrid sterility exists in crosses between distant relatives such as *E. iranica* ($n = 51$) and *E. calcarius* ($n = 8$) where there is no behavioral isolation. These forms have apparently never been in contact; thus the selective basis for the development of isolating mechanisms has never been present.

Differentiation of parasites.   Patterns of differentiation that involve strains preferring different foods or hosts are poorly understood. It has frequently been observed that, in the laboratory, strains of parasitic organisms may be successfully switched from the usual host organism to another by transfer of large numbers. For instance, the human louse (*Pediculus humanus*) can be converted into a rabbit louse. Large numbers are transferred, and the relatively few able to feed survive and reproduce. This process of selection eventually results in a strain that is happy on rabbits but unenthusiastic about men. (Individuals will feed on men, but a colony will not thrive.) A similar selective process may well be responsible for the transformation of head lice into body lice when the former are subjected to the normal environment of the latter. It has been suggested that the genetic structure of louse populations encourages a plasticity that makes the transformation in either direction relatively simple. Needless to say, this structure is itself undoubtedly a result of selection. The *Ascaris* may differ from the *Pediculus* principally in that the Ascari are not so protean genetically.

There can be little doubt about the selective advantage accruing to those cuckoos whose eggs closely match those of the host bird. Rates of desertion of eggs laid in the nests of the usual host are much lower than of those laid in the nests of unusual hosts. However, exactly how gentes develop and are maintained is still a mystery. It seems unlikely that gentes are genetically isolated from each other. There is no sign of differentiation among them *except* in the egg habitus; such differentiation in other characters might be expected if each gens were an isolated evolutionary unit. However, because only superficial characteristics are studied in many ornithological investigations, such differentiation may be present but undetected. On the other hand, gentes might be isolated from each other and kept similar in nonegg characters by winning combination selection. There seems, however, to be no mechanism that could prevent interchange of genetic information in places where several gentes occur together, especially in view of the rather large territories occupied by the males.

The resemblance between the cuckoo egg mimics and the series of mimetic polymorphs found in females of *Papilio dardanus* is largely superficial. The major difference lies in the necessity for the female cuckoos to lay their eggs in the nest of the proper host species. The choice of the proper nests by the female cuckoos may be explained by the phenomenon of imprinting. It is not sufficient for a population to be genetically structured so that a multiplicity of distinct forms is produced; the structure must be such that the proper egg type is laid by a bird which chooses a particular foster parent. Thus if the female cuckoo raised in the nest of a reed warbler were to be inseminated by a male of the gens parasitizing the white wagtail, her eggs might be better mimics of white wagtail eggs than of reed warbler eggs. She would continue to lay her eggs in the "wrong" nest, and in all probability her young would have a relatively low chance of survival. Unfortunately, nothing is known of the genetics of cuckoo egg color, and so the importance of the male contribution is conjectural.

It seems likely that in both cuckoo gentes and *Papilio dardanus* mimetic polymorphs differentiation may have taken place in isolation. When two differentiated isolates of the butterflies came into contact in areas where both models were present, disruptive selection established a polymorphic system (as described in Chap. 7). However, when two cuckoo gentes, each parasitizing a different host, came into contact in an area occupied by both hosts, another factor was probably added. In addition to disruptive selection (lower reproductive success of genotypes producing eggs intermediate between the hosts), there was probably also selection favoring behavioral patterns that encouraged positive assortative mating (pairing of individuals raised by the same foster parents). This seems to have been accomplished most successfully where the host forms have somewhat different ecological requirements, the male cuckoos remaining in the familiar habitat where they were reared and consorting with females of like background.

In places where the habitats of the host forms tend to overlap, intermediate-type eggs are often laid and the mimicry breaks down. The gentes of cuckoos seem to be somewhat intermediate between polymorphic forms and what might be called host races. The degree of perfection of the mimicry apparently depends on the degree of ecological isolation enjoyed by the gens. Where gentes occur together in the same general area, there is no sign of enough genetic differentiation to result in reproductive incompatibility. There is every indication that such gentes are *not* incipient species and that divergence at the species level occurs in geographic isolation. Needless to say, the exact status of the cuckoo gentes deserves much additional study.

Catastrophic selection

An interesting pattern of population differentiation involving chromosome rearrangements is *Clarkia franciscana*, studied by Lewis and Raven. This largely self-pollinated plant is restricted to an area of serpentine rock in the city of San Francisco, which is within the geographic range of the closely related *C. rubicunda*. Apparently *C. franciscana* is also related to *C. amoena*, a northern species. Studies of meiosis in hybrids show that *C. franciscana* differs from *C. amoena* by at least two translocations and two paracentric inversions. It differs from *C. rubicunda* by at least three translocations and four inversions (Fig. 10-18). *Clarkia amoena* and *C. rubicunda* differ by at least three translocations and two inversions. Lewis and Raven have concluded that *C. franciscana* originated relatively recently from *C. rubicunda* and that it may have been the result of a rapid repatterning of the chromosome set, producing many differences in a relatively short time. It also seems likely that *C. rubicunda* may have arisen in a similar manner from *C. amoena* at an earlier time.

Lewis has speculated that chromosomal repatterning, as described above, may be the result of environmental stress, such as drought or a mutator genotype which simultaneously produces a number of gene or chromosome mutations. If after the occurrence of the structural reorganization of the chromosomes the plant for any reason is spatially isolated from the parental population, it can establish a new population; perhaps in an environment unsuitable for the original population. Lewis has called this pattern of population differentiation **catastrophic selection.**

As a result of electrophoretic analysis of eight enzyme systems in *C. franciscana* and *C. rubicunda*, Gottlieb has demonstrated that in six of the eight systems *C. franciscana* has become fixed for an allele not found in *C. rubicunda*. There are two possible interpretations for these remarkable genetic differences between what have been pre-

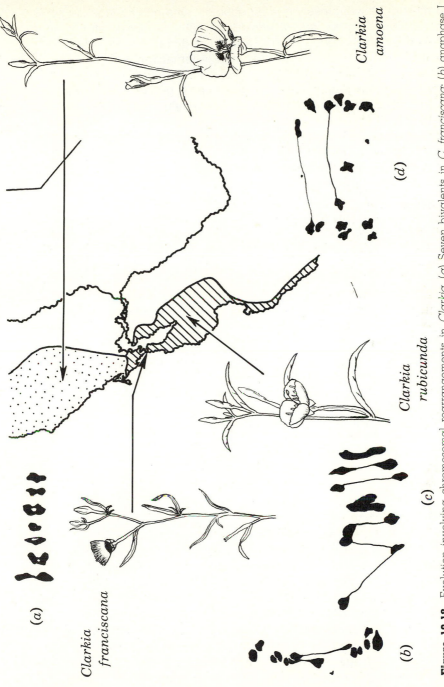

Figure 10-18 Evolution involving chromosomal rearrangements in *Clarkia*. (*a*) Seven bivalents in *C. franciscana*; (*b*) anaphase I showing two bridges and two fragments in cross between *C. rubicunda* and *C. franciscana*; (*c*) anaphase I showing chain of five chromosomes, a chain of three chromosomes, and three bivalents in cross between *C. rubicunda* and *C. franciscana*; (*d*) anaphase I showing two bridges, two fragments, and a lagging chromosome in a cross between *C. amoena* and *C. rubicunda*. Map shows distribution of three species in San Francisco Bay area. [*After H. Lewis and P. H. Raven, Evolution, vol. 12 (1958), Brittonia, vol. 10 (1958); and H. Lewis and M. E. Lewis, Univ. Calif. Publ. Bot., vol. 20 (1955).*]

sumed to be two closely related species. One explanation is that *C. franciscana* was not in fact derived from *C. rubicunda* but instead is the last surviving population of a nearly extinct species ancestral to both *C. franciscana* and *C. rubicunda*. Alternatively it may be postulated that *C. franciscana* was derived from *C. rubicunda* by rapid speciation but farther back in time than suggested by Lewis and Raven. The genetic differences would therefore have accumulated subsequent to the cytogenetic events which lead to the formation of the new species. There is, of course, no way in which either hypothesis can be tested.

Gottlieb has proposed that if a species is accepted as having originated in a rapid series of events involving chromosomal reorganization, it must be substantially similar to its progenitor upon electrophoretic examination. Gottlieb has shown that these criteria are met in *Stephanomeria exigua* subsp. *coronaria* and its derivative species *S. malhurensis*, two species in the sunflower family (Compositae).

Allopatric speciation

When differentiation of physically isolated populations goes to the point that reunion of the populations does not occur if contact is reestablished, it is known as **allopatric speciation.** An abundance of evidence suggests that allopatric speciation is an important cause of organic diversity. It is the splitting mechanism which, coupled with extinction, has been employed to explain the large numbers of relatively distinct kinds of plants, animals, and microorganisms that populate the earth.

Little is known about the time required for populations to differentiate. In any given case, many variables would affect the required time span, including population size, magnitude of selection pressures, degree of isolation, and the genetic system of the organism. Usually speciation seems to be a much more drawn-out process than could be conveniently observed by the evolutionist. However, much recent work, e.g., that on industrial melanic moths and on *Cepaea* and *Natrix*, indicates that selection pressures in nature may be generally higher than once thought; if this is the case, speciation may also occur more rapidly than has been assumed in the past. At any rate, the evidence for the view that allopatric speciation is a primary splitting mechanism in evolution is not direct observation. The presence of patterns of variation that seemingly represent every conceivable stage in the postulated process is the actual basis for this view. Some of these have been discussed in this chapter; the biological literature is replete with others.

Sympatric speciation

Can distinct new kinds of organisms arise in the absence of physical isolation of populations? The answer is certainly yes in the case of alloploidy (discussed in Chap. 9) and for organisms in which sexual processes are absent. Each individual of completely asexual organisms, e.g., some rotifers, is a genetic isolate, and species are clusters of clones that owe their similarity to interclone selection.

One of the enduring controversies among evolutionary theorists concerns the possibility of sympatric speciation in sexual organisms. One argument against sympatric speciation has been that hybridization will swamp out any differences produced by a disruptive selection pressure. However, some provocative work by Thoday and his coworkers and by Streams and Pimentel indicates that this is not necessarily so. In laboratory experiments with *Drosophila* these workers have shown that disruptive selection can produce divergence in the absence of isolation. For instance, Thoday and Gibson subjected a wild-type population to disruptive selection for chaeta number, with both high- and low-selected individuals being placed in a common vial for mating. At the end of 12 generations the original population had split into two populations, which produced few hybrids.

Bush has suggested that the widespread (northeastern and central United States) race of *Rhagoletis pomonella*, an insect which is found on apple trees, was derived from the hawthorn-infesting race of this species. The new race most likely originated in the Hudson River Valley, where it was first reported and where both apples and hawthorns commonly occur. There is a positive correlation between host plant and mate selection in these races. It is believed that host selection has a genetic basis and represents minor genetic changes. Bush has argued that initial differences when reinforced by factors such as different times of emergence on plants with different fruiting times, disruptive selection, conditioning, and semigeographic isolation have led to the sympatric evolution of these races.

Further, Ghent and Wallace have presented evidence that temporal factors may lead to sympatric speciation. Among the sawflies, each of which tends to be associated with one typical host tree, there are several pairs of species on the same host. One overwinters in the egg stage following emergence and mates in the fall. The other overwinters in the pupal stage and then emerges, mates, and oviposits in the spring. Here a simple change—possibly of a single gene—affecting the onset of diapause can produce complete temporal isolation into two groups. The larvae of these species pairs are often found feeding together in common clusters.

The extent to which sympatric speciation has occurred in vertebrates is unknown. As Ghent has pointed out, the actual physical separation of a population into two segments is a positive event, for which actual positive evidence may be found. To prove that a gene pool has never been so divided is a logical impossibility. Field studies will inevitably uncover positive evidence of actual events much more frequently than they will ever find any kind of convincing evidence that something never happened, i.e., a gene pool was never divided. There can be little question that additional data are needed to resolve the sympatric-speciation controversy.

SUMMARY

Available evidence indicates that differentiation in isolation is the major source of organic diversity. Populations physically separated from each other (so that gene flow is minimized or absent) have different evolutionary "experiences" and thus differentiate genetically. If this process proceeds beyond a certain point, the populations will not reunite if contact is once again established. The investigation of this genetic "point of no reunion" and the development of generalizations concerning it are among the most difficult problems confronting evolutionists. Populations that have become so distinct that there is little chance of merger in the future may still exchange genetic information, reciprocally obtaining variation that may stimulate further evolution. It should not be assumed, however, that the "purpose" of evolution is to create diversity or that occasional genetic interchange between nearly isolated groups is in any sense bad or aberrant.

REFERENCES

Bowman, R. I.: Morphological Differentiation and Adaptation in the Galápagos Finches, *Univ. Calif. Publ. Zool.*, vol. 58 (1961). The comprehensive source on the Geospizinae. Includes extensive references to the literature, including Lack's classic 1947 monograph.

Brown, W. L., and E. O. Wilson: Character Displacement, *Syst. Zool.*, **5:**49–65 (1956). A good series of examples of the phenomenon.

Clausen, J.: *Stages in the Evolution of Plant Species*, Cornell University Press, Ithaca, N.Y., 1951. A brief survey of the classic studies of Clausen, Keck, and Hiesey on ecotypic differentiation in plants.

Dobzhansky, T.: *Genetics of the Evolutionary Process*, Columbia University Press, New York, 1970. A fine exposition of the classical view of the differentiation of populations.

Ehrlich, P. R., and P. H. Raven: Differentiation of Populations, *Science*,

165:1228–1232 (1969). A discussion of the role of gene flow in the differentiation of populations, evaluated in the light of current data.

Mayr, E.: *Animal Species and Evolution*, Harvard University Press, Cambridge, Mass., 1963. This scholarly and exhaustive treatise supersedes the author's earlier classic, *Systematics and the Origin of Species*, Columbia University Press, New York, 1942. It will long remain the source book on speciation in animals.

Sibley, C. G. (chairman): Systematic Biology, *Nat. Acad. Sci. Publ.,* 1969. Many examples of differentiation of populations are discussed in these proceedings of an international conference.

Solbrig, O. T.: *Principles and Methods of Plant Biosystematics*, Macmillan, New York, 1970. Many of the topics covered in this chapter are further discussed in this book.

Stebbins, G. L.: *Variation and Evolution in Plants*, Columbia University Press, New York, 1950. Contains many examples of differentiation in plants.

Thoday, J. M., and J. B. Gibson: Isolation by Disruptive Selection, *Nature*, **193:**1164–1166 (1962). This paper and Streams and Pimentel, *Am. Nat.*, **95:**201–210 (1961) are critical references on sympatric divergence in the laboratory.

Woodson, Robert E.: Butterflyweed Revisited, *Evolution*, **16:**168–185 (1962). The resampling of a transect made in 1947 across the United States is described and the results evaluated. Many papers on differentiation of populations are published in *Evolution*, and a survey of the back numbers will be of interest to any student of the subject.

Can the processes that account for the differentiation of populations be the same ones that are responsible for the great diversity of life? The efficacy of mutation, selection, and drift in producing different colors of *Linanthus* flowers, geographic variants among butterflies, or species of birds has been described in the preceding chapter. Now the question may be asked: Is the same constellation of factors also responsible for the differences between flowers, butterflies, and birds? Are these factors responsible for the existence of extremely distinct clusters as well as for those separated by relatively small gaps? Because of the primarily taxonomic orientation of early evolutionary studies, this is often considered to be the problem of the origin of higher taxonomic categories. Some paleontologists and geneticists have felt that higher categories, such as genera, families, and orders, may have resulted from evolutionary processes different from those studied at the species level. It seems clear from the evidence from many fields of biology that the immense amount of time during which evolution has been taking place obviates any need to postulate other processes than those previously discussed.

EXTINCTION AND BIOGEOGRAPHIC PROVINCIALISM

The existence of extremely distinct clusters can be accounted for by extinction or by inadequate geographic sampling. With possibly very rare exceptions, e.g., certain fish species whose entire populations seem to occur in single small springs, even the most distinct clusters are made up of smaller subclusters with some degree of variation among them. If all the near relatives of a group of organisms become extinct in the course of long periods of time, then obviously that group will appear very distinct. Should fossil forms be discovered, the distinctiveness of the group will be less apparent. On the other hand, a newly arisen phylogenetic lineage may evolve in isolation over long periods of time, becoming more distinctive as it becomes more specialized and diversified. The birds as a group are a good example of this phenomenon. The term biogeographic provincialism speaks for itself. What appear to be distinct clusters may prove to be members of a continuum when adequate samples are obtained from all parts of the geographic range.

Extinction

The sole surviving member of the reptilian order Rhynchocephalia, the tuatara (*Sphenodon punctatus*), is found only on about a dozen islets off the coast of New Zealand. The groups of individuals on different islands certainly belong to different mendelian populations, but the degree of genetic divergence among these populations (and the amount of interchange between them) is unknown. Other animal isolates of this sort are numerous. The strange Peruvian butterfly *Styx infernalis* is a member of the family Lycaenidae (related to our common blues and hairstreaks), but its many structural peculiarities clearly set it apart from other lycaenids, and it is placed in a separate subfamily. Again, nothing is known of the degree of differentiation that may exist within the cluster.

If his fossil record is ignored, man is a very distinct organism. There is complete reproductive isolation between *Homo sapiens* and his nearest living relatives (the anthropoid apes). As far as is known, differentiation within the human species has not progressed to the point where segregates within the species are infertile upon crossing. However, in contrast to *Sphenodon* and *Styx*, there is a great deal of geographic variation within *Homo sapiens*. Our concepts of "race" are based primarily on variation in a few conspicuous external characters (skin color, hair type, skull shape), but there is also variation in less obvious characters, notably blood type and hemo-globin type (Chap. 7). The *Homo sapiens* cluster therefore does not fragment easily into distinct subclusters. Virtually all subgroups of man exchange genes to some extent, with the result that patterns of variation are exceedingly complex. Indeed, discordant variation in which patterns for the various characters studied are widely different is very common. A good example of this discordant variation can be seen in the comparison of distributions of blood types in human populations (Fig. 11-1). The distributions of blood-group genes *A* and *B* show little resemblance.

Examples of very distinct forms are also found commonly in the

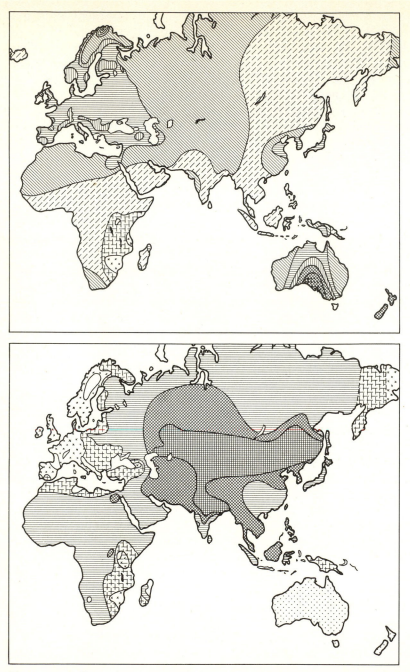

Figure 11-1 Discordant variation in the distribution of blood-group genes (Above, gene *A*; below, gene *B*) in aboriginal human populations. Degree of shading indicates area occupied by populations with approximately the same frequency for the gene in question. (*From A. E. Mourant, The Distribution of Human Blood Groups, Blackwell Scientific Publications, Ltd., 1954, F. A. Davis Company, Philadelphia.*)

plant kingdom. *Ginkgo biloba*, the maidenhair tree, is, like *Sphenodon*, the only living member of its group, a very distinct order of gymnosperms. The phyletic line to which it belongs can be traced far back into the fossil record. During the Mesozoic there were many genera and species of Ginkgoales (Fig. 11-2). For some reason unknown to us, the line became extinct in the Tertiary, with the exception of *Ginkgo biloba*. The exact native habitat of the ginkgo is not known, but for centuries the Chinese have cultivated the species for its edible seeds. In China most of the plants are grown in temple gardens. Since the plant has virtually no insect or fungal pests, and since its leaves turn a striking gold color in autumn, ginkgo has become a common and popular street tree in most temperate parts of the world. It seems to survive in areas where the worst environmental pollutions which man can produce are abundant. *Ginkgo*, like many other isolated gymnosperms, is very diverse genetically, having many horticultural variants.

The dawn redwood, *Metasequoia glyptostroboides*, also apparently is the last survivor of a genus of gymnosperms related to *Sequoia* and *Sequoiadendron*. Unlike *Ginkgo*, *Metasequoia* was first described as a fossil. It was not until some 5 years after its description from leaf impressions in Pliocene deposits in Japan that living plants were found growing in China. Again, this tree apparently has been preserved from extinction at least in part through cultivation by the Chinese.

In addition to such instances of completely isolated species, there are also cases of two or only a few closely related forms which, as a group, are clearly separated from their nearest living relatives. In the butterfly genus *Neophasia* (Pieridae) are found what are considered by all lepidopterists to be two distinct species with no intermediates and little intraspecific variation. One species, *N. menapia*, is widespread in western North America, where it feeds on various coniferous trees. Both sexes are white, with black markings restricted largely to the leading edge and apex of the forewings. *Neophasia terlooti* is found in northern Mexico and Arizona; it also feeds on conifers. Males are similar to those of *N. menapia*, but the dark markings on the forewings are more extensive. The females resemble the males, except that the white ground color has been replaced by orange-red. No red females of *N. menapia* are known, and no white females of *N. terlooti* are found.

Living elephants are an interesting example of two very distinct species as the only survivors of a diversified group. Thus the African elephant (*Loxodonta africana*) clearly is the closest living relative of the Indian elephant (*Elephas maximus*) and vice versa. Both, however, have closer relatives among known extinct forms not directly ancestral to them.

Figure 11-2 *Ginkgo biloba* and extinct relatives. Center, *Ginkgo biloba*; upper left, *Sphenobaiera furcata*, late Triassic; upper right, *Ginkoites lunzensis*, Triassic; lower left, *Baiera longifolia*, Jurassic; lower right, *Ginkgo lepida*, Jurassic. [Adapted from H. N. Andrews, *Studies in Paleobotany, John Wiley and Sons, Inc., 1961, and after R. Kräusel, Paleontographica, vol. 87B (1947), and after O. Heer.]*

Detailed studies of the closed-cone pines of western North America have revealed a similar situation among these plants. *Pinus masoni* is known only from the Pliocene. From study of the abundant fossils, it has been suggested that *P. masoni* became extinct in the Pleistocene after giving rise to two separate lines of development. One of these produced *P. linguiformis*, which also became extinct in the Pleistocene. From *P. linguiformis* arose *P. attenuata*, which has a fossil record extending well back into the Pleistocene and which is today the commonest species of closed-cone pine in California. The second phyletic line deriving from *P. masoni* gave rise to *P. muricata*. This species survives today and is abundant along the Pacific Coast, together with its derivative *P. radiata*. If only modern pines are considered, *P. attenuata* and *P. muricata* are obviously closely related morphologically. Both, however, have close relatives in the fossil forms that preceded and overlapped them in time, only to become extinct during the Pleistocene. These two pines appear to be vigorous expanding species with considerable genetic variability. *Pinus radiata* has suffered range reduction in recent times and perhaps would be on the road to extinction were it not for the fact that the tree is much cultivated and has been introduced into many parts of the world where it is grown for paper pulp.

The situations just described, where a species or small group of species appears to be very distinct from other such groups, all involve **extinction.** The group is distinct because closely related forms have not survived. When the fossil record is considered (if one is available), the distinctness of the species disappears. Similar misconceptions result when forms living in a rather broad geographic area are studied in only a part of their range. Distinct clusters seem to be present when a limited geographic area is considered, but when the world picture is studied, this conclusion no longer obtains. As with the previous examples, the problem is one of sampling, in this instance insufficient sampling in space rather than insufficient sampling in time.

Biogeographic provincialism

In North America, four distinct kinds of animals are included in the butterfly genus *Nymphalis (N. antiopa, N. milberti, N. californica*, and *N. vaualbum*. Individuals may be assigned to one of the four forms with certainty, no intermediates having been discovered. When *Nymphalis* of the Palearctic region are considered as well, the impression of neat packages disappears. The relationship of the species that we call *N. vaualbum* to the Eurasian form of the same name is open to some question. In the few characters studied (size, color pattern, genitalia) in samples from North America and Eurasia there is considerable overlap in variation. As in most other cases, an estimate of the actual degree of differentiation must await thorough studies of large samples of characters. Similarly, the amount of differentiation between *N. milberti* and the Palearctic members of the *N. urticae* complex (Fig. 11-3) and among *N. californica* and the Old World *N. polychloros* and *N. xanthomelas* needs further investigation. There are myriad similar situations.

In Yellowstone National Park there are two kinds of bears: black bears (*Ursus americanus*) and grizzly bears (*U. horribilis*). If the Park were the entire range of the grizzly, then grizzly bears would form a distinct apparently uniform cluster of the same sort as that described for the tuatara (*Sphenodon*). However, the grizzly ranges northward to Alaska, where populations are made up of very large individuals (Alaska brown bears, *U. middendorffi*) and thence into Eurasia (*U. arctos*). It seems likely that these bears form an intergrading series of clusters. The polar bear (*U. maritimus*) is a very closely related cluster adjusted primarily to marine life in the Arctic. The brown and polar

Figure 11-3 Closely related butterflies of the genus *Nymphalis*. Upper left, *N. milberti*, Wyoming; upper right, *N. caschmirenis*, Nepal; lower left, *N. urticae*, Sakhalin; lower right *N. urticae*, Hungary.

bears are largely allopatric, although occasional individuals meet along the arctic coast. To our knowledge, intermediate individuals have not been reported in nature (although zoo hybrids are well known). Thus, although bear taxonomy might seem simple to the inhabitant of Yellowstone Park, consideration of the world picture introduces problems. Are the European bears really the "same species" as the Yellowstone bears? Are the polar bears really a different species? Are these questions biologically important?

This problem is also common in plants. As with animals, it is particularly evident in forms that have a circumpolar or circumboreal distribution. Often the same species may masquerade under different names in Canada, Europe, and Russia until monographic study, bringing together specimens from all areas for the first time, shows the true distribution. It may seem unnecessary to belabor the point in such detail here, but biogeographers often accept the results of local floristic and faunistic studies without considering this particular source of bias.

Reticulate variation

A rather different situation, and one not well understood, occurs in various groups of organisms. Here a species is well marked and easily distinguished. There appears to be little or no morphological intergradation so that each cluster is clear-cut. In the milkweed family (Asclepiadaceae), for example, once the rather complicated floral morphology is understood, a series of nonoverlapping taxa (species) is recognized easily. It is interesting that in this family of flowering plants it is very difficult to delimit clusters of these primary taxa. The pattern of variation is such that genera are virtually impossible to define. The condition in which characters considered important by the taxonomist appear in different combinations in many species is known as **reticulate variation.** Perhaps this is a result of the complex obligate insect-pollination mechanisms found in these plants, but it seems to be common at the species level in various groups of plants and animals.

Reticulate variation must be distinguished from reticulate evolution, which is the joining of phyletic lines (through crossing). It has been demonstrated clearly in many plants, where it usually involves doubling of the chromosome number of the hybrid. Instances of this sort in both plants and animals are discussed in Chap. 9. In the Asclepiadaceae, there is no reason to suspect that union of formerly separate phyletic lines has taken place, and there is no indication that polyploidy has played a role in species formation.

Niche width and variation within taxa

One area in which the Hutchinsonian school of ecology (chap. 5) is contributing to our understanding of evolutionary patterns is in developing models to explain the degrees of difference found between related species. For instance, groups of closely related species presumed to be in competition for a resource (and using the same foraging strategy) are called guilds. Guilds of Caribbean coral-reef fish species of the predator family Serranidae (groupers, hamlets, sea bass) vary a great deal in size and tend to show rather sharp resource partitioning among the member species along the prey-size axis of their niche hypervolume. That is, different species of serranids tend to eat different-sized prey, large species specializing in larger prey, small species in smaller prey. In contrast, guilds of species of the herbivorous families Scaridae (parrot fishes) and Acanthuridae (surgeon fishes) all seem to graze on the same benthic algae. These and similar situations raise a number of important evolutionary questions.

1. Why should some guilds show the kind of partitioning characterized by the serranids and others the close "packing" typified by the scarids and acantharids?
2. Is there a difference in the "invasibility" of such guilds? That is, is there a difference in how easily an immigrant species can become a member of the guild? If so, how is such a difference to be explained?
3. What if any are the limits to phenetic similarity imposed by natural selection on members of closely packed guilds? What are the limits to phenetic differences imposed on members of partitioning guilds?
4. What patterns of interspecies nearest-neighbor distances along a niche axis might be predicted for guilds of varying degrees of species packing?

MacArthur, Levins, Roughgarden, and others have developed mathematical models predicting the evolutionary characteristics of such guilds. The most recent mathematical model by Roughgarden predicts, among other things, that the closely packed scarid-acantharid guild would be more readily invaded than the serranid guild. If true, this should mean that speciation would be occurring more rapidly among scarids and acanthurids than among serranids (since there is more "niche room" for new species). There is, it turns out, some evidence that speciation is occurring more rapidly among fishes feeding at lower trophic levels than among predators. Furthermore, closely related serranids are more often allopatric than closely related scarids and acanthurids, bearing out another prediction of the model.

THE FOSSIL RECORD

In several senses, time is the crux of the problem of macroevolution as opposed to microevolution. The past history of life can be studied only by examination of fossils and comparison with extant forms. Opportunities for checking these observations and the conclusions drawn from them are few indeed. The discovery of "living fossils" such as *Latimeria* and *Metasequoia* provides such a check. *Latimeria* is a living representative of the coelacanth fishes, a group once thought to have become extinct in the Cretaceous. *Metasequoia*, the dawn redwood, was first described as a fossil from the Pliocene, only to be found extant in China some years later. Another check of our inferences drawn from comparing recent and fossil forms is the discovery of previously postulated "missing links." The most famous such find was perhaps Dubois' discovery of the remains of *Homo erectus* (*Pithecanthropus*), a fossil man with ape-like characteristics (see Chap. 12).

Fossils must be viewed as a very biased sample of the remains of past life. Disturbances in the earth's crust have all but wiped out critical parts of the fossil record, and the most interesting and vital earlier portion is the most distorted. Parts of some organisms, such as vertebrate bones, shells of mollusks and Foraminifera, woody portions of plants, and cuticles of leaves, are more readily fossilized than others. Organisms less well endowed with hard or resistant parts, e.g., worms, larval insects, most algae and fungi, are less often preserved. The habitat of the organism also has a large effect on the probability of its becoming a fossil. A steppe-dwelling animal or plant stood a much smaller chance of being preserved than one from a swamp. Factors favorable for the preservation of plant remains often are not suitable for animal preservation. For example, acid conditions that might preserve leaves would dissolve shells and bones. Finally, certain specialized habitats, such as tar pits, mineral water bogs, and volcanic slopes, may be preserved almost *in toto* at times with whatever assortment of organisms might happen to be present or attracted to them.

Nevertheless, the fossil record is sufficiently ample and diverse to enable paleontologists to recognize many patterns and describe processes thought to have been responsible for their development. Many analytical problems are peculiar to paleontology. For example, in addition to the taxonomic difficulties resulting from diversification across a heterogeneous environment (*horizontal speciation*), one must contend at the same time with differentiation through time (*vertical speciation*).

Modes of evolution

Simpson and others have summarized what is known of the major patterns of evolutionary change, particularly with respect to animals. Simpson divides these patterns into three modes. Many instances of the first mode—the **splitting** of phylogenetic lineages—may be traced in the fossil record. These range from the division of the archosaurian reptiles into terrestrial and aerial lines (represented now by the crocodiles and birds) to the formation of species of mammals. On the other hand, some lineages did not split for long periods of time but underwent such gradual morphological change that taxonomic distinction anywhere along the continuum is virtually impossible, except at the beginning and the end. This evolutionary mode is called **phyletic evolution.** Simpson recognized also a third mode, **quantum evolution,** which is rapid change involving the acquisition of characteristics for a completely different way of life. The partial or complete joining of phylogenetic lineages is difficult to recognize in the fossil record, although presumably it played the important role in the past that it does today, e.g., in alloploidy.

One of the most interesting problems that can be studied with the fossil record is the **rate** at which evolution has occurred. There are many different ways of measuring rate of evolution. Although none of them is entirely satisfactory, they still are of great interest. In many of the well-worked-out lineages of vertebrates and mollusks, rate of change in particular characters can be studied with time as a dimension. For example, the change in size of a particular organ can be studied over time. Such rates are called **morphological rates.** The unit of morphological rates is a **darwin,** which is a change in a dimension of 1 percent in 10,000 years. A good example of the study of morphological rates has been provided by Simpson for fossil horses. In two main lines of horse evolution there has been an increase in height of cheek-tooth crowns relative to their horizontal dimensions from the Eocene to the Pliocene. In neither line was the rate of change constant. Both show acceleration toward the end of the Oligocene and a deceleration in the late Miocene. The average rate of change in horse-tooth dimensions was about 40 millidarwins.

It is also possible to measure the rate of "formation" of taxonomic categories, assuming that these are proportional to some stage of morphological distinctness or some degree of genetic isolation. The rate of extinction of taxonomic groupings, as expressed in survivorship curves, also may be of use. Simpson has called them **taxonomic rates.** It must be remembered, of course, that comparisons based upon taxonomic distinctions between groups must be interpreted with great caution. There is ample evidence from the study of living and fossil forms that what is called a genus, say, in one group may not be equivalent "biologically" to what is so designated in another. Nevertheless, in many well-understood major groups, where the taxonomy is generally agreed upon, the **duration** of genera may be a valid measure of rates of evolutionary change, as discussed below.

The multivariate-correlation analyses now being employed in numerical taxonomy seem to offer real possibilities for the development of more satisfactory methods of measuring evolutionary rates. By studying changes through time in a large sample of characters, objective estimates of the total amount of structural divergence can be obtained. The smaller the degree of correlation between successive samples, the greater the amount of evolution. In such an analysis, arbitrary taxonomic judgments can be avoided.

The measurement of time is also beset with difficulties. While there are modern methods of dating which presumably give reliable results, there is no general agreement on the beginning or the duration of the major geologic periods. Radiometric age determinations have the same sampling problems as studies of fossils and must be correlated with the occurrence of known fossils. Experimental error may be relatively large. In general, establishment of an absolute time scale is most useful for rather long periods of time. For shorter spans of geologic time, it is often necessary to use a relative scale, placing organisms earlier or later in time, according to their relative positions in layers of sediment, the level of their pollen in varves, etc. For the most part, the evolutionist must be content with an approximate radiometric age for fossils, even though he may be almost certain of the sequence of appearance of a given series of forms.

Rates of evolution

Rates of phyletic evolution and of splitting vary greatly. Evolutionary rates are complex functions of many factors in the environment and in the organism and of interactions between these factors. Changes in genetic systems, the disappearance or appearance of predators or parasites, changes in temperature, increasing salinity of the sea, rise in sea level, mountain building, and change in intensity of solar radiation are a random sample of the factors that may act and interact to affect the rate of evolution.

If the rates of evolution for the various members of a major taxonomic grouping are plotted as a frequency diagram, they form an asymmetrical distribution with the modal rate nearer the maximum rate. Simpson has called this distribution **horotelic,** and the rates falling within it are horotelic rates. In most groups it will be found that some organisms apparently evolved at rates either slower or faster than those making up the horotelic distribution. These have been termed **bradytelic** and **tachytelic rates,** respectively.

The land carnivores and the pelecypod mollusks (bivalves) may be compared to illustrate horotelic rates. Simpson has calculated survivorship curves for these groups. Survivorship curves are a measure of the duration of a genus in the fossil record (Fig. 11-4, continuous lines). Plotted against time is the percentage of genera that survived a particular length of time. Here duration of a genus is used as a measure of rate of evolution, i.e., the turnover time of genera. The fact that the mean survivorship for genera of mollusks is 78 million years in comparison with 8 million years for a genus of mammals suggests that the carnivores have evolved perhaps 10 times faster than pelecypods, in terms of duration of arbitrarily established genera.

If it is postulated that correlation of survivorship and rate of evolution is perfect and negative, then a frequency distribution of rates can be plotted (Fig. 11-5). Along the abscissa the total range of rates for mollusks and carnivores is divided into deciles (tenths) so that the two groups can be compared even though the carnivores have faster absolute rates. The percent of genera in each decile is shown on the ordinate. The distributions of rates in the two groups are quite similar despite the differing absolute rates. These distributions show the standard or horotelic rates for the Pelecypoda and Carnivora.

If the pelecypods are now examined in greater detail, with survivorship of exclusively fossil genera compared with genera arising in the past and still existing, the phenomenon of **bradytely** is revealed. Again using Simpson's calculations and graph (Fig. 11-4), it can be seen that actual survival is greater than expected survival. *Some living genera have had distinctly longer spans than genera that have become extinct.* The longest-lived extinct genera lasted 275 million years, but extant genera that have survived for 400 million years are also known. According to Simpson, then, any living pelecypod genus older than about 250 million years represents a bradytelic genus. Its rate of evolution, measured in this fashion, is outside the horotelic distribution, toward the slow end of the scale.

One can easily think of many groups of plants and animals in which the rate of evolution appears to be very slow. It must be remembered, however, that bradytely is a statistical effect, detected

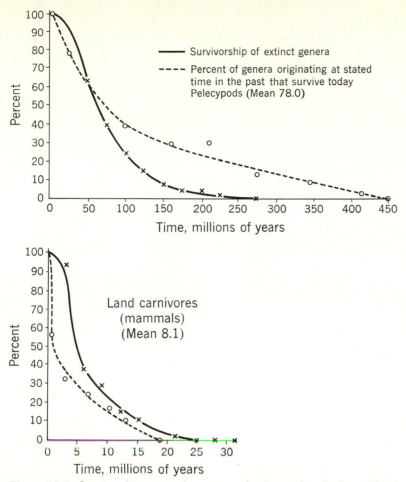

Figure 11-4 Survivorship curves for genera of pelecypod mollusks and land carnivores. Continuous lines, survival of genera now extinct; broken lines, survival of genera known as fossils that are still living. Ordinate, percent of genera surviving; abscissa (continuous lines), duration of extinct genera in the fossil record; abscissa (broken lines), time elapsed since appearance in fossil record of genera now living. (*From G. G. Simpson, The Major Features of Evolution, Columbia University Press, 1953.*)

by special means of analysis. So to label contemporary organisms that have changed little for long periods of time is thus not strictly correct. Organisms of many different kinds, such as opossums, giant sequoias, crocodiles, and club mosses, appear to have evolved very slowly. Whether they can be classified as bradytelic is another matter. Nor is it possible, with the present state of our knowledge of these or

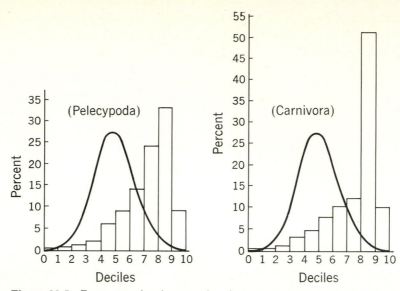

Figure 11-5 Frequency distributions of evolutionary rates in genera of pelecy-pod mollusks and land carnivores. Histograms are based on survivorship of extinct genera (see Fig. 11-4) and on the assumption of perfect negative correlation of survivorship and rates of evolution. Ordinate, percent of genera in each class; abscissa, rate of evolution expressed in deciles so that histograms can be compared despite differing absolute rates. Normal curves equal in area to the histograms are drawn for comparison. (*From G. G. Simpson, The Major Features of Evolution, Columbia University Press, 1953.*)

of groups such as the pelecypods, to determine why evolution appears to have been at least partly arrested. There is no reason to believe that such organisms are more "primitive," have a low mutation rate, or are depauperate genetically and less variable.

The method of study using survivorship curves, which makes bradytely apparent, cannot show the presence of rates faster than those of the horotelic distribution. Organisms with the very rapid rates of evolution referred to as tachytely exhibit this phenomenon presumably for relatively short periods of time and then slow to horotelic or bradytelic rates if they do not become extinct. Rapid evolution, or tachytely, often is associated with a new mode of existence, what Simpson has called a new **adaptive zone,** and is characteristic of quantum evolution. By adaptive zone is meant here not simply the place where the organisms live but the mode of life in such a place as well. It has been suggested that tachytely is one of the various possible reasons for the gaps in the fossil record which often occur at the apparent origination of a new phyletic line. If evolution took place at an unusually rapid rate, and particularly if the ancestors of the new line were distributed in partially isolated, small populations, the chances of their becoming fossilized would be very small. Simpson's view of a possible relationship between horotely, bradytely, and tachytely is shown in Fig. 11-6. Here a bradytelic line produces a tachytelic branch which becomes horotelic.

MAJOR EVOLUTIONARY PATTERNS

Adaptive Radiation

The birds perhaps represent a group that experienced an initial period of rapid evolution. They split rapidly from the archosaurian reptiles in the Mesozoic, apparently perfecting flight through selection for increased flight range and agility in a group of climbing-gliding reptiles. Once true flight was attained, its advantages (for escape, dispersal, food seeking, etc.) led to rapid proliferation of the group. Such proliferation is sometimes called **adaptive radiation.** Apparently there was sufficient overlap between the flying and gliding adaptive zones for the more successful birds to replace the group from which they arose. It is interesting to note that the gliding zone still remains largely unoccupied (except for a few fish, lizards, snakes, and mammals), although the mammals have successfully entered the

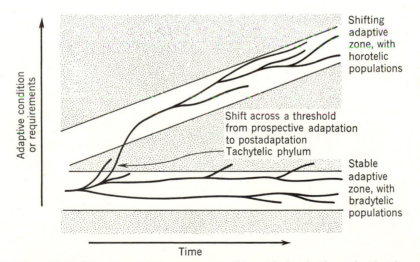

Figure 11-6 Diagrammatic representation of horotely, bradytely, and tachytely (see text). (*From G. G. Simpson, The Major Features of Evolution, Columbia University Press, 1953.*)

aerial zone (or a specialized part thereof), in part at least, by specializing in night flying and sonar hunting.

An outstanding example of adaptive radiation in animals is the tremendous diversity of the marsupials in Australia. In the absence of placental mammals, marsupials developed grazing forms (some kangaroos), burrowing forms (marsupial moles), forms resembling tree and flying squirrels (phalangers), rabbit-like forms (hare wallabies), wolf-like carnivores (Tasmanian wolves), badger-like carnivores (Tasmanian devil), ant-eating carnivores (banded anteater), and arboreal forms with no obvious placental equivalent (koalas, tree kangaroos). Unfortunately for the fields of comparative psychology, linguistics, and sociology, no marsupial equivalents of the primates appeared. The poisonous snakes of the family Elapidae (cobras and their relatives) similarly show adaptive radiation in Australia. They have produced, among other types, forms superficially resembling the vipers or pit vipers (Viperidae or Crotalidae). There are, of course, many other examples of adaptive radiation. The Galápagos finches and African lake cichlids are good examples of such radiation on a scale less spectacular than that of the Australian mammals and reptiles.

There are also instances of adaptive radiation by plants. Within most of the large plant families there has been radiation into trees, shrubs, lianas, and various types of herbs, including aquatics. A number of families have also independently produced saprophytic and insectivorous derivatives. Specialized types of adaptive radiation are concerned with particular mechanisms in plants. A remarkable series of pollination systems exists within the phlox family (Polemoniaceae); these are diagrammed in Fig. 11–7. The basic floral type with five corolla lobes, five stamens, and a superior ovary (usually with three stigma lobes) has become modified so that bee flowers, hummingbird flowers, butterfly flowers, hawkmoth flowers, bee fly flowers, beetle flowers, and even bat flowers occur. There are also flowers that are regularly self-pollinated (autogamous) and flowers that are self-pollinated in the bud (cleistogamous). It seems likely that selective forces leading to developmental modifications have resulted in this diversification which takes advantage of most types of available pollinators. Specialization in the Polemoniaceae has taken place at the generic and subgeneric taxonomic levels. In other angiosperms, very nearly an entire family may be specialized for one pollinating agent, usually wind, as in the grasses (Gramineae). In the almost exclusively wind-pollinated sedge family (Cyperaceae), there are one or more insect-pollinated forms, e.g., *Dichromena*, in which clusters of flowers and associated colored leaves together simulate large flowers.

These examples indicate that the concept of adaptive radiation has been applied to changes involving both relatively long and short periods of time. It may refer to diversification at what taxonomists would call the specific level, as well as to proliferation of phyla within very broad adaptive zones. It may even be used to describe the evolution of special relationships such as those between flowering plants and their pollinators. It is clear that the meaning of the term adaptive radiation must, in large part, be determined from the context in which it is employed.

Differing Rates of Evolution and Adaptive Zones

It is unwise to speculate too specifically about the causes of differing rates of evolution. Possibly organisms in ecological situations that change only slowly in the course of geologic time show bradytely. It has been suggested that the world's oceans represent such a situation, but we really know relatively much less about marine habitats than land ones. Other organisms in the same general habitat may not be bradytelic. Similarly, the "cause" of an instance of tachytely is usually impossible to determine. Entry into a new adaptive zone clearly is involved. It is almost impossible to avoid visualizing adaptive zones and thinking of discontinuities (nonadaptive zones) between them even though they represent an abstract space whose many dimensions are all the factors of the environment, including other organisms. In the past, as today, the situation must be exceedingly intricate, involving complexly (and often cyclically) shifting gradients.

In some way, it would appear that certain lineages gain access to a new adaptive zone that is constantly changing. This new zone must be contiguous in a multidimensional sense. The organisms must be physically near it, happen to be provided (by the mechanisms of population genetics) with genotypes that will survive and function in it, and be able to exist with the organisms already present or possibly to replace them. Having successfully entered an adaptive zone, the group may then diversify to occupy subzones of the original zone or spread further into contiguous zones. The amount of proliferation of adaptive subtypes seems to be primarily a function of the extent (diversity) of the adaptive zone. The Strepsiptera, curious, highly modified insects which are endoparasitic in bees, wasps, and other insects, seem to have entered a rather narrow adaptive zone. The birds apparently have found a relatively wide one. It is important also to remember that what today appears to be a relatively unstable (and relatively unoccupied) intermediate zone may at one time have been

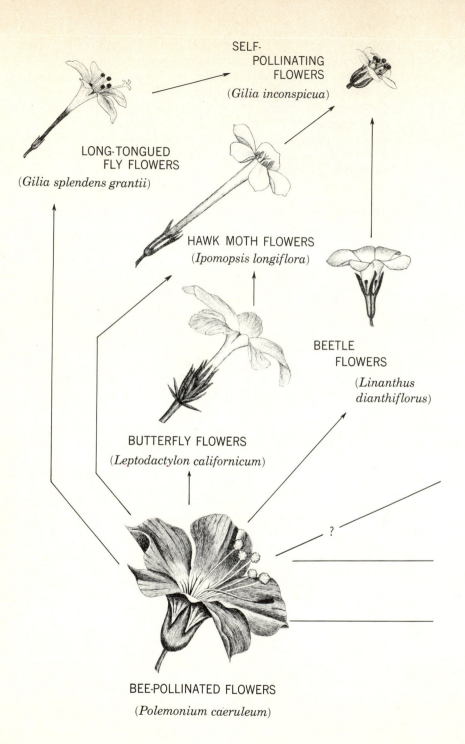

SELF-
POLLINATING
FLOWERS

(Gilia inconspicua)

LONG-TONGUED
FLY FLOWERS

(Gilia splendens grantii)

HAWK MOTH FLOWERS

(Ipomopsis longiflora)

BEETLE
FLOWERS

*(Linanthus
dianthiflorus)*

BUTTERFLY FLOWERS

(Leptodactylon californicum)

?

BEE-POLLINATED FLOWERS

(Polemonium caeruleum)

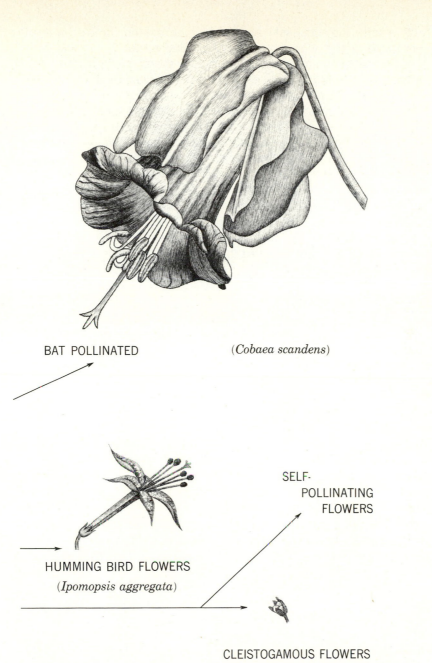

BAT POLLINATED   (*Cobaea scandens*)

SELF-
POLLINATING
FLOWERS

HUMMING BIRD FLOWERS
(*Ipomopsis aggregata*)

CLEISTOGAMOUS FLOWERS
(*Polemonium micranthum*)

Figure 11-7   Diversity of pollination types in the Polemoniaceae. Flowers exemplify structural types associated with the various systems of pollination; arrows indicate possible directions of evolution of pollination systems. (*After V. Grant.*)

quite stable. The successful entry of reptiles into the completely aerial zone quite possibly unstabilized the gliding zone.

As has been emphasized in Chap. 5, there is a strong tendency for biologists to fragment the environment to serve their own analytical purposes. While the concept adaptive zone refers to a way of life rather than a physical area, it nevertheless represents an artificial fragment of the space-time continuum in which the organisms are found. In the final analysis, it is the total situation that evolves, even though we may choose to separate organic evolution from environmental change. The results of our study depend to a very large extent on the choice of artificial distinctions. Science progresses by comparing the results of study of similar situations from different points of view until a relatively unvarying answer is obtained. It seems clear that the study of the evolution of organisms together with changes in "their habitat" has not reached the point where answers are unvarying.

Competition

The same problems arise with respect to the consequences of two organisms meeting in the same adaptive zone. It is usually thought, or at least stated, that the inevitable result is some sort of **competition.** The term has been used in so many different ways that its general unqualified use conceals rather than reveals information (see Chap. 13). Furthermore, with respect to plants, the term usually must have a meaning rather different from its possible meanings when applied to animals. By extension, the word competition has been used to refer to situations where major groupings of organisms have replaced others in adaptive zones of great breadth. Thus it is often said that placental mammals have eliminated, through competition, the marsupials and monotremes over much of the earth's surface or that mammals as a group have completely exterminated the mammal-like reptiles. The angiosperms appear to have replaced the gymnosperms in most parts of the world also; this too is described as the result of competition somehow extended to the level of higher taxonomic categories. In retrospect, one can look at any one of these situations and say that, by definition, the surviving organisms were "better adapted" than those which became extinct.

An example of competition has been reported by Jaeger. Two species of salamander are found on the north-facing slope of Hawksbill Mountain, Shenandoah National Park, Virginia. *Plethodon richmondi* primarily is limited to infrequent pockets of soil within areas of talus. *Plethodon cinereus* is found in the soil surrounding these talus areas. In another location, Blackrock Mountain, *P. cinereus* is found only in the soil surrounding the talus. *Plethodon richmondi* has not been observed on the mountain. Where the talus is disintegrating and soil erodes into the rocks, *P. cinereus* populations are expanding their territory at the expense of the *P. richmondi* populations. Jaeger has predicted that these populations, which apparently cannot compete with *P. cinereus,* will eventually become extinct because of the erosion of the talus which at present for unknown reasons prohibits the entry of *P. cinereus* and its subsequent expansion into these regions. Extinction of one group, however, is not the only possible outcome of "competitive occupation" of the same adaptive zone. One or both of the occupants may become more specialized, thus restricting or eliminating the interaction between the types. Cockroaches, at one time the dominant group of insects, have changed from what was once probably a generalized herbivorous or omnivorous zone and now make their living mainly as scavengers. Their former zone seems to have been highly "fragmented" among a great many insect specialists, particularly of the orders Hemiptera, Homoptera, Orthoptera, Coleoptera, Lepidoptera, Diptera, and Hymenoptera—mostly relative newcomers on the evolutionary scene.

Convergence

Sometimes organisms that are not closely related enter similar adaptive zones and as a result of selection come to bear a superficial resemblance to each other. This phenemenon is called evolutionary **convergence.** Classic examples are those of whales and fish and of desert succulent plants of various families. In the aquatic environment, the selective advantage of genotypes producing phenotypes with a certain type of streamlining is obvious. Desert plants (Fig. 11–8) often resemble each other in having a thick cuticle and low surface-volume ration, both of which reduce water loss; and they are often armed with spines which tend to discourage desert animals from eating them. The three plants of Fig. 11–8 belong to three separate families, Euphorbiaceae and Asclepiadaceae from the Old World, and Cactaceae from the New World. The spines in all three are derived from different structures.

Convergence of vegetation types has also been intensively studied. Mooney and Dunn have characterized the numerous selective forces which have been operative in the formation of the highly convergent vegetation in the Mediterranean climate regions of the world, i.e., California (*chaparral*), South Africa (*fynbosch*), central Chile (*matorral*), and southwestern Australia, as well as in the Mediterranean region (*maquis*). The climate is characterized by summer drought and winter rain. The vegetation has a similar

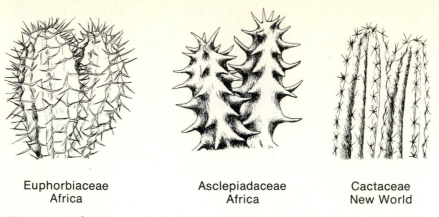

Euphorbiaceae	Asclepiadaceae	Cactaceae
Africa	Africa	New World

Figure 11-8 Convergence in desert plants. Left, *Euphorbia* (Euphorbiaceae), Africa; center, *Huernia* (Asclepiadaceae), Africa; right, *Cereus* (Cactaceae), Latin America.

appearance, a dense scrub dominated by woody evergreen species. The great species differences between these regions implies that they have had distinct evolutionary histories. On the other hand, the average similarity of vegetative form suggests that the same selective forces such as fire, drought, high temperature, rainfall unpredictability, and mineral deficiencies have been operative to limit the various combination of forces and thereby lead to convergence in form and function.

More complex are the various patterns of mimicry in which selection favors genotypes that are least likely to be destroyed by predators. An example of mimetic convergence is shown in Fig. 11–9. In this case, *Alcidis agathyrsus* (moth) and *Ideopsis daos* (butterfly) are presumably distasteful to predators, whereas *Papilio laglaizei* (butterfly) and *Cyclosia hestinioides* (moth) are thought to be palatable. Selection has apparently favored any genotypes within populations of the latter two that produced phenotypes resembling in any degree the protected forms.

Convergence among relatively closely related forms is sometimes called **parallelism.** For example, several lines of mammal-like reptiles probably independently acquired characteristics by which we define mammals. There is a continuum between convergence in the strict sense and parallelism, and the difference between the two is unimportant.

Coevolution

As has already been made clear, different groups of organisms which have close ecological relationships exert strong reciprocal evolutionary pressures on each other. This type of interaction has been called **coevolution** by Ehrlich and Raven, who studied the relationships between butterflies and their larval food plants. Plant-herbivore interactions are just one of many sets of coevolutionary interactions; others include the coevolution of predators and prey, parasites and hosts, mimics and models, and pollinators and plants.

People have long recognized that insects and other herbivores had evolved varied responses to the plants upon which they feed. The appearance, smell, or taste of the leaf, flower, or fruit very often provides a feeding stimulus for the herbivore, and its feeding apparatus and digestive tract are specialized for dealing with the appropriate plant materials. One need only think of the response of certain birds and men to brightly colored fruits or of the grinding teeth and special digestive system of ruminants to see that plants have had a profound evolutionary effect on the animals which devour them. The

Figure 11-9 Convergence presumably due to mimicry. Upper row: left, the moth *Alcidis agathyrsus*; right, the moth *Cyclosia hestinioides*. Lower row: left, the swallowtail butterfly *Papilio laglaizei*; right, the danaine butterfly *Ideopsis daos*. All from New Guinea. (*After R. C. Punnett, Mimicry in Butterflies, Cambridge University Press, 1915.*)

degree to which this relationship was reciprocal has not, however, been as widely recognized. Most biologists have been aware that spines on plants evolved as a defense against herbivores, but only recently has it been realized that many **secondary plant compounds** also evolved as defense mechanisms against plant predators (including fungi). These compounds include alkaloids, quinones, essential oils (including cyanogenic substances and saponins), flavonoids, and raphides (needle-like calcium oxalate crystals).

Long before man began to produce synthetic pesticides, he used some of these plant substances for the same purpose for which the plant evolved them—as insecticides. Examples are nicotine (from tobacco) and pyrethrum (from chrysanthemum-like plants). Many other of these substances are familiar as the active agents in spices.

Prominent among the plant biochemicals are the **alkaloids,** a chemically diverse group which includes nicotine, caffeine, quinine, and the active substances in marijuana and peyote. The hallucinogenic properties of some of these may well not be fortuitous. Herbivores with altered states of perception or reaction time undoubtedly are easy prey for carnivores, and there is reason to believe that the plants evolved molecules with such properties in response to herbivore selection pressure.

Of course, if an animal evolves the ability to detoxify alkaloids, then an unoccupied adaptive zone may become available to it. Kircher and his associates have studied the role that alkaloids play in the desert ecology of species of the genus *Drosophila*. The senita cactus (*Lophocereus schotti*) produces the alkaloids pilocereine and lophocereine which are toxic to eight species of *Drosophila*. On the other hand, these alkaloids have no effect on *Drosophila pachea*, whose sole breeding sites are the rotting stems of this cactus.

By studying the pattern of food-plant utilization by butterflies Ehrlich and Raven came to the conclusion that the development of secondary plant substances has been a major factor in the evolution of butterflies. For instance, the breaching of various plant biochemical defenses by butterfly groups has led to adaptive radiation on plant groups possessing those defenses. Examples are the diversification of one group of swallowtails on plants of the dutchman's pipe family Aristolochiaceae and the monarch butterflies on the milkweeds (Asclepiadaceae). In some cases the repellent substances in the course of the coevolutionary interaction have actually become attractants to the insects.

That herbivores can exert enormous selection pressure on plant populations has been demonstrated repeatedly. Larvae of a single small species of butterfly have been found to be capable of destroying almost 50 percent of the potential seed production of a lupine population, and weevils can kill up to 100 percent of the seed production of a large woody legume vine (*Dioclea megacarpa*) in Costa Rica. Formerly huge populations of *Opuntia* cactus in Australia were decimated by larvae of the moth *Cactoblastis cactorum*, and the cactus population is effectively controlled at a low level by the moth. Predators on seeds and seedlings clearly constitute the most potent agents of selection operating on plant populations.

The foliage of many acacias (trees of the Leguminosae) contain bitter-tasting chemicals which serve as protection against many groups of herbivores. Other species do not have such substances. Janzen has done a brilliant study of the swollen-thorn (bull-horn) acacias, which lack these chemicals. The function of the chemicals has been replaced by the activities of ants that live in the hollow swollen thorns and feed on Beltian bodies (specially modified leaflet tips) and nectar from foliar nectaries. The ants attack most herbivores and prune away competing plants, thereby serving as an effective defense system for the acacia. If the ants are killed with a nonpersistent insecticide, the acacias are killed by herbivorous insects. Since acacias without the ant association have bitter-tasting leaves, which the ant acacias do not, the protection provided by the ants apparently relaxes the selection pressures for the production of chemical defenses. This permits the plants to invest energy elsewhere, presumably in such things as Beltian bodies.

The potential complexity of coevolutionary interactions can be seen in the acacia-ant example, for in addition to the close relationship between the plant and its guardians these are probably coevolutionary interactions between the two species and various herbivores.

Perhaps the most complex known coevolutionary interactions are those involving **mimicry.** Two kinds of mimicry are generally recognized. In **Batesian mimicry** a harmless or tasty mimic organism superficially resembles a dangerous or distasteful model organism. This is a case of false advertising resulting from a selective advantage presumably occurring to individuals which bear some resemblance to the model. In **Müllerian mimicry** a series of dangerous or distasteful models resemble each another. This presumably evolved because few individuals were lost from a population of any member of the Müllerian complex in the course of training predators. An unpleasant experience with any model serves as a stimulus for avoiding them all.

Presumably, in any Batesian complex, the model and mimic are involved in a constant coevolutionary race. While it is to the mimic's advantage to evolve toward the model, it is to the model's advantage to move away from the mimic. After all, each pleasant experience a predator has with a mimic increases the threat to the model: it creates a credibility gap. The situation is even more complex because dangerous or distasteful models have usually evolved striking warning coloration (aposematic coloration) making them very conspicuous. It is this aposematic coloration which normally is copied by the mimic. But if mimicry becomes very close and the mimic becomes very common, most predators will have pleasant experiences with the conspicuous pattern and both model and mimic will suffer. This may lead to a partial divergence from the model pattern by the mimic (cases often described as imperfect mimicry).

Experimental evidence makes it clear that there is a continuum between Batesian and Müllerian mimicry. For instance few, if any, butterflies are totally tasteful. Whether or not a mimic moves towards the Müllerian end of the spectrum probably depends on its foodplant and the ease with which it can master the physiological problems of becoming poisonous. Indeed it seems likely that many butterflies not involved in mimicry complexes are sufficiently distasteful to be able to evolve directly into Müllerian complexes when appropriate partners become available. One might well wonder why all butterflies, indeed all organisms, are not evolving into one vast Müllerian complex. The answer would seem to be that the more species of prey that are recruited into a mimicry complex, the more profit accrues to the predator that evolves successful methods of attacking any or all members.

The complexity of coevolutionary interactions surrounding mimicry grows when it is realized that the members of the complex coevolve with their predators. For instance, selective advantages obviously accrue to those predators capable of discriminating model from mimic in Batesian complexes. In addition butterflies in mimicry complexes are also coevolving with their larval foodplants. Indeed in some cases the chemicals which make the butterflies distasteful to predators are picked up directly from the foodplant, which in turn had evolved them as defenses against herbivores.

Interesting coevolutionary interactions can also be traced in predator-prey, parasite-host, and pollinator-flower complexes. The principles are the same as those involved in plant herbivore and mimicry complexes. One such interesting coevolutionary relationship is hemiparasitism, well documented in the plant genus *Orthocarpus* in California. Atsatt has established that if plants from a single population of *O. purpurascens* are grown individually in pots under uniform garden conditions, less than 50 percent of the plants reach sexual maturity. However, when plants are grown together in clusters of three or, even more dramatically, with other species typically found in *O. purpurascens* environments, there is a significant increase in the survival rate. It has been suggested that survival-rate increase is the result of a genetic complementation. *Orthocarpus* plants readily form attachments to plants with which they grow. By this means the genetically impoverished hemiparasites obtain from the host plants physiologically essential materials that they themselves are incapable of producing.

Orthocarpus purpurascens is often associated as a hemiparasite with the composite plant *Hypochoeris glabra*, introduced from Europe. This dandelion-like plant has seed-like fruits with five spines, which aid in their dispersal (Fig. 11-10). The seeds of *Orthocarpus* develop a loose, net-like outer layer which frequently gets hooked upon the spines of the *Hypochoeris* fruits. Thus when dispersal of the host plant occurs, its hemiparsite is dispersed along with it. Native genera of composites with fruits similar to those of *Hypochoeris* occur

Figure 11-10 A fruit of the plant *Hypochoeris* showing the spines which serve for dispersal. Caught in the spines are two seeds of the hemiparasite *Orthocarpus*. (*Courtesy of P. Atsatt.*)

in California also. Undoubtedly hemiparasitism in *Orthocarpus* has had a selective impact on its hosts, but this remains to be documented.

An example of a coevolved predator-protection mechanism has recently been described by Ehrlich and Ehrlich in coral-reef fishes. Fishes of different species but of the same size, class, and general color pattern form resting schools over reefs in the daytime (Fig. 11-11). There is evidence that a fish in a large school is somewhat less vulnerable to predation than a lone fish. By schooling together, the different species achieve a larger school size than would be possible if each species schooled separately, and they therefore gain greater freedom from predation. Countervailing selective forces which prevent perfection of color-pattern mimicry may include the need for species-recognition patterns involved in reproduction. Different grunts, members of the genus *Haemulon* (family Pomadasidae) which school together are extremely similar yet can be distinguished

Figure 11-11 A portion of a heterotypic school of fishes on a coral reef at St. Croix, Virgin Islands. Three species are present: *Haemulon flavolineatum* (French grunt), with diagonal stripes on lower portion of body, *H. chrysargyreum* (smallmouth grunt), with even parallel stripes over entire body, and *Mulloidichthys martinicus* (yellow goatfish), with a single stripe running through the eye. (*Photograph by P. R. Ehrlich.*)

underwater by an experienced diver. They often school with goat-fishes (*Mulloidichthys*, family Mullidae) which appear very similar to the grunts from a distance of a few feet.

Using coevolution as an approach to community ecology should help clear up some areas of past disagreement. For instance, it is often said that the range of species *A* is limited by competition from species *B*. If this is true, then *A* and *B* probably make up a coevolutionary system. This means that questions about selection pressures must be posed, and field and laboratory investigations carried out to find the answers.

EVOLUTIONARY TRENDS

In retrospect one can observe evolutionary trends of virtually any desired degree of generality, ranging from a trend toward increased (or possibly now decreased) melanism in a British moth to a trend toward increasing complexity in the biosphere. Unfortunately such trends observed *a posteriori* have all too often been interpreted as if the evolutionary process were a means to a predetermined end. The fallacy of this view can be exposed with a very crude analogy. A driver leaves New York City and at each intersection takes the fork that seems to promise the easiest going based upon what he can observe at the intersection about the amount of traffic and condition of the road. At the end of a week of this whimsical travel he is found on the Indiana toll road. A teleologically oriented observer would doubtless claim that someone had planned a trip to Chicago for him. Readers interested in the sorts of trends that have been observed in the fossil record are referred to the works of Rensch and Simpson. These writers are highly successful in interpreting these trends in the light of the basic mechanisms of mutation, recombination, selection, and drift.

Increase in size

Two very general trends are considered briefly here and in the following section. In many phyletic lines of animals, there is a tendency toward increase in size. This has been called Cope's rule. The validity of this generality, in the broadest terms, can be question-ed only if one is willing to assume that the earliest living organisms were above or near the middle of the present size spectrum. From the discussion of the origin of life (Chap. 1) it will be seen that this is hardly likely. Possible selective advantages of being large can be found in virtually every physiological function of organisms: maximum exposure of leaves to photosynthesis (trees), heat conservation (whales), room for complex brains (pigs), or resistance to desiccation by a low surface-volume relation (barrel cacti).

The fossil record shows many trends toward larger size. A well-documented instance is that of the phylogeny of horses, with dog-sized Eocene horses giving rise to, among others, the contemporary work horse. The mammal record is replete with similar stories, although exceptions are known; e.g., modern marsupials and sloths are smaller than their Pleistocene counterparts.

Large size seems to have been selected for in some animals because of concomitant benefits derived through **allometry.** Allometry refers to differential growth of body parts. The horns and antlers of ruminants, for example, often show a strong positive allometry. As body size increases, they grow proportionately much larger. (The head of a postnatal human being grows less, proportionately, than the rest of the organism as a whole; it shows negative allometry.) Horns and antlers are important both in defense and in battles between males for access to females. It does not seem unreasonable, therefore, to assume that the large size of many ruminants is accounted for in part by selection for larger weapons. This is an example where selection for proportionately larger appendages may have produced, through positive allometry, a correlated increase in body size.

It is also possible that, in some cases, selection for large body size on the basis of increased efficiency of arrangement of internal organs, say, could have produced appendages overlarge for their supposed function. Such might be the case with the evolution of *Megaceros*, the Pleistocene Irish elk which had the largest antlers of any known deer. As long as the disadvantage in large antlers did not counterbalance the advantage of large overall size, the trend would continue. The discussion is perhaps overspeculative, but it shows that there is no reason to view the great size of these antlers (or the giant mandibles of stag beetles, or many other such instances) as the result of some sort of "momentum" that carried once-adaptive trends to nonadaptive extremes. As should be obvious from a consideration of population genetics (see Chap. 6), this would be a meaningless analogy. Selection against extremes would bring a trend to an abrupt halt as soon as the adverse effects counterbalanced the beneficial effects of the trend. As Simpson has pointed out, organisms that have extreme characteristics may seem bizarre to human eyes, but this is no reason to infer that they are inadaptive. Indeed, the thriving possessors of many bizarre features (male peacocks, bottlebrush weevils, narwhals, plants with intricate flowers or such complicated modes of pollination as pseudocopulation) hardly support the idea that these characteristics are liabilities.

Increase in complexity

Perhaps the broadest and most general trend that has been postulated is that toward increased complexity. Difficult as it is to specify level of organization within groups of organisms, few biologists would deny that an aardvark is more complex or highly organized than a coacervate droplet. A monarch butterfly is a more complicated apparatus than an amoeba or even a fern. Although relatively simple organisms still are present in some adaptive zones (more conspicuously perhaps in aquatic ones), the superiority of more highly organized types in many situations apparently is attested by their seeming dominance. However, dominance in an ecological situation is an exceedingly difficult thing to measure. Do the pine trees in a coniferous forest "dominate" the fungus that inhabits their roots and without which they could not survive? Do mammals "dominate" the bacteria and other organisms in their intestinal flora that synthesize vitamins necessary for their life? Indeed, careful study of any ecological situation reveals interactions among organisms that make comparisons of "superiority" and "dominance" exceedingly difficult.

Nevertheless an aardvark cannot be produced by amassing coacervate droplets. Many complex problems of structure and function had to be solved before DNA could use aardvarks or man to make more DNA. These problems apparently were solved slowly, one or a few at a time, over long periods of time. The solutions may well have spelled the doom of less efficient types, and the visible record of life becomes one of increasing complexity or organization. Some more simple organisms also have survived, presumably to become integrated with highly complex ones into complex ecosystems. The basic question remains unanswered: Why should DNA have evolved aardvarks and men for making more DNA when bacteria and other simpler organisms apparently can serve just as well? Perhaps the answer lies in the complexity of ecosystems whose cybernetic mechanisms result in great stability. But this is a difficult problem to study in the fossil record. A major evolutionary trend intimately associated with increase in complexity is the trend toward increased homeostasis in the individual, population, and ecosystem.

SUMMARY

There is no evidence to justify the assumption of an essential difference between the mechanisms that produce gene-frequency changes within populations and those which account for the differences between men and microbes. In view of the time available for the evolutionary process,

mutation, recombination, selection, and drift quite adequately account for the diversity of life. The discontinuities in variation patterns appear to be the result largely of extinction. The discontinuities in the taxonomic description of variation patterns reflect extinction plus generous sampling errors. Examination of the fossil record has resulted in a classification of the major patterns of phylogenetic change as splitting of lineages, phyletic evolution, and quantum evolution. Rates of evolution appear to have fluctuated widely in the course of time. In addition to the standard rates for a group, there often have been unusually slow and unusually rapid rates.

REFERENCES

Chambers, K. L. (ed.): *Biochemical Coevolution*, Oregon State University Press, Corvallis, Oregon, 1970. A series of symposium papers which cover several important aspects of coevolution.

Ehrlich, P. R., and P. H. Raven: Butterflies and Plants, *Sci. Am.*, **216**(6):104–113 (1967). Examples of plant and animal coevolution.

Raup, D. M., and S. M. Stanley: *Principles of Paleontology*, Freeman, San Francisco, 1970. A summary of current and classical views of patterns of evolution from the paleontologist's perspective.

Rensch, B.: *Evolution above the Species Level*, Columbia University Press, New York, 1960. An interesting analysis of evolutionary patterns, primarily in animals, with detailed consideration of trends in evolution and of the evolution of the nervous system.

Simpson, G. G.: *The Major Features of Evolution*, Columbia University Press, New York, 1953. The most thorough and general account for zoology of the patterns of phyletic change and their explanation in modern biological terms. Simpson's more recent chapter in Sol Tax (ed.), *Evolution after Darwin*, vol. 1, pp. 117–180, University of Chicago Press, Chicago, 1960, is an excellent brief consideration of the problems.

Stebbins, G. L.: *Variation and Evolution in Plants*, Columbia University Press, New York, 1950. The best source for an overall view with genetical orientation of the record of past plant life. The chapter The Evolution of Flowering Plants, by D. I. Axelrod in *Evolution after Darwin*, pp. 227–305, contains more recent work on the Angiospermae.

Weller, J. M.: *The Course of Evolution*, McGraw-Hill, New York, 1969. A comprehensive survey of the evolutionary history of the major plant and animal groups.

HUMAN EVOLUTION:
PHYSICAL AND CULTURAL

The phenomenon of man has been explained in many ways by man. Our view is that the theory explaining the evolution of other organisms is necessary and sufficient for man as well. Cultural evolution is an inevitable consequence of man's biological evolution.

In this section the evolution of man and of his culture is considered from various aspects. Chapter 12 places in biological perspective a very brief description of what is known of the evolutionary history of Homo sapiens. No special processes appear to be responsible for the origin of man, but with the development of culture and the extragenetic transmission of information, a complication appears: evolution within culture itself. The evolution of man now seems to be a resultant of the interactions between biological evolution, in the usual sense, and this psychosocial evolution.

One aspect of man's culture is his attempt to deal with the physical universe he perceives around him. His techniques for handling reality determine what reality he sees. Man's understanding of the process responsible for his coming into existence is an aspect of this more general problem and is discussed in the last chapter.

CHAPTER
TWELVE

THE
EVOLUTION
OF MAN

From the point of view of processes, the evolution of *Homo sapiens* is unique. The evolutionary forces described thus far have played and still play an extremely important part in human evolution. However, in addition to these forces, an entirely new kind of evolution, that of **culture,** has entered the picture. This *nongenetic body of information* is, like genetic information, transmitted from generation to generation. The evolution of all organisms except man depends with very minor exceptions, upon the information stored in the nucleotide code and upon its expansion and rearrangement through mutation and recombination. In man there is superimposed upon this a large body of extrinsic information which, at least in theory, is potentially available to all members of the species. Through the utilization and manipulation of this body of information, man has evolved devices (both mental and physical) that have given him a unique ability to modify his environment and, indeed, to influence the evolution of all organisms on the face of the earth.

MAN'S EVOLUTIONARY HISTORY

Man shares a vast inheritance with all mammals. Those interested in the long story of the attainment of mammality are urged to consult textbooks on vertebrate paleontology. The conquest of land by vertebrates and the eventual appearance during the age of the dinosaur of our inconspicuous warm-blooded ancestors with differentiated teeth and highly developed devices for fetal nourishment make a fascinating story. However, it is a story which sheds little additional light on the processes of evolution. The genotypes that, through recombination, provided more efficient tetrapodal locomotion, eggs resistant to desiccation, insulating body hair, metabolic control of body temperature, faster conducting neurons, and all the other "inventions" on the road to the status of mammal contributed their information differentially to the gene pool of subsequent generations. At some point early in mammalian history, quite likely in the Paleocene (70 million years ago), came the first evolutionary step that was to lead eventually to the differentiation of man from the rest of the mammals through the possession of culture.

We can only guess at the reason or reasons for this step; perhaps it was the presence of efficient terrestrial predators, perhaps an abundance of fruit. But, for whatever reason, in one group of mammals, individuals living in the branches of trees and shrubs started to leave more offspring than their cohorts who preferred a strictly terrestrial life. Ascent into the trees meant the penetration of an entirely new environment, in which the requirements for survival were strikingly different from those met by terrestrial animals. Many of the trends caused by selection for an arboreal creature need little explanation. Flexible grasping organs at the ends of the limbs are useful devices for remaining in trees. When leaping from branch to branch it obviously is necessary to be able to judge distances, and organisms with genotypes that tended toward good binocular vision were likely to reproduce themselves better than their less fortunate relatives. Those forms with genotypes in which the eyes rotated toward the front of the head tended to have binocular vision only if a large long snout did not interfere. Both binocular vision and grasping hands and feet lessened the need for a long snout for investigation and manipulation. Shortening of the snout and the resultant reduction of olfactory membranes would be a handicap for a ground-dwelling animal, and genotypes having it would probably be selected against. However, in the treetops the loss of sense of smell was less serious and was more than compensated for by improved vision.

Stereoscopic vision in itself would be of little use without the neural mechanism necessary to evaluate the sensory input and translate it into highly coordinated voluntary movements. Thus, in the arboreal primates, selection resulted in a trend toward high development of the cerebral cortex as a center for evaluation of sensory input and the formulation and initiation of responses to the environmental stimuli received. Because of the arboreal habitat, sight and touch came to override smell and hearing as sources of information.

Living in trees presents some serious problems in the care of offspring. In primates this gave a selective advantage to individuals that had smaller litters but gave them a high level of care. In many mammals, sexual activity is confined to a single season of the year, and the young are born at a time when a suitable food supply is available. In the tropical-forest environment of our distant ancestors, the food supply of fruits and insects was presumably relatively more constant than in temperate zones. Thus there was probably little selection in favor of a single period of sexual activity. In the absence of this factor, selection favored development of social bonds and year-round sexuality important to the continuing presence of males, the establishment of the family group, and increased protection for the smaller litter of helpless young.

Man is not the only descendant of the shrew-like animals that originally invaded the trees. Baboons have returned to a largely terrestrial life, while chimpanzees and gorillas spend most, but not all, of their time on the ground. Remaining in the arboreal habitat is an array of forms including such diverse primates as orangutans and gibbons, as well as monkeys, marmosets, tarsiers, and tree shrews.

It is important that our early ancestors lived in the trees, but it is also very important that they left them. It is difficult to see how a culture even vaguely resembling ours could have been developed by tree-dwelling organisms, if for no other reason than that there was an almost complete lack of raw materials for even primitive technology in the arboreal habitat. Although the reasons why our ancestors left the trees are obscure, perhaps one was that some of the larger primates found competition from smaller more agile ones too severe. Because of their increased size and intelligence, they were better able than their ancestors to cope with the problems of terrestrial living. Almost certainly the return was gradual, and at one time our ancestors must have lived much as modern-day chimpanzees do, foraging on the ground in the daytime but retreating to the trees at night. It seems likely that the efficient bipedal posture so characteristic of man was achieved after early prehominids left the shelter of the trees and began to forage in bands out on grassy savannahs. Fossil evidence indicates that an abundance of food was available in the game animals that roamed the open spaces, and selection probably favored any protohuman genotypes permitting, by whatever means, utilization of this food resource. An upright posture, providing reasonably rapid locomotion while at the same time freeing the hands to grip stones or clubs, would have been at a premium. Intelligence and social organization would also have had their reward in food.

It is important to remember, when one is casually discussing our family tree in this manner, that the evolutionary processes discussed are no different in principle from those accounting for bandless water snakes or banded snails. To say that our ancestors moved out of the trees to escape the competition of more agile foragers is merely a shorthand for the following. At one point in our evolutionary history any recombinant that had the slightest behavioral tendency to descend from the trees and forage on the ground had a better chance of contributing to the gene pool of the following generations than other genotypes lacking this tendency. The frequency of the kind of genetic information producing this sort of behavior therefore increased in the populations concerned, and the behavioral norm was slowly shifted. A great many generations after the first pioneer individuals foraged briefly on the ground, the behavior of all individuals in the populations concerned became terrestrial.

The details of the human fossil record are not particularly pertinent to our theme. A brief review of the salient features of the record (as we interpret it) is given here for the convenience of those not familiar with them. The earliest fossil organisms generally conceded to be "men" are the Australopithecines (members of the genus *Australopithecus*). These men, who may have made their appearance in the upper Pliocene (possibly as early as 5 million years ago) and who disappeared some 500,000 to 600,000 years ago, were fully erect and bipedal and made and used stone tools. The time of emergence of *Australopithecus* is doubtful because of uncertainties in the dating of recent finds, such as those in ancient deposits of Lothagam Hill in northern Kenya, which have been reported to be 5 million years old. The primate fossil record prior to *Australopithecus* sheds little light on the line leading to man, although one fossil from the upper Miocene (some 14 million years old) known as *Ramapithecus* seems to be more related to our distant ancestors than to those of monkeys or apes and probably represents the departure point leading to the hominid line.

Direct descendants of one group of the Australopithecines probably are the Java and Peking men and their relatives, *Homo erectus*. The first fossil *H. erectus* has an estimated age of 600,000 years, indicating a possible overlapping in time with some of the Australopithecines. Most *Homo erectus* probably disappeared around the time of the Riss glaciation, some 200,000 to 250,000 years ago. The last remnants of this species may well have persisted in geographic isolation after selection had transformed other groups of *H. erectus* into what we now call *Homo sapiens*. Other less well-documented remains indicate that *H. erectus* was widespread and quite variable and had achieved a culture involving the use of fire and tools.

It seems likely that much of the confusion regarding the emergence of *Homo sapiens* has been caused by reticulate evolution. About the time of the last interglacial (100,000 to 200,000 years ago), various populations of the geographically variable *H. erectus* probably gave rise to numerous populations of *H. sapiens*. These populations had varying fates; some died out, others met and fused through interbreeding; and some may have persisted and evolved in relative isolation until after the last glaciation. The famous Neanderthal man seems to have been a geographic variant of *H. sapiens*, one which disappeared, in all probability, from different combinations of causes in different areas (interbreeding with more modern *H. sapiens*, competition from more modern *H. sapiens*, starvation due to disappearance of game, etc.).

Once our predecessors became upright, there remained only one major physical change to convert them into modern men. This change was a great increase in brain size and skull volume. *Australopithecus*, one of our earliest upright ancestors, had a brain volume of 450 to 600 cm³ (about that of a large ape). The cranial capacity of *H. erectus* bridges the gap between the largest great apes and modern man, the smallest skull on record having a capacity of 775 cm³ and the largest 1,200 cm³, well within the range of present *H. sapiens*. The Java men tended to have a slightly smaller brain than Peking men. Modern men (*Homo sapiens*) average about 1,450 cm³. Because of the physical limitations on pelvic expansion in anthropoid females, most of the growth resulting in large brain size is postnatal. This great postnatal growth in skull capacity results in a very long period of helplessness in the infants of the larger-skulled forms, creating a mother-offspring relationship that has left a considerable mark upon our present-day culture.

In summary, then, man owes many of his most characteristic features to an ancestral sojourn in the trees. This was responsible for the well-developed association centers of his brain and the skillful manipulating devices on his forelimbs. It also gave him his family association with its social bonds, year-round sexuality, and mother-offspring relationship. All these were instrumental in the development of culture, which will be considered next.

CULTURE

At one time it was commonly thought that man's large brain made it possible for him to invent culture. It now seems that possibly the reverse was true. The earliest presumed ancestors (or near ancestors) of modern man, the australopithecines, were erect creatures with brains not differing appreciably in size from those of modern anthropoid apes. They almost certainly made tools. The evidence indicates that the australopithecines were animals of the plains and that they were primarily vegetarian. There is also evidence that their diet was supplemented somewhat with the meat of small animals. Little is known about the tools used by these protohumans, but it seems unlikely that they could have left the shelter of the forests before they acquired a reasonable security by employing rocks and clubs in their own defense. A variety of stone tools have been found which are believed to have been used by *Homo habilis*, a more advanced hominid which lived alongside the australopithecine *Zinjanthropus* 1.75 million years ago. This discovery is *prima facie* evidence that *H. habilis* had at least a rudimentary culture, like modern chimpanzees, which have recently been shown by van Lawick–Goodall to make tools. They also learn toolmaking from each other. The stone tools of *H. habilis* are much more complex artifacts than the stick and leaf tools of chimpanzees.

There can be little doubt that an ape-brained anthropoid, quite possibly our own direct ancestor, was the possessor of a complex body of information that passed from generation to generation non-genetically. It also seems highly likely that these protohumans utilized a reasonably complex system of verbal communication. Making stone implements is not as simple as the twentieth-century armchair observer might believe. While it is conceivable that the young of *H. habilis* learned to do this merely by careful observation and mimicry, it seems more likely that a certain amount of spoken instruction went along with the demonstration. The possession of culture, and perhaps of speech, by these long-extinct, very small-brained anthropoids clearly outlines the probable solution of one of the most vexing problems in human evolution, the "cause" of the roughly threefold increase in brain size between that of earliest fossil man and modern man.

As culture became important in prehuman society, genotypes with the mental characteristics permitting optimal utilization of this extragenetic information were more successful reproductively than their less-well-endowed cohorts. Genotypes were favored that produced brains with the highest ability to associate, integrate, and store incoming sensory data and to utilize these data in a manner that enhanced the survivability of the genotype. This selection pressure resulted in a trend toward great expansion of the cerebral cortex and an increase in the number and complexity of the neuronal systems necessary for "thinking and speech." It is not unreasonable to assume that much of this increased volume is the result of a premium being placed on storage capacity. Man's tremendous neopallium is relatively more free from commitment to special sensory and motor functions than that of other mammals. These "uncommitted" areas may be presumed to be concerned with association and memory. This presumption is supported by results obtained from electrical stimulation of the brain in conscious patients undergoing brain surgery. For example, stimulation of the temporal lobe may lead to the patient's rehearing a complete symphony or reliving an event of the distant past. When a human being is subjected to a frontal lobotomy, his sensory and motor functions are relatively unimpaired, but he becomes "irresponsible."

There is a considerable body of literature on the reasoning power of chimpanzees. On certain types of tests designed primarily to evaluate human reasoning power, some "chimps" score higher than many human adults. Indeed, as Harlow succinctly puts it, if man is defined as the possessor of mental abilities that occur in other animals only in the most rudimentary forms, if at all, we "must of necessity disenfranchise many millions . . . from the society of *Homo sapiens.*" Chimpanzees may lack a complex culture not because of any great lack of reasoning power but because of some other factor that inhibited the development of speech or the regular utilization of tools, or the reduction of intermale aggressiveness. The "invention" of rudimentary culture started a selective trend that led eventually to man's large brain; the large brain did not just mysteriously develop and then discover culture.

Cultural and biological evolution cannot proceed independently. Indeed, from the very beginning of culture, man's evolution has been characterized by the interactions of biological and cultural evolution. The existence of culture put a selective premium on certain types of brains; the evolution of the brain permitted an expansion and enrichment of the culture. Such interactions were certainly very important during the transition period from the early hominids to *Homo sapiens*, but they are still very much with us. Before going further into such interactions, however, we shall consider some characteristics of culture and some features of cultural evolution.

One of the outstanding characteristics of human cultures is their tremendous diversity. Human beings speak some 2,800 different languages, describe their genetic relationships with each other with myriad complex kinship systems, believe in a great diversity of gods and spirits, are organized into groups which practice every degree and kind of governmental control, and fill their everyday lives with galaxies of taboos concerning everything from forms of salutation to shapes of wine glasses. This cultural diversity is by no means superficial; indeed, people of different cultures often have very basically different world views. This difference is frequently reflected in the language of a culture, and in a very real sense, as discussed in Chap. 13, the language creates the world view.

Language differences are among the most important of all cultural isolating mechanisms. Communication of information about complex phenomena may be exceedingly difficult within a culture—as almost any teacher will gladly testify; but between cultures with widely different languages the problems are immense. The reader is invited to contemplate the difficulties of explaining even a simple word such as "also" to an Eskimo if there are no dictionaries, no shared third language, nor even the certain knowledge that an equivalent concept exists in his mind. The intricate and highly developed language of the Eskimos does not have a structure congruent with that of our language.

In spite of their great diversity, however, many similarities may be observed among cultures. Some form of religious belief is virtually universal. It has been suggested that these beliefs are based on observed differences between living and dead human beings, the assumption being that the absence of breathing and lowering of the body temperature result from the desertion of the body by a spirit. This seemingly logical assumption combined with, among other things, dreams and ignorance of the forces of the physical world, may well be the basis of all religions. The elaboration of these or other simple ideas into the complex pattern of religions that we have today was a long and complicated process. It seems eminently fair to say that, even with the flourishing of science in the last few centuries, man's creation of spirits, gods, and the related paraphernalia of religion has had the most far-reaching effects of any cultural phenomenon. In most societies of the past and in the majority of societies today, organized systems of religion provide the principal means for the individual to orient intellectually to his physical and cultural surroundings. Among other things, such systems of orientation make it very difficult for the individual to appreciate the outlook of members of other societies.

Most people in every culture believe that their own way of doing things is, in some absolute sense, *right*. They are unaware, or at most only dimly aware, of their own biological and cultural history. They do not understand why they love their children, why the sun comes up in the morning, or why they must hate their country's enemies. Such cultural chauvinism can create considerable amusement for those conscious of it. For instance, in a 1971 television broadcast about Laotian tribesmen suffering in the Indochina war, the commentator intoned: "They are very primitive; they just pray to spirits and don't have a God." In short, people generally accept the dicta of their culture without question. The acceptance of these dicta at one point in time and space may well have added to the viability of the culture. At another point the same set of values may be suicidal.

Many of the important rules for living in our culture are believed by some to have been handed down from heaven a mere few thousand years ago. Most of them probably trace to the time when human beings gave up a nomadic hunting and food-gathering way of life and, with the invention of agriculture, began to settle down in rather large organized groups. There were numerous advantages to living in such groups, among them cooperative defense, the ability to carry out projects requiring a great deal of manpower, and the opportunity for specialization into various trades and professions. For groups of any size, from family on up, to enjoy the fruits of cooperation, internal strife must be kept at a minimum. Thus, for instance, intergroup selection favored those groups which suppressed killing within the group. One logical way to do this was for the elders to tell the young that killing a member of the in-group would offend the spirits, and indeed that is essentially the way it is done today in our own society. It should be noted carefully that, in spite of constant reiteration of "Thou shalt not kill," our society allows certain kinds of killing of socially sanctioned types. Thus a society may kill its internal enemies (assorted "criminals"), and killing its external enemies may be encouraged. The social approval of killing outsiders at one time doubtless had considerable selective advantage for the society as a whole, but improved weapons make this no longer true. Certain types of murders within our society, although technically illegal, are approved socially.

To recapitulate, cultures are extremely diverse and are separated by, among other things, language barriers. In spite of this, there are common threads running through most, if not all, cultures that make it rather obvious that major features of human culture have proliferated from a common source. People are generally unaware of their biological and cultural history, and most assume that they were, miraculously, born into the culture which has "The True Word."

A major interaction between cultural and biological evolution has been in the change of selection pressures. The development of modern medical techniques; the elimination of many large predators; the increase of the food supply through improved agricultural methods; control and evaluation of the environment with furnaces, air conditioners, dams, radar weather-warning systems, and the like have permitted many individuals with otherwise nonviable genotypes to persist. The diabetic controls his disease with insulin and reproduces. There is time to lead the congenitally blind man to the storm cellar because of the tornado warning over the radio. Thus some differentials in reproduction have been ironed out by cultural factors in *some* societies. However, some recently introduced cultural factors have imposed other selective pressures by favoring those with genotypes that are relatively immune to insecticides in their food, air pollution, nervous tension, heart disease, and cancer. In the last two cases differential reproduction is increased because, with increasing life spans, reproduction is carried into the years when these diseases are prevalent.

The patterns of gene flow in human populations have also been tremendously changed by cultural developments. Systems of transportation have steadily improved, moving the entire human population more and more in a direction of panmixia. However, this trend has been countered to some degree by immigration quotas and other cultural barriers to random mating.

Other interactions between cultural and biological evolution are obvious. Incest taboos tend to lower the coefficient of inbreeding. Social disapproval of interracial, interreligious, and interclass marriages tend to keep the population in many parts of the world divided into relatively small partially interbreeding groups (somewhat analogous to Wright's model for a population with a structure favorable for evolutionary progress).

The influence of our primate background on our culture has been profound. The loss of a sharply defined oestrus period in the female, the general helplessness of the primate infant, and the concomitant establishment of a family group stable throughout the year, has led to systems of interpersonal relationships that have been vastly elaborated in cultural evolution. On top of the relatively simple male-female and female-offspring relationships of prehuman family groups, cultural evolution has produced the monstrously complex set of phenomena usually included under such topics as love, sex, and kinship. That these phenomena have become deeply and basically interwoven into the entire fabric of our behavior has been amply demonstrated by anthropologists and psychologists. These phenomena enter into choices of political systems and political leaders, legal systems, the characteristics of the deities that men have devised, and even into choices of designs for automobiles.

Unfortunately, very little is known in detail about the ways in which culture evolves. Some similarities with biological evolution are obvious but the value of the following analogies is open to considerable doubt. They are given here more as food for thought than as established fact.

Many apparent parallelisms may be detected. Virgin births are found in the mythologies of many different cultures. Complex puberty rites are also widespread, ranging from severe tests of manhood involving torture and genital mutilation to ceremonies such as Christian confirmation and Jewish Bar Mitzvah. Ceremonial appeals to spirits are nearly ubiquitous—*vide* Navajo dancers and San Francisco ministers appealing to their gods for rain. Some examples seem closer to the biological phenomenon of convergence: the appearance of functionally similar structures in very dissimilar entities. An example of this might be the military dictatorships which sprang up in both Germany and Japan between World Wars I and II. The histories of the two cultures in which these phenomena appeared were widely divergent, and yet in many superficial aspects the dictatorships were similar. The stories of virgin births in different mythologies may be a better example of convergence than parallelism; the line is difficult to draw in any case. Similarly, the idea that one may achieve a desired condition by eating a portion of a cadaver is widely and spottily distributed through human cultures. The bodies of victims put to death in Aztec religious ceremonies were devoured so that the eater could establish close contact with his god. As Linton has said, "It was a religious concept not unlike that of the Christian communion except that the Aztecs were painfully literal about it."

An obvious cultural analog of natural selection can be found in the differential reproduction of entire cultures. The body of information making up some cultures has become more and more widespread, i.e., possessed by more and more individuals, while others have decreased or become extinct. The cultures of some small groups have doubtless disappeared without a trace. However, cultural evolution is obviously more reticulate than biological evolution, and large cultures virtually never disappear without transferring some of their information to other cultures. Although American Indian cultures have been badly swamped and in some cases completely destroyed by the spread of western Europeans, some of their elements have been transferred into western European culture. The use of tobacco is a good example.

The ascendance of individuals with novel ideas may be a random phenomenon in human society. Thus one could consider the advent of Aristotle, Darwin, Hitler, Buddha, Tecumseh, etc., to be a sort of cultural analog of mutation. Men with their proclivities are doubtless present from time to time in all cultures, but, as with gene mutations, the proper environment is necessary for them to gain prominence.

An interesting analogy can be drawn between genetic homeostasis and cultural integration. As will be remembered from Chap. 7, when strong selection is applied to a single character, only a certain amount of progress can be made before the effects of unbalancing a well-organized genotype counterbalance the selective pressure on the character. In other words, natural selection preserves a well-balanced genotype. It is possible that overdevelopment of some feature of a society may lead to the destruction of the integrated properties of the society. One might view the development of an extreme military dictatorship in Japan in this light. Another example was the promulgation of a fantastic taboo system by the Hawaiian priesthood, which among other things led to the destruction of the entire religious system.

Today this imbalance can be seen in the overdevelopment of the physical and biological sciences relative to our other disciplines. For many centuries systems of ethics (the time-tested rules under which a society operates) have been taught and maintained by church and state. It is either too much trouble or impossible to orient most people to the evolutionary basis for ethics, and so these institutions control human behavior through a combination of force, social pressure, and promised supernatural punishment. As scientific knowledge increased over in the last two centuries, governments and churches slowly changed their ideas so that to a large degree they did not conflict directly with the findings of science. In the past few decades, however, the advances in science and technology have far outstripped the ability of conservative religious and governmental systems to adjust to the changes.

Medical advances and public health programs have permitted an unprecedented surge in population size, a surge which the ill-informed portion of our cultural structure will not permit us to counteract. Increased population pressures increase the danger of war, but our most enlightened political and religious leaders and, indeed, many of our scientists are just beginning to have the vaguest grasp of the possible consequences of another world conflict. Even if war is avoided, the growth of population in a world already malnourished will have devastating consequences on the ecology of our planet. Irreplaceable natural resources, such as the fossil fuels and minerals are rapidly being used up; even water will be in short supply in a relatively few years. At a time when we should be finding ecologically sound ways of producing more food, we seem to be dedicated to a galloping antiecology technology which is drastically affecting the very life-support systems of the earth.

The natural sciences have helped to initiate and support the population explosion and have produced thermonuclear weapons, nerve gases, and agents of biological warfare, together with more subtly destructive products of technology. In a sense they are in the same position as parents who permit their children to play with loaded guns. One of the possible consequences of applying too strong a selection pressure against a single character in a *Drosophila* experiment is that the line will become extinct. Perhaps in permitting this tremendous gap to develop between the scientists and the laymen, we have doomed our line to the same fate. Extinction may well be the ultimate interaction between cultural and biological evolution.

SUMMARY

Homo sapiens is the product of biological and cultural evolution. The processes of his biological evolution do not differ in kind from those of other diploid, outcrossing organisms. Cultural evolution, change in the mass of nongenetic information shared by human beings, is easily recognized but poorly understood. The two kinds of evolution are inextricably bound in a complex of interactions. *Homo sapiens* is the only organism to have become aware of its origins and of the possible evolutionary consequences of its actions. It remains to be seen what the consequences of this knowledge will be.

REFERENCES

Dobzhansky, T.: *Mankind Evolving*, Yale University Press, New Haven, Conn., 1962. Introduction to the entire literature on human evolution may be gained from the extensive bibliography of this scholarly and interesting book.

Ehrlich, P. R., and A. H. Ehrlich: *Population, Resources, Environment*, 2d ed., Freeman, San Francisco, 1972. A comprehensive examination of human ecology.

———, J. P. Holdren, and R. W. Holm (eds.): *Man and the Ecosphere*, Freeman, San Francisco, 1971. A collection of readings from *Scientific American* concerning the population-resource-energy crisis.

Huxley, J. S.: Evolution, Cultural and Biological, in *Yearbook of Anthropology, 1955*, pp. 3–25, Wenner-Gren Foundation, New York. An interpretation of cultural evolution quite contrary to that of the authors.

Linton, R.: *The Tree of Culture*, Knopf, New York, 1955. A very readable account of the diversity of cultures.

Montagu, Ashley (ed.): *Culture and the Evolution of Man*, Oxford University Press, New York, 1962. A series of stimulating essays on the interaction of organic and psychosocial evolution.

Peacock, J. I., and A. T. Kirsch: *The Human Direction*, 2d ed., Appleton-Century-Crofts, New York, 1973. An evolutionary approach to social and cultural anthropology.

Smith, H. W.: *Man and His Gods*, Grossett & Dunlap, New York, 1956. A physiologist looks at religions.

Washburn, S. L., and P. C. Jay (eds.): *Perspectives on Human Evolution*, vol. 1, Holt, New York, 1968. A well-selected series of readings on the evolution of man and his culture.

Man is the product of biological and cultural evolution, and man has developed the explanation of evolutionary processes put forth in this book. It is therefore inevitable that our view of evolution has been colored by the biological and cultural history of *Homo sapiens* in general and by that of western science in particular. What is the true (or objective) explanation of organic diversity? To ask this question is to fall into the error of assuming the existence of absolutes. It is not possible temporarily to renounce our membership in the human race and view the available data with Jovian detachment. It is possible, however, to point out some of the obvious sources of bias, some of the weak points in the story, some current points of controversy, and some possible ways of increasing our understanding. Complete objectivity may be unattainable, but perhaps we can hope to approach it asymptotically.

CHAPTER THIRTEEN
EVOLUTIONARY THEORY

ANTHROPOCENTRISM

Perhaps the most obvious effect of being human observers is the nearly ubiquitous tendency to use *Homo sapiens* as a standard. This species chauvinism is manifest in many ways. It is easily recognized in its most naive form when the assumption is made that evolution has always worked toward man as its ultimate goal. Thus Lecomte du Noüy states:

> Evolution begins with amorphous living matter or beings such as the Coenocytes, still without cell structure, and ends in thinking Man, endowed with a conscience. It is concerned *only* with the principal lines thus defined. It represents *only* those living beings which constitute this unique line zigzagging intelligently through the colossal number of living forms. [Lecomte du Noüy (1947), p. 66; emphasis his.]

In more subtle forms, such ideas persist

in diverse ways in modern evolutionary literature. Often the term "higher animals" is used to refer to those more like *Homo sapiens* than "lower animals"; the implications of the "higher" and "lower" are quite clear. When standards of "success" in evolution are selected, by some odd coincidence the winning characteristic usually seems to be one in which man excels: ability to control the environment, intelligence, possession of culture, retention of "generalized" characteristics, etc. Some other rather obvious standards are often glossed over, for example, persistence through time (cockroaches excel here), total numbers of individuals (the forte of microorganisms), and reproductive potential (many candidates, such as the housefly and beef tapeworm). People just assume that man *must* be the most successful organism and define success accordingly. Since man invented the language, he most certainly has a right to do so. However, the threat of thermonuclear extinction hangs more heavily over our heads than over the cockroaches, and we may yet envy them their ability to crawl under rocks and their relative immunity to radiation.

Since most scientific work is done by individuals who are reproductive (or, more rarely, postreproductive) members of *Homo sapiens*, the "adult" stage of the life cycle has acquired a certain prestige relative to other stages. For instance, illustrations of phylogenetic trees almost invariably depict series of adult organisms, and the taxonomy of almost all groups is based primarily on this stage of the developmental sequence. How different would be our view of evolution if we were intelligent mayflies, with a long nymphal life in which to ruminate over nature, followed by a few frantic, flapping days as reproductives? Similarly, mosses might have a rather different view of the relative importance of the haplophase and the diplophase; if dandelions were authors, one might find sexual reproduction discussed in the literature as a rare and imprudent luxury.

Needless to say, there is not a shred of evidence to suggest that man is the ultimate goal of evolution; indeed, there is none to indicate that, barring catastrophe, he is even the terminus of his own lineage. If there were purposive forces guiding evolution, we would expect to find traces of them in the process; in the absence of such evidence, it must be assumed that such forces do not exist.

CULTURAL BIAS

Evolutionary theory has been almost exclusively the product of Western minds—minds that think in terms of the Indo-European languages. The structure of these languages has acted to mold our view of nature into a form easily handled by the language. Ideas such as that an effect implies a cause or a creation a creator have, since Aristotle, been considered to be immutable laws of logic. It is interesting that Oriental religions have emphasized the artificiality of the subject-object dichotomy. Their philosophies aim to eliminate this division, the supposed result being similar in many ways to the professed goals of certain psychotherapies in Western cultures. A linguistic need for a doer and the done, for objects and relationships among them, may have deeper and more damaging effects than are presently realized. Our language requires us to put things into various relationships even when it is patent nonsense to do so. For instance, we often say "It is snowing," although the "it" is a meaningless word which soothes our sense of syntactic aesthetics only. Similarly we tend to think of natural selection as *something* that somehow changes a population.

People of other cultures order natural phenomena in ways quite different from those we consider natural and proper. For instance, Eskimos have no generic term for water, but they have a detailed and useful terminology describing the various kinds of frozen and liquid water. Gauchos have some 200 terms for horse colors, but they divide the vegetable world into four species: *pasta*, fodder; *paja*, bedding; *cardo*, woody materials; and *yuyos*, all other plants. As a language system develops, the effects of its structure seem to be invasive and widespread. All aspects of the culture eventually are involved, and a network develops that is difficult to escape. It has been suggested that the person most nearly free to describe nature impartially would be a linguist familiar with many widely different linguistic systems.

Some impression of the relation of language to behavior and to the description of nature can be gained by comparing even super-ficially the basic aspects of Indo-European languages with a very different language. The language of the Hopi Indians has been studied in considerable detail by Whorf and offers revealing com-parisons. It is difficult to describe the differences in English, for the languages are scarcely congruent. For example, our concept of plurality causes us to use cardinal numbers in referring both to real and imaginary pluralities. We count 10 objects and regard them as a group. However, we say that there are "10 at a time," introducing the concept of time into group perception. When we refer to 10 hours or to any other cyclic sequence, actually only one item is experienced at a time; the others are remembered or predicted. We think of time in such a way as to "know" that there was a day yesterday and that there will be a day tomorrow. We can actually quantify "tomorrow" quite "precisely" in minutes, hours, days, months, and years. The Hopi Indian, on the other hand, would not think of using numbers for entities that do not form an objective group. He recognizes a group of 10

Indians. But, if they stay for a visit, he reports that they "left *after* the tenth day," not that "they stayed 10 days."

It is interesting and important to realize that similar differences between the languages are manifest when physical quantity, phases of cycles, and other aspects of time such as duration are investigated. Our mass nouns, which we use to refer to unbounded homogeneous phenomena, imply, besides indefiniteness, lack of outline or size. When we particularize, we often must say "body of water," "dish of food," or "bag of oats." The relator "of" denotes or suggests *contents*; we must have a "container" for the "portion" of matter described. In Hopi, mass nouns also imply indefiniteness but not lack of outline and size. "Water" always means a specific mass or quantity of water. No "container" is implied. One could give examples almost without end. In Hopi, there is no basis for a formless item such as our "time"; our structuring of time with three verb tenses does not occur. Metaphors involving an imaginary "space" ("this discussion is *over* my head") are lacking.

It seems clear that such concepts as newtonian time, space, and matter are inherent in the language of the newtonian physicist. A scientist working in a language with very different structure conceivably might have been compelled to describe nature in, say, relativistic terms. It also seems clear that much of what we think of as "real," "commonsense," and "beyond doubt" in biology are recepts from our language and culture. Biologists have much to learn from the study of the ways other cultures with different languages view nature. Biologists may also benefit from using what is perhaps the only less-biased language presently available to them for describing nature. That language is mathematics.

SCIENTIFIC BIAS

Good examples of the effects of language on the biologist's view of nature are not difficult to discover. Biologists have long believed that sexually reproducing organisms occur in distinct clusters or kinds, commonly called **species.** In the 20-year period between 1937 and 1957, outstanding evolutionists such as Theodosius Dobzhansky and Ernst Mayr, taking cognizance of the evolutionary importance of isolation, attempted to develop definitions of the concept *species*, using genetic criteria. Mayr's short definition, as given in his classic *Systematics and the Origin of Species*, is the one still employed in essence by the majority of modern evolutionists:

Species are groups of actually or potentially interbreeding natural

populations, which are reproductively isolated from other such groups.

This definition is a description of what is known as the **biological-species concept.** Its acceptance was responsible for a strong shift of emphasis toward studies of evolving entities in nature rather than the application of static concepts to dried insects and stuffed birds. It was recognized that all other taxonomic categories, such as genera and families, could be adjusted arbitrarily in size. If a worker thought that *Homo sapiens* and *Homo erectus* were too different to be placed in the same genus, he could place the Java Man in the genus *Pithecanthropus*. Others might disagree with him but could not demonstrate that he was wrong. Such disagreements rested only on opinion. With the biological concept at the species level, things were different; one had only to find out if two populations were actually interbreeding or in the absence of the physical contact necessary to permit this gene exchange were at least *potentially* able to interbreed. Partially differentiated segregates within a species were called subspecies and were considered, with considerable justification, to represent the early stages of species formation. In borderline cases, such as the *Drosophila paulistorum* example discussed in Chap. 10, where information derived from laboratory crosses did not permit a clear species or subspecies decision, the description "species in *statu nascendi*" has been used.

In recent years a group of biologists has questioned the continuing usefulness of the biological-species concept. They regard the generality that organisms occur as discrete units—species—as an artifact of the procedures of taxonomy. These procedures decree a hierarchic structure in which every entity to be recognized formally with a scientific name must be assigned to some level in the hierarchy. In other words, taxonomists were required to find distinct entities, whether or not any existed.

Doubt has also recently been thrown on the notion that species are evolutionary units, their component populations tied together by gene flow. Soon after the idea of biological species was promulgated, it became obvious that the great multiplicity of genetic systems found in plants, as discussed earlier, resulted in a very limited applicability of the concept in botany. The concept was never intended to apply to asexual organisms, but it has also proved difficult to apply in the many groups of plants where sexual reproduction makes up only one part of the genetic system, where alloploidy has produced a reticulate phylogeny, or where the "organism" is multiple (as in lichens). As more knowledge of the invertebrates is uncovered, the inadequacy of the biological-species definition has also been apparent.

The last stronghold of the utility of the concept was in the insects and vertebrates. However, there seems to be only one entity, a vertebrate, about which we have sufficient information to be reasonably safe in assuming that it is a "biological species." This vertebrate is *Homo sapiens*. We have sufficient information about interbreeding within the group to be relatively sure that all subdivisions can exchange genes with all others (the sort of information that is usually lacking). In addition, man's nearest living relatives are so different from him that the possibility of gene flow is discounted. Investigations of insects, *which did not start from the premise that organisms must (in most instances) occur in distinct clusters*, have indicated that the ease with which various groups of insects may be fragmented into distinct biological species has indeed been overestimated. A study of North American butterflies showed that less than one-half of the genera could be divided neatly into groups of entities that might correspond to biological species.

In retrospect, these conclusions are not very surprising. The very nature of the biological-species definition makes its use in practice impossible. The accepted test of conspecificity, or lack of it, is what happens or would happen when two forms occur together in nature. Even when, conveniently, two forms are sympatric, studies may reveal some level of partial interbreeding and make a firm decision impossible. In such a situation, the question revolves around whether or not swamping will overcome any tendency toward isolation caused by a selective disadvantage of the hybrids. This problem has been discussed in the chapter on differentiation of populations. Since the environmental conditions are certain to change, the only way to answer the question of what will happen is to wait and see.

In allopatric entities the problem is complicated by the idea of potential interbreeding. One must predict the courses of populations if they approach each other and the events at their hypothetical meeting under assumed conditions. Then, if the formation of hybrids is postulated, the fitness or viability of the hybrid population must be estimated. Unfortunately, fitness and viability are very difficult to estimate for a population of *Drosophila* in a bottle, let alone for a hypothetical population resulting from presumed hybridization of populations that have traveled unknown routes to an assumed locality where postulated conditions prevail.

Laboratory tests of interbreeding potential, while yielding valuable information, are not considered definitive. Two kinds of mice, *Peromyscus leucopus* and *P. gossypianus*, will hybridize in the laboratory, but where they occur together naturally in the Great Dismal Swamp of Virginia they remain distinct. Laboratory hybrids between northern and southern populations of *Rana pipiens* do not develop properly. As discussed in Chap. 10, these northern and southern frogs are connected by a long chain of intermediate populations. Should these intermediate populations suddenly become extinct, would the two terminal groups represent two "good" biological species? The question cannot be answered because the biological-species concept has no operational definition. There is no set of operations by which a value can be assigned to each entity: in this case, either "good species" or "not good species." It is conceivable that if the two terminal populations approached one another naturally, selection might alter them so that they would interbreed freely on meeting; then again it might not.

The biological-species concept might be redefined in a way that would make laboratory tests conclusive. Nevertheless, the problem of cutting the continuum of possible degrees of success in hybridization would remain. In addition, the amount of work involved in delimiting a single species would be staggering, and one would still be faced with the omnipresent problem of the experimental biologist: deciding what the experimental results tell him about situations in nature.

Considering the patent difficulties of dealing with the concept of the species as an evolutionary unit, how can one account for its great tenure in the biological literature? The idea seems to have had its origins in western European parochialism. As in the bear example in Chap. 11, the apparent distinctness of clusters is much enhanced when only a small geographic area is known. Equally, inadequate sampling over a large geographic area will increase the impression of distinctness. If an expedition from Mars collected samples of man by shooting a few inhabitants of Norway and one or two African pygmies, the Martians might easily come to the conclusion that there were two distinct species of man.

The appearance of Darwin's *Origin of Species* signaled the end of the idea that species were eternal and unchanging but did relatively little to dispel the notion of prevalence of distinct clusters. However, the idea that scientists, in expecting to find discrete species, were imposing their own prejudices on nature was not entirely lost on biologists of the last century. J. Victor Carus (*Geschichte der Zoologie*, 1872) stated:

> It is of interest to note that in Aristotle the difference between plants and animals is already touched upon. . . . Regarding the nature of some marine growths one may be in doubt whether they are plants or animals. . . . Even the ascidians, says Aristotle, properly may be called plants since they give off no excrement. . . . One sees that Aristotle fell into the same error as almost all moderns. The term "plant," which came to us as part of our language, was interpreted as a term that must correspond to a class of naturally occurring entities. The same thing has happened to later workers with respect to the term

"species." Instead of investigating whether there exists in nature anything that is unchangeable and circumscribed and that corresponds to this term, and then, in the absence of such, to allow nature her liberty and only artificially to assign a meaning to it that corresponds to the current state of knowledge, one simply assumed that one was compelled to consider the words as a symbol for one of nature's secrets, a secret that one might still hope to unveil.[1]

In the century following the writing of this passage biologists have been rather well disabused of the idea that species are unchangeable, but the idea that they are "things" lying in wait to be discovered in nature lingers on.

There may be an additional cultural factor tending to prevent recognition of anything but neatly segregated animal entities in nature. There is a revulsion against so-called "miscegenation" in many parts of Western culture. It has been suggested that although it is considered permissible for plants to engage in illicit activities (hybridization), such behavior could not be recognized in animals. The validity of this interesting speculation is, to say the least, difficult to evaluate. There can be little doubt, however, that cultural factors strongly influence the biologist's view of the structure of nature.

If the generalization that species are units in nature is largely invalid, does this mean that there is no practical way of investigating the diversity of nature? At first glance, it might seem that the current nomenclatural structure would be an insuperable obstacle to reasonably objective description of the patterns of variation found in the earth's biota. However, as long as names are viewed merely as convenient landmarks in the continuum of life, rather than as possessing some deep, if obscure, meaning in themselves, the problems are not too serious.

In recent years taxonomists have been investigating multivariate methods of assaying similarities among organisms. These techniques rely on high-speed, automatic, data-processing equipment to compare simultaneously many features of organisms or groups of organisms and to express numerically the degree of similarity of each entity in the study with every other entity. For instance, it is easily within the capabilities of modern digital computers to compare 200 different butterflies with each other, each of the 19,900 individual comparisons being based on 100 or more attributes of the individuals compared. Such computing systems obviously have a capacity that transcends that of the human mind for making relatively objective multiple comparisons. The details of the techniques are beyond the scope of this book, but an introduction to the field of numerical taxonomy can be found in references at the end of this chapter.

[1]   We are indebted to the late Prof. R. G. Schmieder for translating this passage and bringing it to our attention.

Development of these techniques for comparing organisms shows promise of being of considerable help in solving some of the more vexing questions about the evolutionary process. For instance, it led to the birth of the field of population phenetics in the last decade. Relative to more classic taxonomic questions, methods have been developed that mathematically and pictorially express the patterns of relationship found in numerical taxonomic studies. In this way, a portion of the cultural bias may be removed; at least the computer presumably does not have a deep fear of miscegenation. One of the important consequences of the numerical revolution in taxonomy has been the realization that there is no one "real" set of taxonomic relationships among a series of organisms but rather an infinite number of such sets based on different arrays of characters or standards of relationship. It is now clear that the taxonomic relationships of the larvae of five different insects may well be different from those of the adults if the taxonomy is designed to show ecological or genetic resemblance. The taxonomies would have to be identical if the purpose of the taxonomy were to show phylogenetic relationships.

Needless to say, an accurate and repeatable description of patterns of variation is of extreme importance in the study of evolutionary processes. Evolutionary theory is an explanation of the origin of organic diversity, both neontological and paleontological. Therefore, agreement on the nature of this diversity and unambiguous methods of describing it are obvious prerequisites to achieving maximum coherence in our theory. Otherwise we shall always be somewhat in the position of the blind men formulating a theory of elephant morphology.

It is *not* being suggested here that human and cultural biases and subjective taxonomy have conspired to give an entirely fallacious picture of the structure of the biotic world. It *is* being suggested that in several areas distorted ideas may be impeding progress toward a more thorough understanding of evolutionary processes. For instance, it seems very likely that the required taxonomic structure, as well as the cultural factors discussed, resulted in misapprehensions about the role of gene flow in "holding species together." Unfortunately, most taxonomic studies do not provide the kinds of data that are helpful in establishing just what patterns are present; all too often the results merely present nature tied up in neat packages. This became especially obvious during the writing of the chapter of this book dealing with differentiation of populations. North American butterflies are considered by most taxonomists to consist of an array of some 400 "species" and many more "subspecies." Unfortunately many, probably most, of these entities exist only in the cabinets of collectors, not in nature. The appalling complexity of interrelationships is now slowly being elucidated and the inadequacy of current taxonomic systems for describing observed situations is very striking. It is in the area of description that numerical taxonomy is superior to classical taxonomy and in some cases of special evolutionary interest numerical techniques should be employed, e.g., in comparing larval and adult taxonomies in insects in order better to understand the action of evolutionary forces on different stages of the life cycle. Numerical techniques, however, are too time-consuming in data gathering to be worthwhile for the first-order structuring that is the goal of most taxonomic work.

It would, of course, be folly to assume that the situation in butterflies (or birds, or any other group) is representative of that in all animals. A series of studies of patterns of differentiation in all groups is now being done with sampling and analyses designed to describe the patterns present regardless of their configurations, not to apportion the variation into prepared compartments.

More objective and quantitative methods of describing relationships will also aid in the study of the evolution of developmental systems. Can the larvae and adults of holometabolous insects evolve semi-independently? Can the gametophytic and sporophytic generations of the same plant follow different evolutionary pathways? Growing evidence from numerical taxonomy indicates that the answers to both questions are yes. Indeed, numerical taxonomic studies, based on different sets of adult characters, give systems of relationship that are not entirely congruent. To some degree the head can evolve in different directions from the feet. It is all too common to think of evolution as operating on adult forms only, whereas actually the entire course of development is under selective control. The study of interactions of evolutionary pressures at various stages of development has barely been started.

In this section, the so-called species problem has been used as representative of a whole class of problems because it holds great interest (often mainly practical) for most evolutionists. Ideas such as niche, community, and climax, which are ecological concepts having much in common with the biological species, could be treated similarly.

EVOLUTIONARY BIOLOGY

In the past two centuries, few, if any, scientific ideas have been subjected to as many vicissitudes as the idea of evolution through natural selection. The theory was violently opposed by many clergymen when it was first proposed by Darwin and is still anathema to many religious groups. Many of the clergy today may have serious reservations, particularly in regard to man's place in the evolutionary scheme. Evolution has the distinction of having been legislated against by some of our states; indeed, teaching evolution is still technically illegal in Tennessee. More than 100 years after the idea was first put forth, few laymen have any real idea of what it is all about. In many high school biology classes and textbooks the subject is not touched on at all or is dealt with only in euphemisms. A high school teacher was quoted in the *Palo Alto Times* (California) to the effect that he had been experiencing little difficulty in teaching evolution in his local school system, although in Arizona he had met with less success because many students and other teachers had refused to accept his statements on the subject. There followed a flurry of letters to the editor, condemning the teacher for daring to express such dangerous and heretical ideas.

Recently, the California State Board of Education passed a set of guidelines requiring all elementary school science textbooks adopted by the state to emphasize the notion that evolution is a theory not a fact and that not all scientists believe in the theory. Although the American scientific community was shocked, a large segment of the public praised the action of the Board of Education.

Evolutionists have not met with smooth sailing within biology either. Many outstanding scientists of Darwin's day vigorously opposed the idea, and until the science of genetics began to mature in the 1920s and 1930s the theory of natural selection led a tenuous existence. Today the vast majority of biologists accept the theory of evolution, although a great many of them have only a passing acquaintance with recent thinking on the process itself. In addition, there has always persisted a small but determined group of scientists and pseudo scientists who use distorted versions of evolutionary ideas to support their own social theories. Those interested in this aspect of evolution should consult Dobzhansky's excellent *Mankind Evolving*. It is perhaps only natural that the massive, and often ignorant, opposition presented to evolutionary theory should have left its mark on the theory itself. There can be little question that it has.

The most obvious aspect of evolutionary theory that can be at least partially explained as reaction to the Bishop Wilberforce approach has been the development of a rather stringent orthodoxy. This orthodoxy is easily detected in the compulsion of biologists to affirm *belief* in evolution rather than to accept it as a highly satisfactory theory and in their tendency to list *proofs* that evolution has occurred. It is, of course, a matter of debate where healthy conservatism leaves off and dogma begins. Suffice it to say that the discipline is at least close enough to the danger area to call for some critical reexamination of basic tenets.

By the standards of science, the idea that all modern organisms are the modified descendants of organisms that lived in the very distant past is almost, but not quite, a "fact." Scientific facts are percepts that people easily agree on: the height of a column of mercury, the number of rats in a litter, the length of a femur, etc. They are mostly things that can be counted or measured, where repetition of the operation gives the same results. Perhaps the best description of this idea of descent with modification is that it is a theory that seems to be overwhelmingly supported by the available evidence. To a lesser degree, the same statement can be made for the broad outlines of the evolutionary process as currently understood.

The strong urge to believe in present evolutionary theory, which is so evident among workers in the field, seems to stem partly from a very common human error, the idea that *one* of a number of current explanations *must* be correct. Usually the theory of evolution is contrasted with that of special creation, a one-sided contest, to say the least. The demonstration that the idea of special creation is scientifically meaningless does not, however, "prove" that the theory of evolution is correct. Current faith in the theory is reminiscent of many other ideas which at one time were thought to be self-evidently true and supported by all available data—the flat earth, the geocentric universe, the sum of all the angles of a triangle equaling 180 degrees. It is conceivable, even likely, that what might facetiously be called a noneuclidean theory of evolution will be developed. Perpetuation of today's theory as dogma will not encourage progress toward more satisfactory explanations of observed phenomena. As Hardin puts it:

> There is always a considerable lag in teaching. Many years ago it was remarked that the Military Academy of St. Cyr in France trained its students splendidly to fight the battles of the *last war*. So it is in science teaching; we too often train our students to fight battles already won, or equip them with weapons that no longer fire.[2]

Is our current explanation of evolutionary processes without flaws? Hardly; even the most sanguine evolutionist would admit that there is much to learn. The fine theoretical structure of **population genetics** has not been thoroughly tested in natural populations. Indeed, it is only recently that biologists have begun to realize that very large selection coefficients may be the rule rather than the exception in nature. One of the triumphs of theoretical population genetics was to show that in view of the long periods of time available,

2   Garrett Hardin, Meaninglessness of the Word Protoplasm, *Sci. Mon.*, **82:**119 (1956).

very small selection coefficients could account for the observed diversity of life. Similarly, a whole complex of questions about the relationship of genetic and phenetic heterogeneity and about the nature of genotype "integration" remain unsolved.

Much work needs to be done to clarify and make operational (if possible) related concepts such as **population fitness** and **adaptation.** The former concept is involved in such questions as: Which is more fit, a population showing a high degree of polymorphism or one that is homozygous for superior alleles at most loci? This may be like asking: Which are better, apples or oranges? Certainly the question cannot be answered until and unless a satisfactory definition of population fitness can be formulated and the current data on genetic polymorphism in natural populations are better understood.

Adaptation is one of the most overused terms in biology. Natural selection has become recognized as an *a posteriori* description of events, but adaptation, the result of selection, has been relatively tenacious of its status as a *thing*. It is difficult to see much merit in the term, as all known organisms are the result of more than a billion years of selection and are therefore "adapted." Often adaptation is used in vague comparisons of an organism's way of life with the extent of usable habitat (parasites are more "narrowly adapted" than omnivores). At worst, it is used for inciting wonder at the diversity of vertebrate forelimbs, bird beaks, or pollination mechanisms. One is reminded of Lincoln's remark that his legs were, miraculously, just long enough to reach the ground. In the former instance, once the relationships are described (preferably quantified), the comment on adaptation seems extraneous. Under present conditions, elephants cannot survive in as many places as human beings. Does it really help to add that elephants are more narrowly adapted than people? The continuing idea that adaptation is some phlogiston-like beneficial substance that a population may possess in varying quantities has been at least partly responsible for the difficulties that theorists have had in coming to grips with the problems attendant to the question of population fitness.

Because of the extremely loose application of the term adaptation in the biological literature, it might be wise to drop it completely. The many fine studies of microevolution discussed earlier, e.g., those of Dobzhansky on *Drosophila*, Kettlewell on *Biston*, Clausen et al. on *Achillea*, are best viewed as investigations of natural selection, not as studies of progressive adaptation. The populations concerned are always "adapted"; at different points in time they are "adapted" to different conditions. Although the broad outlines of the splitting processes in evolution seem to be understood adequately, no general mathematical treatment has been possible and many of the details are obscure. For instance, the kinds of interactions that lead to patterns of differentiation such as are shown by the Galápagos finches (Chap. 10) have not been satisfactorily analyzed. To say that "competition between *Geospiza magnirostris*, *G. fortis*, and *G. fuliginosa* on Indefatigable Island caused them to adapt to more specialized feeding niches" sounds very impressive and scientific but is almost meaningless. The word **competition,** for instance, has many meanings and very unfortunate connotations. In this case, it might imply that at first all three kinds of birds utilized the same limited food supply. Competition is not between kinds but between individuals. If we assume that there was at least a slight differentiation in food preference in isolation, individuals of the same species would "compete" with each other more strongly than with the individuals of other species. Presumably, for food specialization to be enhanced, some selective advantage would have to accrue to variants that tend to restrict their diets to particular types of food. It is very difficult to see the selective advantage of such restriction, but it must be tied in with the presence of several different species because where a species occurs alone, its members tend to be more general feeders. The problem may involve complex interactions among such factors as feeding efficiency, recognition characteristics, psychological reactions to individuals of other species, changes in the food supply in lean years, the general structure of the ecosystem, and so forth. Whatever the factors, more study is needed.

There are many other points of disagreement among evolutionists, some more important than others. Some biologists still think that under certain circumstances populations of diploid outcrossing organisms may differentiate without the intervention of physical isolation, through disruptive selection or positive assortative mating. Others cite cogent reasons why this so-called **sympatric speciation** is unlikely. In a more general context, the question of the degree to which gene flow prevents differentiation remains unsettled, as does the related question of whether or not species ordinarily are evolutionary units. As mentioned in Chap. 7, there is still considerable disagreement on the role of **drift** in evolution, and the disagreement will doubtless persist until more is known about natural populations.

EPILOGUE

Many scientists like to feel that they approach the world in a completely objective manner. For instance, a biologist may scoff at the religious, saying that they accept on faith a system of beliefs that cannot be put to a rational test. The scientist all too often overlooks the articles of his own faith, a faith that almost always includes a belief in a real world in which a sort of statistical order exists. He believes that internal consistency of a theoretical construct is "good," that quantification is "good," that curiosity is "good," and that certain kinds of logic are pertinent to his real world. He may even resort to appeal to authority. Virtually all scientists dogmatically accept as fact that there has been a past and there will be a future, although neither concept is readily amenable to operational analysis. Finally (although not exhaustively), subjective aesthetic standards such as "beauty" are part of the criteria used in judging many scientific theories.

In this book, we have accepted most of these standards and have clung to other dogmas, *faute de mieux*. Our reasons are manifold and subjective; in brief, they are that, by accepting these standards, we gain insight which we find pleasing. It is important for the student of science to remember, however, that other methods of gaining insight may be equally or even more satisfying to others. We have, for instance, assumed that events in long time stretches can be induced from knowledge of short time stretches. It is, of course, conceivable that we are doomed to the sort of disappointment that awaited the physicists who thought that the "laws" which apply to baseball could be applied to subatomic particles.

Being scientists does not, however, absolve us of all responsibility as human beings. One must stand ready to make value judgments, and if necessary to implement these judgments, when scientific decisions in the usual sense are not possible. An ardent believer may state that the human population can never be large enough, since for him each soul reflects the glory of a god; a politician may claim that the destruction of a continent is preferable to a change in political or economic philosophy. We could not contend that either is wrong in an absolute sense, but we can easily evaluate both views as inimical to goals that we cherish. The scientist may abdicate political or moral responsibility because the problems involved cannot at present (or may never) be analyzed by his rules. In our opinion, he is as culpable as the preacher or senator who does not attempt to appreciate what men can gain from science. Science is a human activity; where it becomes divorced from man, it loses meaning.

The biologist, and in particular the evolutionist, seems to be in an enviable position to help integrate science with other human thought patterns and activities. Physicists and mathematicians have contributed so much of the literature on the philosophy of science that an important fact often has been ignored. Unless our sense data have completely misled us, physicists and mathematicians are themselves products of evolution.

Man is a curious animal, and his curiosity has led to such

questions as whether or not there is a reality divorced from the human mind. Indeed, the most important philosophical questions still hover, as they have for centuries, in the area of the mind-matter duality and the nature of percepts. It may be that a thorough understanding of the processes that led to man and his curious mind will help us to decide whether such questions have any meaning. It is our guess that they will be found meaningless and that when this is widely recognized both science and man will have reached a new level of maturity.

REFERENCES

Dobzhansky, T.: *Mankind Evolving*, Yale University Press, New Haven, Conn., 1962. See especially the discussion of Social Darwinism and Racism, pp. 10–15.

Ehrlich, P. R., and R. W. Holm: Patterns and Populations, *Science*, **137:**652–657 (1962). This paper expounds the prejudices of two of the authors of this book and cites other writers with similar views.

————, and Peter H. Raven: Differentiation of Populations, *Science*, **165:** 1228–1232 (1969).

Frank, Philipp: *Philosophy of Science*, Prentice-Hall, Englewood Cliffs, N.J., 1957. A good discussion of the link between science and philosophy as seen by a physicist.

Hennig, W.: *Grundzüge einer Theorie der phylogenetischen Systematik*, Deutscher Zentralverlag, Berlin, 1950. By far the best account of the concepts and views of the phylogenetic school of taxonomists.

Lecomte, du Noüy, P.: *Human Destiny*, Longmans, New York, 1947. Very little in this volume would be acceptable to a biologist acquainted with evolutionary processes.

Sneath, P. H. A., and R. R. Sokal: Numerical Taxonomy, *Nature*, **193:**855–858 (1962). A fine brief summary of the philosophy, methodology, and literature of this field.

Sokal, R. R., and P. H. A. Sneath: *Principles of Numerical Taxonomy*, Freeman, San Francisco, 1963. A rigorous and scholarly treatment of the scientific aspects of modern taxonomy.

Teilhard de Chardin, P.: *The Phenomenon of Man*, Harper & Row, New York, 1959. An unorthodox view and discussion of the meaning of evolution which may be uncomprehensible to the typical scientist.

West, D. A.: Hybridization in Grosbeaks (*Pheucticus*) of the Great Plains, *Auk*, **79:**399–424 (1962). The bibliography of this paper is a guide to the literature on hybridization in birds.

Whorf, B. L.: *Language, Thought and Reality*, Wiley, New York, 1956. An excellent source of Whorf's ideas on the relationship of language to behavior and *Weltanschauung*.

Abiogenesis the origin of life from nonliving systems.

Acentric fragment a fragment of a chromosome lacking a centromere.

Acridine a heterocyclic organic molecule which can induce frame-shift mutations by permitting the introduction of an extra base into the DNA molecule at the time of replication.

Acrocentric chromosome a chromosome with a terminal or nearly terminal centromere.

Adaptive radiation evolutionary diversification, often over a relatively short period of time, of a group of organisms, presumably following their entry into a new adaptive zone; also used with reference to structures.

Adaptive value the survival value and reproductive capability of a given genotype relative to other genotypes in the population.

Adaptive zone the "way of life" of a taxonomic group of organisms, in a broad sense; may be subdivided into adaptive subzones.

Adenine one of the two purines (6-aminopurine) involved in the structure of DNA and RNA.

Adenosine diphosphate (ADP) a compound made up of adenine, the five-carbon sugar ribose, and two phosphate groups, which is involved in the mobilization of energy in cellular metabolism.

Adenosine triphosphate (ATP) ADP with an additional phosphate group attached by a high-energy bond; decomposition of ATP to ADP makes energy available for other reactions.

Adventitious embryony production of an embryonic sporophyte by mitotic divisions from tissues of another sporophyte without an intervening gametophytic generation.

Agamic without gametes; used with reference to complexes of organisms in which all individuals reproduce asexually.

Agamospermy formation of seeds without fertilization, with male gametes, if present, serving only to stimulate division of the embryo and produce endosperm.

Aggregate any grouping of more than one object.

Alkaloids nitrogenous organic compounds produced by plants, some of which are known to have physiological activity in animals, e.g.,

GLOSSARY

morphine, caffeine, and nicotine.

Allele one of the several alternative states of a functional gene unit.

Allometry different growth rates in different parts of the same organism.

Allopatric with nonoverlapping geographic ranges.

Allopatric speciation differentiation of populations in geographic isolation to the point where taxonomists recognize them as separate species.

Alloploid a polyploid formed by increase of chromosome number in an individual with more than one type of genome.

Allotetraploid a polyploid formed by doubling of chromosome number in a diploid hybrid between two organisms with different genomes or by the fusion of diploid gametes of such organisms.

Allozyme a form of an allele separable from another allele at the same locus by differences in electrophoretic mobility. Allozymes are genetically segregating isozymes, q.v.

Amino acids the chemical building blocks of proteins; on hydrolysis, proteins yield amino acids.

Amphidiploid a tetraploid organism that is diploid for two genomes, usually from different species.

Anaphase the stage of nuclear division during which the chromatids of the chromosomes separate and move toward opposite poles of the spindle.

Aneuploidy increase or decrease of chromosome number by values less than whole genomes.

Apogamety agamospermy in which cells other than the egg form the new embryonic sporophyte.

Apomixis all forms of reproduction in which meiosis and syngamy are partially or completely circumvented.

Apospory agamospermy in which somatic cells of the sporophyte produce a diploid gametophyte through mitotic divisions.

Arrhenotoky parthenogenesis resulting in the formation of haploid male offspring, the females being diploid.

Articular a bone found in the jaw of many vertebrates which presumably has been modified into the malleus of the mammalian ear.

Artificial selection the choosing by man, as far as possible, of the genotypes contributing to the gene pool of succeeding generations.

Asexual reproduction apomixis, q.v.; often used to refer to vegetative propagation.

Autocatalysis promotion of a reaction by its end products.

Autoploid a polyploid formed by increase in the number of identical genomes in the same organism or by the fusion of diploid or polyploid gametes from organisms with essentially identical genomes.

Autosome a largely euchromatic chromosome that is not a sex chromosome.

Autotetraploid an autopolyploid with four similar genomes.

Autotriploid an autopolyploid with three similar genomes.

B chromosome a supernumerary heterochromatic chromosome found in varying quantities in some plants and animals (not necessarily in all cells).

Backcross the mating of an offspring to one of its parents or to a parental type.

Bacteriophage   virus attacking bacterial cells, often resulting in their lysis.

Balanced polymorphism   polymorphism maintained in a population usually by the heterozygotes, at the locus under consideration, having a higher adaptive value than either homozygote.

Barrier   a physical or behavioral condition that hinders or prevents the movement of individuals (and thus of genetic information).

Basal granules   cytoplasmic organelles of characteristic fine structure to which cilia and flagella are attached.

Batesian mimicry   mimicry in which a harmless or tasty species superficially resembles a dangerous or distasteful model species.

Bimodal population   a population with measurements of a given character clustered around two values.

Bivalent   the synapsed pair of chromosomes in the meiotic prophase.

Blastopore   an invagination in the surface of an embryo in the gastrula stage.

Blastula   an early stage in embryonic development in which the cells are often arranged in a hollow sphere.

Bradytelic evolution   evolution at a much slower rate than horotelic evolution.

Breeding size   the number of individuals in a population that are actually involved in reproduction in a given generation.

Buffering   protection of a system from change by outside forces.

Bulbil   a gemma or miniature plantlet which is produced asexually on a plant and which can develop into an adult organism.

Canalization   a description of developmental pathways, usually thought of as resulting from buffering.

Canals   enclosed regions within the tube-like portions of the interconnecting channels of the lipoprotein membranes which form the endoplasmic reticulum.

Carbohydrate   an organic compound consisting of a chain of carbon atoms to which hydrogen and oxygen are bound in a ratio of 2:1; this class of compounds is the primary source of metabolic energy.

Carotenoids   fat-soluble pigments ranging in color from yellow to red; most are tetraterpenes.

Cellulose   a primary constituent of the cell wall in most plants; an insoluble carbohydrate which forms microfibrils from polysaccharide chains of glucose.

Cell wall   the outermost layer of plant cells, consisting primarily of cellulose and other polysaccharides.

Centric fragment   a portion of a chromosome that contains a centromere.

Centriole   an organelle generally found in pairs and functioning in the origin and motion of cilia and flagella, as well as in nuclear and cell division; usually lacking in higher plants.

Centromere   the portion of a chromosome, usually greatly restricted in length, to which spindle fibers are attached in cell division. Also called kinetochore.

Centrosome   the area surrounding the centriole in many cells, presumably associated with the organization of protein fibers.

Character   any feature that varies within the group of items under study.

Character displacement condition in which sympatric populations of two species are more dissimilar than allopatric populations of the same two species.

Chemosynthetic able to elaborate carbohydrate substances of relatively high complexity with only inorganic substances, e.g., sulfur, iron, as a source of energy.

Chiasma, chiasmata a cross-shaped configuration of the chromosomes in a bivalent in the first meiotic prophase, usually the visible result of prior cytological crossing-over.

Chlorophyll magnesium complexes of porphyrins found as green pigments in the cells of photosynthetic organisms, where they absorb light energy.

Chloroplast membrane-bound, chlorophyll-containing organelle of most eukaryotic plant cells, the site of photosynthesis.

Chromatid a visible subdivision of a chromosome having but one centromere (or behaving as if the centromere were undivided); a half chromosome.

Chromatography the separation of chemical compounds as a result of their differential migration on or through various substances, e.g., gels or paper.

Chromonema the finest visible longitudinal morphological subdivision of a chromosome.

Chromosome the cell organelle with which most of the nuclear genetic information is associated and which contains the centromere or spindle attachment point.

Chromosome complement the set of chromosomes included in a nucleus which may include one or more genomes, depending upon the cell and its state.

Cilium, cilia a short protoplasmic extension with characteristic fine structure, projecting from a cell and moving with a characteristic pattern and beat; usually present in large numbers.

Cisternae apparent spaces within the endoplasmic reticulum whose profiles in electron micrographs suggest that they are flattened vesicles.

Cis state two mutational alterations in the same cistron or on the same chromosome, as opposed to the trans state, q.v.

Cistron an operational unit, equivalent to or smaller than a genetic region, controlling a specific protein.

Cleavage division one of the early divisions of the zygote that leads to the formation of an embryo in animals.

Cleistogamous producing inconspicuous flowers that never open and are thus self-pollinated.

Climax community the relatively stable community following a successional series; usually thought of as characteristic of a particular climatic zone.

Cline a gradient of variation in the measurement of a character of a population or in a complex of characters of a population, the gradient often varying in steepness along its length or being stepped.

Clone a population derived by asexual reproduction of a single individual.

Coacervates aggregates of varying degrees of complexity, resulting from the interaction of two or more colloids.

Codons nonoverlapping triplets of DNA which code for particular amino acids.

Coenocyte multinucleate plant body which lacks cross walls.

Coenzymes substances which combine with specific proteins to form complexes that are catalytically active.

Coevolution the patterns of evolutionary interaction between major groups of organisms with a close and evident ecological relationship, e.g. plants and herbivores.

Colloid a dispersion of one substance within another, having properties of both a solution and a suspension.

Community the group of organisms found in a particular place.

Competition use of the same limited resources by two or more organisms.

Conjugation union of cells or organisms during which all or part of the genetic information of one individual is transferred to another.

Complementary genes genes whose products must interact in order to produce a particular phenotypic effect.

Conspecific considered by taxonomists to belong to the same species.

Convergence superficial resemblance resulting from occupancy of similar adaptive zones.

Cretaceous the last period of the Mesozoic; beginning about 130 million years ago.

Crossing-over the exchange of usually corresponding segments between chromatids of homologous chromosomes; results in chiasmata and gene recombination.

Culture a body of nongenetic information transmitted from generation to generation or within generations.

Cyanogenic organic compounds which contain the cyanogen (CN) radical.

Cybernetic involving a control or governing mechanism operated by feedback from the process.

Cytoplasm the portion of the cell within the cell membrane and exclusive of the nucleus.

Cytoplasmic inheritance inheritance of traits whose determinants are not located on the chromosomes.

Cytosine the pyrimidine (2-4-aminopyrimidine) involved in the structure of both DNA and RNA.

Decay of variability the reduction of heterozygosity because of genetic drift leading to loss and fixation of alleles at various loci.

Deletion the loss of a segment of a chromosome.

Denatured protein a structurally modified protein showing decrease in solubility and change in biological activity.

Density-dependent selection selection in which the fitness of a genotype depends upon the density of the population in which it occurs.

Deoxyribonucleic acid (DNA) giant molecules which in most organisms are the nuclear repository for the genetic information and which are replicated and transmitted equationally to daughter nuclei; see nucleic acid.

Dicentric having two centromeres.

Differential segments   portions of chromosomes that do not pair in meiosis.

Differentiation   the changes observed in development as a zygote becomes a multicellular entity in which many diverse kinds of cells, tissues, and organs are found.

Dikaryosis   the condition of possessing two nuclei in each cell.

Dimer   a compound formed as the result of the polymerization of two like molecules.

Diploid   having the zygotic number of chromosomes, *or* having two genomes, in reproductive cells other than gametes.

Diplophase   that part of the life cycle in which the zygotic chromosome number is found in reproductive cells other than gametes.

Diplosis   establishment of the zygotic chromosome number, usually by syngamy and karyogamy.

Diplospory   a mode of apomixis in plants in which a diploid gametophyte is formed after mitotic or partly meiotic divisions of the spore-forming cells.

Directed alternate disjunction   regular movement of alternate centromeres to the same pole of the spindle in nuclei heterozygous for one or more reciprocal translocations.

Directional selection   selection resulting in a shift in the population mean for the character considered.

Disjunction   movement of the centromeres of a bivalent to opposite poles of the spindle during the first meiotic anaphase.

Disruptive selection   selection in which two or more different genotypes are at an advantage and intermediate types are at a disadvantage.

Disulfide linkage   a covalent bonding of cysteine residues (S—S) in or between protein molecules; thought to be responsible for spindle formation.

DNA   see deoxyribonucleic acid.

Dominance   the effect of the phenotype of a heterozygote for a given locus being more similar to one homozygote than it is to another; the allele involved in the proximate homozygote is referred to as the dominant allele; in complete dominance the heterozygote phenotype is identical to that of the dominant homozygote.

Duplication   the presence of a section of chromosome more than once in a genome.

Ecosystem   the thermodynamically interrelated set of organisms in a particular environment.

Effective breeding size   the breeding size mathematically adjusted so that populations with varying sex ratios, degrees of inbreeding, etc., can be compared.

Effectors   small molecules (metabolites) that combine with repressor molecules, activating or inactivating them with respect to their ability to combine with an operator.

Electrophoresis   a procedure in which molecules with different charges are separated by their different rates of migration in an electric field.

Emigrantes   the parthenogenetically reproduced offspring of fundatrix generations of the aphid *Tetraneura ulmi*, which develop wings and fly away to feed on roots of grasses.

Endoplasmic reticulum   the system of tubes and vesicles in the cytoplasm of plant and animal cells, the membranes bounding which are usually associated on the outside with ribosomes and are continuous with the plasma membrane and the outer membrane of the nuclear envelope.

Endoploidy   division of the chromosomes without division of the nucleus, resulting in higher than zygotic chromosome number.

Endosymbionts   organisms which live within other organisms in a mutually beneficial relationship.

Enucleated   having the nucleus removed.

Environmental variance   that portion of the phenotypic variance caused by differences in the environments to which the individuals in a population have been exposed.

Enzymes   proteinaceous catalysts of cellular reactions.

Epigenotype   the series of interrelated developmental pathways through which the adult form is realized.

Epistasis   interaction of nonallelic genes.

Equational division   division of the chromosomes so that daughter cells are alike genetically.

Erythrocyte   the type of hemoglobin-containing cell found in the blood of vertebrates.

Euchromatic   describing those portions of the chromosomes that manifest the usual prophase-telophase transformations and contain those genes with major phenotypic effect.

Eukaryotic   cells having a well-defined nucleus and cellular organelles.

Euploidy   increase in chromosome number by whole genomes.

Expressivity   a measure of the uniformity of the phenotypic expression of a gene in a particular environment.

Exules   parthenogenetically reproduced generations of females by the emigrantes generation of the aphid *Tetraneura ulmi.*

Fat   an organic compound which contains carbon, hydrogen, and oxygen; two basic subunits of fat are glycerol and fatty acids.

Fatty acid   an organic acid composed of long carbon chains.

Feedback   the influence of the result of a process upon the functioning of the process.

Fitness   the survival value and reproductive capability of a given genotype relative to other genotypes in a population.

Flagellum   long thread-like locomotory or feeding structure.

Flavonoid   a class of organic compounds often considered to be metabolic by-products of plants; they include many characteristic plant pigments.

Founder principle   the principle which states that when a new population is established in isolation, its gene pool is not identical with that of the parent population because of sampling error; these differences are enhanced as different evolutionary pressures in the areas occupied by the two populations will then also be operating in different population genetic environments; and the result is increased divergence.

Frame-shift mutation a transposition in the reading of the triplet code along the DNA molecule because of the insertion of an extra base into the DNA molecule at the time of replication.

Frequency-dependent selection selection in which the fitness of a genotype depends upon its frequency in the population.

Fundatrix the all-female sexually reproduced generation of the aphid *Tetranuera ulmi* which reaches maturity within the elm leaf galls.

Gametocyte a cell which, through division, will form gametes.

Gametogenesis the combined process of cell division and differentiation which results in the production of gametes.

Gametophyte that phase of the life cycle in many plants in which the gametes are produced, usually by mitosis.

Gap a discontinuity in variation.

Gastrula the stage in development following the blastula in which the cells at the surface move to form an invagination (the archenteron), which has an opening to the outside known as a blastopore.

Gemma, gemmae bud-like outgrowth or fragment of a plant which can asexually develop into a new individual.

Gene the segment of a chromosome between two closest points of crossing-over; a hereditary unit having more than one state and whose different states produce differences in the phenotype; a segment of genetic material that bears the information specifying the structure of a single protein or polypeptide chain. See cistron, recon.

Gene flow the movement of genetic information within and between populations.

Gene frequency the number of loci at which a given allele is found within a population divided by the total number of loci at which it could occur.

Gene pool the total genetic information possessed by a population.

Genetic assimilation the incorporation into the genotype, by a selective process, of characteristics appearing in ontogeny as a response to the environment.

Genetic drift random fluctuation of gene frequency (usually due to sampling error inherent in the genetic mechanism); present in all populations, its effects are most evident in very small populations.

Genetic homeostasis the tendency of populations under selection to regress toward the original mean.

Genetic load in a population, the average number of potential deaths per individual due to genetic causes (lethals, semilethals, etc.).

Genetic system all the factors, internal and external, that affect recombination in an organism.

Genetic variance that portion of the phenotypic variance caused by variation in the genetic environment of the individuals in a population.

Genome the minimum set of nonhomologous chromosomes all of which must be present to ensure the proper functioning of a cell.

Genotype the totality of the genetic material of a cell (usually restricted to nuclear genetic material); the total genetic endowment of an individual; the genetic endowment of an individual at a given locus.

Gens, gentes   a subset of cuckoos tending to parasitize one kind of bird but not necessarily geographically isolated from other such subsets.

Glucose   a six-carbon sugar which is a primary source of energy for most organisms.

Golgi complex   a cell organelle with secretory role found in most plant and animal cells, often in proximity to the nucleus, and comprising a series of concentrically arranged cisternae without ribosomes, as seen in electron micrographs.

Grana   pigment-containing structures within a plastid, usually appearing as many disks stacked in series.

Guanine   one of the two purines (2-amino-6-oxypurine) involved in the structure of DNA and RNA.

Guild   a group of closely related species presumed to be in competition for the same resource and using the same foraging strategy.

Gynogenesis   reproduction by parthenogenesis in which stimulation by sperm of another species is necessary for development.

Habitat   the place where an organism lives. In the felicitious analogy of ecologist E. P. Odum the habitat is an organism's "address," the niche its "profession" (see niche).

Haplodiploidy   a genetic system found in some animals in which males develop from unfertilized eggs and are haploid, the females being diploid also called arrhenotoky.

Haploid   having the gametic number of chromosomes, *or* having a single genome, in reproductive cells other than the gametes.

Haplophase   that part of the life cycle in which the gametic chromosome number is found in reproductive cells other than gametes.

Haplosis   establishment of the gametic chromosome number, usually by meiosis.

Hardy-Weinberg law   the law stating that in a large panmictic population, in the absence of mutation, selection, and differential migration, the frequency of autosomal genes at a given locus remains constant and that after one generation the frequency of genotypes at the locus reaches equilibrium.

Hemiparasitic   photosynthetic plants partially dependent on host plants for nutritional needs and joined to them by root grafts.

Heritability   the genetic variance divided by the phenotypic variance; an estimator of the degree of resemblance between offspring and parent.

Hermaphroditic   possessing the phenotype of both sexes in organisms with male-female sex differentiation.

Heterochromatic   describing entire chromosomes or portions of the chromosomes that do not manifest the usual prophase-telophase transformations and appear to lack genes with major phenotypic effect.

Heterogametic   producing gametes with differing chromosome complements.

Heterokaryotic   containing within one cell or coenocyte nuclei of different genotype.

Heterotrophic requiring organic carbon, which has originated in other organisms, for nutrition.

Heterozygote an organism in which the alleles of a gene are different.

Higher categories taxonomic categories above the level of genus in the established hierarchy.

Holarctic the land mass of the entire northern region of the continents of the Old and New World.

Holometabolous having a pattern of development that includes distinctly different egg, larval, pupal, and adult stages.

Homolog one of the set of two or more chromosomes which are identical with respect to their constituent loci.

Homozygote an organism in which the alleles of a gene are the same.

Horotelic evolution rates of evolution falling within the distribution (asymmetrical with a mode nearer the upper than the lower end) most commonly found when evolutionary rates are plotted in a frequency distribution.

Hybridization gene flow between populations; usually restricted to populations recognized as distinct by taxonomists.

Hydrated protein a protein molecule about which water molecules are held.

Hypha a tube or thread-like fungal filament which forms a mycelium.

Hypostatic preventing or masking the action of one gene by the action of another gene.

Imprinting the imposition of a stable behavior pattern by exposure, during a particular period in development, to one of a restricted set of stimuli.

Inducer a small molecule found in the cytoplasm which eliminates the effect of a repressor molecule and permits the transcription of the structural genes of the operon.

Induction determination of the developmental fate of one cell mass by another.

Interstitial segment that portion of a chromosome between a translocated segment and the centromere.

Introgression incorporation of genetic material from one population into that of another by repeated backcrossings; usually restricted to populations regarded as distinct by taxonomists.

Inversion reversal of the sequence of a portion of a chromosome.

Isozyme a form of an enzyme which has a catalytic function similar or identical to that of another form of the enzyme but is structurally different.

Kappa particles particles of DNA within the cytoplasm of certain individuals of some species of *Paramecium* that result in the death of other individuals with certain genotypes.

Karyotype the characteristic phenotype of the chromosomes of an organism; usually used with respect to the chromosomes at mitotic metaphase.

K selection selection favoring genotypes that enhance competitive ability at densities near the carrying capacity.

Kin selection selection favoring behavioral traits which result in an individual's helping another individual, to an extent which reflects their family relationship (family relationship itself reflects genetic similarity).

L chromosomes   see limited chromosomes.

Limited chromosomes   heterochromatic chromosomes restricted to the germ line in some species of the fungus gnat *Sciara*.

Linkage   the association of genes as a result of their occurrence in the same chromosome.

Lipid   an organic compound which is fatty or fat-like in composition and insoluble in water but soluble in organic solvents.

Locus   the position of a gene on a chromosome; the location of equivalent genes (alleles) on chromosomes of homologous sets.

Lysogenic   referring to those bacteria that carry temperate phages.

Lysosome   vesicles in animal cells which function in the storage and transport of enzymes or may be the site of intracellular digestion.

Macroevolution   evolutionary events usually viewed through the perspective of geologic time, such as the evolution of the horse from a dog-sized mammal to *Equus caballus*.

Macromolecular   having a very high molecular weight, particularly referring to the molecules of the biologically important classes of organic compounds, e.g. proteins, carbohydrates, nucleic acids, and lipids.

Maxilliped   a jointed appendage of an arthropod.

Megaspores   those spores which, in vascular plants, divide to produce the female gametophytic generation.

Meiocyte   a cell of which the nucleus divides by meiosis.

Meiosis   nuclear division associated with cell division which results (potentially) in four daughter nuclei containing one-half the number of chromosomes of the parental nucleus.

Meiotic drive   a higher probability that one allele rather than others at a locus will be included in the gametes.

Melanic   black, dark brown, dark-colored.

Mendelian population   a reproductive group sharing a common gene pool.

Mesozoic   the age of reptiles; beginning about 200 million years ago and ending about 70 million years ago.

Metacentric chromosome   a chromosome in which the centromere is about midway in its length, the arms therefore equal.

Metalloenzyme   a protein associated with a metal atom or complexes of metal atoms and functioning as an enzyme.

Metaphase   the stage of nuclear division during which the chromosomes become attached to, and oriented on, the spindle at the equatorial plate of the cell.

Microevolution   evolutionary events usually viewed over a short period of time, such as changes in gene frequency within a population over a few generations.

Microsome   a cytoplasmic constituent, obtained upon centrifugation of homogenized cells, thought to consist of ribosomes and portions of the endoplasmic reticulum (now largely replaced by the latter terms).

Migration   the transfer of genetic information between populations; *or* the dispersal and establishment of organisms beyond their place of origin; *or* a periodic movement of individuals.

Mimicry the superficial resemblance of one organism by another, presumably affecting the actions of predators.

Mitochondria cytoplasmic organelles, filamentous or spherical, composed of two membranes, the innermost of which is convoluted, that are the site of many reactions of cellular respiration.

Miocene a middle epoch of the Tertiary period, beginning about 26 million years ago and ending about 11 million years ago.

Mitosis nuclear division associated with cell division in which the chromosomes divide longitudinally, separate, and form genetically identical daughter nuclei.

Modifier genes genes whose major obvious phenotypic effect is to modify the expression of other genes.

Monomer the basic unit of a compound formed by polymerization.

Morphogenesis the process leading to the development of the characteristic mature form of an organism.

Müllerian mimicry mimicry in which a series of dangerous or distasteful species resemble one another.

Multiple allelomorph an allele occurring at a locus at which at least two other alleles are known to occur.

Multivalent an association of more than two chromosomes whose homologous regions are synapsed by pairs.

Mutagenic agent an agent leading to an increase in mutation rate above the spontaneous level at a locus.

Mutation a change in a gene; sometimes applied also to chromosomal changes.

Muton a mutational site within a gene.

Mycelium the plant body of a fungus composed of a mass of hyphae.

Natural selection nonrandom (differential) reproduction of genotypes without the conscious intervention of man.

Neopallium the evolutionarily recent expanded surface layer of the cerebral cortex which is the primary coordination center for motor and sensory function involving all senses and all parts of the body.

Nephric groove external groove leading away from the excretory pore of certain crustaceans.

Neurula the stage in a vertebrate embryo following the gastrula, characterized by rapid differentiation.

Niche in broadest terms the "way of life" of an organism. Various more refined definitions have been constructed (see habitat).

Nondisjunction failure of the chromosomes in a bivalent or multivalent to separate at the first meiotic anaphase.

Nucellus the tissue within an ovule which is functionally the megasporangium.

Nuclear envelope the outer boundary of the nucleus, composed of two perforate unit membranes, the outer of which is continuous with the endoplasmic reticulum.

Nucleic acids complex acids composed of nucleotides.

Nucleolus a spherical body within the nucleus composed primarily of RNA and protein; the site of ribosome production.

Nucleotide a phosphate ester of the *N*-glycoside of a nitrogenous base; chemical building block of a nucleic acid.

Nucleus a membrane-bound organelle of the eukaryotic cell which is the primary site of DNA.

Oligogenes those genes with major obvious phenotypic effect; switch genes.

Ontogeny the development of an individual.

Oocyte cells which have the potential to develop into a mature ovum.

Operator a genetic element that acts as a receiver of specific cytoplasmic signals in the form of repressor-substance molecules and controls an operon.

Operon a unit of linked cistrons, the expression of which is controlled by an operator; a genetic unit of transcription of the DNA code.

Organelle a structure of characteristic morphology and function within the cytoplast; the unicell analog of an organ in a multicellular organism.

Organizer a portion of an embryo that determines the developmental fate of the cell masses with which it comes into contact.

Oriented meiotic divisions meiotic divisions with the spindle oriented so that a particular group of chromosomes always enters a polar body rather than the egg.

Overdominance the result of the heterozygote's being more extreme than either homozygote.

Oxidation a reaction in which oxygen is acquired or hydrogen is lost by a compound or the valence of the metallic element is raised.

Pairing segments the euchromatic regions of chromosomes that synapse with a homolog in the first meiotic prophase.

Palaearctic the Old World subregion of the Holarctic region; comprises generally Europe, North Africa, and Asia to the Pacific Ocean at latitudes north of the Himalayas.

Panmixis random mating.

Paracentric applied to inversions that do not include the centromere.

Parallelism convergence among closely related forms.

Parthenogenesis the development of an individual from an unfertilized gamete.

Pectate a compound of pectic substances (acid polysaccharide amorphous carbohydrates).

Penetrance a measure of the proportion of individuals homozygous for a gene that show its phenotypic effect.

Pericentric applied to inversions that include the centromere.

Phage a virus that reproduces in bacteria.

Phenetic phenotypic, used especially to describe taxonomic relationships based on phenotypic similarity in contrast to phyletic (phylogenetic) relationships based on recency of common ancestor. Crocodiles and lizards are more closely related phenetically than crocodiles and birds are; crocodiles and birds are more closely related phylogenetically than crocodiles and lizards are.

Phenocopy an organism in which a phenotypic change simulates a genotypic change.

Phenotype the resultant of the interaction of the genetic information with the environment; loosely, the characteristics of an organism exclusive of its genetic endowment.

Phenotypic variance the total variance observed in a character.

Phosphate bond (high-energy) usually the anhydride linkage between phosphate ions which may be an important source of energy.

Phosphorylation the addition of a phosphate to an organic compound.

Photolysis splitting water with light as the energy source.

Photophosphorylation the addition of phosphate to organic compounds with light as the source of energy.

Photosynthesis the processes involved in the elaboration of organic compounds from inorganic compounds with light as the energy source.

Phyletic evolution any change occurring sequentially in a single line of descent.

Phylogenetic pertaining to phylogeny, the evolutionary history of an organism.

Pilus, pili hair-like projection on the surface of a bacterial cell which may serve as connecting tube between bacterial cells during conjugation.

Plankton the population of floating organisms in a body of water.

Plasma membrane the outer boundary of the cytoplast, continuous with the endoplasmic reticulum and surrounded by a cell pellicle or wall.

Plasmodesma, plasmodesmata cytoplasmic connection that extends through pores in plant cell walls and connects the protoplasts of adjacent cells.

Plastids organelles of plant cells, usually containing pigments and usually synthesizing soluble or insoluble carbohydrates.

Pleiotropy the influence upon two or more characters, not obviously related, by one or more alleles.

Pleistocene the major epoch of the Quaternary period, beginning about 1 million years ago and ending about 10,000 years ago.

Pliocene the last epoch of the Tertiary period, beginning about 11 million years ago and ending about 1 million years ago.

Ploidy the number of genomes in the zygote of an organism.

Poisson distribution a frequency distribution of the number of times a rare event takes place assuming these ocurrences are independent; approximation to a Poisson distribution of events in nature implies independence of these events.

Polar bodies the two or three meiotic products that do not develop into eggs in oogenesis in animals.

Polygenes those genes without obvious major phenotypic effect; numerous factors affecting a characteristic.

Polyhaploid a haploid sporophyte plant which has been derived from a polyploid and which may contain duplicated or largely homologous genomes; high degree of bivalent formation may be present at meiotic metaphase I.

Polymerase an enzyme which catalyzes the process of polymerization.

Polymerization the formation of large organic molecules from many like subunits.

Polymorphism the presence in the same population and at the same stage of development of two or more conspicuously different forms of an organism in such proportions that the rarest of them could not be accounted for by recurrent mutation alone.

Polyp the sedentary form of a coelenterate.

Polypeptide the substance resulting from the formation of long amino acid chains, the acid group of one molecule combining with the base of another molecule with the loss of a molecule of water.

Polyploid containing more than two haploid chromosome complements in the reproductive cells other than gametes, or presumed to contain more than two genomes in such cells.

Polysaccharide large carbohydrate molecules resulting from the combination of many smaller carbohydrate and possibly other molecules.

Polytene chromosome a giant chromosome consisting of many chromosome strands closely associated along their lengths.

Population a set of items in which one is interested; the individuals in a given area at a given time; in many biological discussions, a synonym of mendelian population, although this is often unstated.

Population biology the study of the pattern in which organisms are related in space and time and including such disciplines as ecology, taxonomy, behavior, population genetics, and others that deal primarily with the interactions of entire organisms.

Population structure the sum of all the factors that govern the pattern in which gametes from various individuals unite with each other.

Porphyrins a class of compounds derived from pyrrole nuclei which, in metal chelate complexes, form pigments important in many biological processes.

Primitive an organism or character judged to be less changed from a presumed common ancestral state than another with which it is compared.

Probability density function a function that describes how probability is distributed among the possible events, i.e., the expected relative frequencies of the events.

Prokaryotic cells without a well-defined nucleus, plastids, Golgi bodies, or mitochondria.

Propagule a portion of a plant which can produce a new plant.

Prophage the noninfectious stage of bacteria phages capable of establishing a sort of symbiotic relationship which need not result in lysis of the host bacteria.

Prophase the stage of nuclear division during which the chromosomes become visible, shorten, and move to the equatorial plate.

Protein a class of compounds of high molecular weight formed by the combination of amino acids with or without other molecules.

Proteinoid a synthetic polypeptide.

Protist unicellular; or having unicellular reproductive organs.

Protobiont coacervate droplets which selectively absorb and concentrate small molecules from the environment and change and incorporate them once inside.

Pseudoalleles   genes which behave like alleles when tested against one another but which can be separated by crossing-over.

Pseudocopulation   a mode of pollination in some orchids in which structures of the flower resemble a female insect and male insects attempting copulation transfer pollen from one flower to another.

Pseudogamy   parthenogenetic development of an ovum after stimulation (but not fertilization) by a male gamete or gametophyte.

Puffing   the extreme uncoiling of a region of a chromosome which facilitates the transcription of ribosomal RNA; easily visible in polytene chromosomes or those of some oocytes.

Purine   an organic base made up of two condensed heterocyclic rings, a pyrimidine ring and an imidazole ring.

Pyrimidine   an organic base consisting of a six-membered ring system with nitrogen atoms at the 1 and 3 positions.

Quadrate   a bone found in the jaw of many vertebrates which presumably has been modified into the incus of the mammalian ear.

Quantum evolution   rapid evolutionary change resulting in what a taxonomist would regard as a new higher taxon.

Quinine   an alkaloid which has been used in the treatment of malaria.

r selection   selection favoring genotypes that result in rapid population growth at low densities.

Random sample   a subset of a population selected so that all items in the population are equally likely to be included in the sample.

Recapitulation   the idea that in the course of development an individual passes through stages similar in form to adults of its presumptive ancestors; often stated as "ontogeny recapitulates phylogeny."

Recessiveness   the converse of dominance; a completely recessive allele is expressed only in the homozygous state.

Reciprocal translocation   transposition of two segments between nonhomologous chromosomes.

Recombination   the formation of gene combinations not present in the parental types.

Recombination index   a measure of the number of new gene combinations that can be produced in a given time; of Darlington: the average number of chiasmata plus the gametic chromosome number.

Recon   the smallest unit whose interchange between chromosomes can be detected by the techniques of the fine-structure geneticist.

Reduction   a reaction in which hydrogen is acquired or oxygen lost by a compound or the valence of the metallic element is lowered.

Regulator gene   the DNA transmitter of specific cytoplasmic signals in the form of repressor-substance molecules.

Replicate   to duplicate or copy.

Repressors   molecules with the property of combining with a particular metabolite (which may either activate or inhibit it) and a particular operator whose action is then blocked.

Reproductive cells   the gametes and their immediate predecessors from which they are produced by division.

Reticulate evolution   evolution in which lines combine as well as split.

Rho   a protein whose presence is necessary for the termination of RNA synthesis.

Ribonucleic acid (RNA)   giant molecules of various types, some of which, in most organisms, copy the genetic code from the DNA and carry it to the sites of protein synthesis (messenger RNA) and some aid in the organization of the synthesis (transfer RNA); see nucleic acid.

Ribosomes   granules associated with the boundary of the endoplasmic reticulum; the presumptive sites of protein synthesis; see microsome.

RNA   see ribonucleic acid.

Sample   a subset of a population; a group of items drawn by some procedure from a population and from which one hopes to learn certain things about the population.

Sampling error   variation due to random elements in the sampling process.

Segmental alloploid   an alloploid in which the combined genomes are homologous in many small segments throughout the complement; crossing-over may recombine material from different genomes.

Selection   differential reproduction of genotypes.

Selection coefficient   the measure of the disadvantage of a given genotype relative to other genotypes in the population.

Sex chromosome   usually a largely heterochromatic chromosome functioning at least in part in sex determination.

Sexual reproduction   reproduction involving the fusion of gametes produced by a prior meiotic process.

Sexuales   males and females parthenogenetically reproduced by the sexuparae which reproduce and give rise to fundatrices of the aphid *Tetraneura ulmi.*

Sexuparae   winged females which were parthenogenetically reproduced by the exules of the aphid *Tetraneura ulmi.*

Sigma   one of the five polypeptide chains of RNA polymerase, responsible for the recognition of initiation sites along the DNA molecule.

Specialized   applying to an organism or character judged to be more changed from a presumed common ancestral state than another with which it is compared.

Speciation   the splitting process of evolution, responsible for the existence of different kinds of organisms that are classified as species by taxonomists.

Species   a group of organisms judged by taxonomists (by diverse criteria) to be worthy of formal recognition as a distinct kind.

Spherosomes   vesicles in plant cells which appear to perform functions similar to those of lysosomes in animal cells.

Spindle   the ellipsoidal mass of protein fibers visible during cell division and thought to play a role in chromosome movement and division of the cytoplasm; actually microtubules.

Spontaneous generation   see abiogenesis.

Stabilizing selection   selection in which genotypes closer to the mean for a character have an advantage over those at the extremes.

Standard deviation   the square root of the variance of a distribution where the variance measures the spread of the distribution about the mean.

Stationary frequency distribution   the representation of a probability density function showing the way the probability in a given situation is distributed over the possible events.

Structural gene   the DNA code unit whose primary product is messenger RNA and which thus controls synthesis of a particular protein or polypeptide.

Structural hybridity   heterozygosity for a chromosomal rearrangement.

Sublethal   causing death later than in the embryonic or infantile stages.

Subspecies   a geographic subdivision of a species deemed worthy of formal recognition by a taxonomist.

Succession   the sequence of transient communities occurring in an area before the climax community, q.v.

Supergene   a group of genes inherited as a unit.

Supernumerary chromosomes   chromosomes present, often in varying numbers, in addition to the characteristic relatively invariable complement.

Switch gene   see oligogenes.

Sympatric   occurring in the same geographic area.

Sympatric speciation   speciation without geographic isolation.

Synapsis   the association of homologous chromosomes in the meiotic prophase.

Syngamy   union of gametes to form a zygote.

Systematic pressure   one of the nonrandom evolutionary pressures: selection, mutation, or migration.

Tachytelic evolution   evolution at a much faster rate than horotelic evolution.

Telophase   the last stage of nuclear division during which the chromosomes lose their definition and the nuclear envelopes of the daughter nuclei form.

Terminalization   the movement of chiasmata to the ends of the chromosomes during the later stages of the first meiotic prophase.

Territoriality   the defense of an area against organisms of the same or similar kind.

Tetraploid   containing four genomes in reproductive cells other than the gametes.

Tetrasomic   describing an organism with one chromosome in the complement represented four times.

Thallus   a plant body not sharply differentiated into root and shoot.

Thelytoky   a form of apomixis in animals in which diploid females are produced from unfertilized eggs.

Thymine   5-methyl-2,4-dioxypyrimidine, the pyrimidine involved in the structure of DNA but not RNA.

Totipotency   possession by cells of the developmental potential which would lead to the formation of a fully mature individual.

Transduction a mechanism of gene transfer in bacteria in which genetic material is transferred from one cell to another by bacteriophage.

Transformation a change in the genetic information of a bacterium as a result of exposure to a DNA extract of a bacterium of different genetic constitution.

Transhydrogenation the sequential transfer of electrons in cellular oxidation cycles.

Translocation see reciprocal translocation.

Trans state two mutational alterations in cistrons on different homologous chromosomes; see cis state.

Trisomic describing an organism with one chromosome in the complement represented three times.

Univalent a chromosome lacking a homolog and therefore unsynapsed at the first meiotic prophase.

Uracil 2,4-dioxypyrimidine, the pyrimidine involved in the structure of RNA but not DNA.

Variance a measure of dispersion obtained by squaring the deviations of the individuals from the mean and dividing the result by the number of individuals; the square of the standard deviation.

Vesicle a spherical intracellular structure which is bound by a membrane and which may function in storage and transport.

Viability the capability for living or continuing to develop; often incorrectly used as synonymous with fitness.

Vitamins organic compounds required in small amounts by heterotrophic organisms and essential to metabolism; often part of an enzyme system.

Vivipary in plants, vegetative reproduction in which propagules replace flowers in the inflorescence; in animals, the production of living young rather than eggs.

Wild type the most frequently observed phenotype or the phenotype arbitrarily designated as "normal."

Zygote a cell formed by the fusion of gametes.

Zygotene the stage in the first meiotic prophase during which homologous chromosomes pair.

INDEX

Acacias:
  ant-dwelling, 301
  swollen thorn, 301
Acanthuridae, 284–285
Acetic acid, 5
Achillea, 233–234, 255, 337
  lanulosa, 168–170
Acridine, 40
Adaptation, 173, 337–338
Adaptive radiation, 291–293
Adaptive zone, 290, 293, 296–297
Adenine, 9, 11, 14
Adenosine diphosphate (ADP), 7
Adenosine phosphate, 11
Adenosine triphosphate (*see* ATP)
Adenylic acid, 11
ADP, 7, 22
Adventitious embryony, 216, 217
African lake cichlids, 243, 244, 267, 292
Agamic complexes, 220–221
Agamospermy, 216–217
Aggregating behavior, 81
Agriculture, invention of, 319
Agrostis tenuis, 234
Alanine, 4
Alcidis agathyrsus, 298, 299
Algae:
  blue green, 182
  brown, 184
  chloroplasts in, 181–182
  diatoms, 184
  diploidy in, 184
  Euglena, 181
  filamentous, 183
  golden, 181
  green, 181, 184
  haploidy in, 183
  and lichens, 73
  -like fossils, 3
  polyploidy in, 212
  red, 181
  and sloth moth, 85
Aliphatic acids, 3
Alkaloids, 116, 300

Alleles, 111, 114, 115
  and continuous variation, 52
  defined, 34
  multiple, 39
  (*See also* Genes)
Allen's rule, 232, 255–256
Allison, A. C., 159
Allium, 28
  cepa, 208
  fistulosum, 208
  hybridization in, 208
  vivipary in, 74, 216
Allometry, 306
Allopatric speciation, 272, 330
Alloploidy, 207–209, 214, 225, 273, 286, 329
  segmental, 209
Allozymes, 114–115
Alternation of generations, 184–185, 217
Altruism, 119
Amazilia beryllina, 251
Ambystoma, 213, 222–223
American Indian culture, 321, 322, 327–328
American Naturalist, 129
Amino acids, 3–6, 11, 13–16, 40
Aminopurines, 41
Ammonia in atmosphere of primitive earth, 3
Amphibia, polyploidy in, 213
Amphidiploid, 213
Anaphase, 24–26, 29, 31
Anchovy, 213
Anderson, Edgar, 264
Andrewartha, H. G., 84, 88
Andrews, H. N., 280
Aneuploidy, 201–205
Angiosperms, pollination of, 85
  (*See also* specific name)
Animal cell, 18–21, 25
  differentiation of, 63
  DNA in, 213
  and polyploidy, 206, 208, 212, 213, 220, 222–223
Animals:
  adaptive radiation in, 292
  and alternation of generations, 184

DNA:
    double helix of, 11, 12
    kappa particles of, 181
    and mutation, 40
    and organelle synthesis, 23
    and polyploidy, 213
    structure of, 11–12
    and transformation, 178
    and viruses, 178–179
Dobzhansky, Theodosius, 176, 258, 259, 328, 335
    and *Drosophila* research, 150–156, 241, 337
Dolinger, P. M., 116
Dominance, genetic, 36
    evolution of, 41–43
Dowdeswell, W. , 138, 144
Dragonfly, differentiation in, 62
Drift, genetic, 98–100, 159, 256, 338
    defined, 99
    founder principle, 147, 266
Drosophila, 43, 51, 163, 337
    and alkaloids, 300
    bristle number in, 161, 165–167
    carcinophilia, 78–79
    centric fusion in, 205
    chromosome structure changes in, 192, 197
    crossing-over suppression in, 30, 37
    eye color in, 113
    gene assimilation in, 170–171
    genetic variation in, 114
    geographic variation in, 241
    hybridization in, 265–266
    melanogaster, 158, 165–166, 170–171, 191
    miranda, 152, 197
    pachea, 300
    paulistorum, 241, 262, 329
    persimilis, 152, 157, 265–266
    polymorphism in, 150–159
    polytene chromosomes in, 26, 43, 45, 46, 151
    population fusion in, 262
    pseudoobscura, 114, 151–155, 157, 265–266
    salivary gland chromosomes of, 26, 43, 45, 46, 151
    segregation-distorter factor (SD), 191
    sex chromosomes in, 44
    and sympatric speciation, 273
    tropicalis, 157
    virilis, 197
    wild type, 58
Drought, 260
Dubois, R., 285
Dunker religious community, 159
Dunn, L., 296

Earth:
    atmosphere of, 3–4, 6
    primitive, 3–4
Earthworms, 221
Ecological communities:
    closed, 187
    open, 187
Ecotype, 233
Edaphic ecotypes, 233
Eggs:
    formation of, 29
    mimicry of, 254, 268–270
Ehrlich, A. H., 304
Ehrlich, P. R., 14, 67, 74, 76, 79, 139
    on coevolution, 299, 300, 304
    on gene flow, 257
    on water snakes, 147, 149
Elapidae, 292
Electrical discharge and origin of life, 3–4
Electrons, 2
Electrophoresis, gel, 113–114
Elephants, 279
Elephas maximus, 279
Elk, Irish, 306
Elodea, 214
Embryo and recapitulation, 66–67
Embryoids, 63
Emerita analoga, 231
Emigrantes, 224
Emigration, 104
Empidonax, 263
Endoplasmic reticulum, 19–22
Endopolyploidy, 26, 28
Endosperm, 217
Endosymbionts, 23
Energy:
    cycling in communities, 88–90
    sources, 6–9
    transformations, 2
England, industrial melanism in, 132–137
English language, 327
Ensatina:
    geographic variation in, 241–242
    ring-of-races pattern, 262
Environment, 83–87, 105
    adjustment to, 173–174
    coarse-grained, 126
    components of, 84
    fine-grained, 126
    niches, 89
    selective agents of, 104
    stress, 270–272
    variable, 123–129
    variance of, 51–54
Enzymes, 42, 113–115
    digestive, 22
    metallo, 8
    primitive, 11
    and protein synthesis, 6–7, 14, 16

Water:
   as resource, 323
   vapor, 3, 5
Water striders, 26
Water vacuole, 19
Watson-Crick model, 12
Weevils, 221, 301
   boll, 164
Weisz, P. B., 16, 19
Whales and convergence, 297
White, M. J. D., 219, 223
Whorf, B. L., 327
Wiener, Norbert, 71
Wilberforce, Bishop, 335
Wild-type phenotype, 58
Wilson, E. O., 89
Wolves, 88
Woman (*see* Human beings)
Woodson, R. E., 234–236
Worms, parasitic, cyclical thelytoky in, 218

Wright, S., 42, 102, 121, 122, 124, 125, 214, 258, 321

X chromosome (*see* Sex chromosomes)
Xenopus, 64

Y chromosome (*see* Sex chromosomes)
Yanofsky, Charles, 40
Yarrow (see *Achillea*)
Yeast, 185
   colony of, 73
   life cycle of, 60
Young, W. J., 47, 48, 194, 199

Zea, 43
Zinjanthropus, 317
Zweifel, R., 223–224